Forestry Research Handbook

Volume I

Forestry Research Handbook
Volume I

Edited by **Pierre Simone**

R CALLISTO
REFERENCE

New York

Published by Callisto Reference,
106 Park Avenue, Suite 200,
New York, NY 10016, USA
www.callistoreference.com

Forestry Research Handbook: Volume I
Edited by Pierre Simone

© 2015 Callisto Reference

International Standard Book Number: 978-1-63239-344-9 (Hardback)

Contents

Preface

The tradition of formal forestry practices can be traced to the 7th century, when it was developed by the Visigoths. They instituted a code for the preservation of oak and pine forests, when they were faced with an ever increasing shortage of wood. Today, forestry is considered one of the three primary land-use sciences, along with agriculture and agroforestry.

Forestry is an interdisciplinary field which requires an extensive knowledge of a variety of subjects like genetics, soil science, biology, botany among others. Silviculture is also an important aspect of Forestry. The main objective of forestry is to manage forests in such a way that they are preserved and yet continue to provide environmental supplies and services. This involves development of better methods for the planting, felling, extracting and controlled burning. In fact, one of the applications of modern forestry happens to be reforestation.

The main objective of Forestry is also its challenge because it is difficult to sustain a resource while ensuring that other resources remain unaffected. Today, a strong body of research exists regarding the management of forest ecosystems.

I'd like to thank all the researchers for sharing their studies in this field with us. I would also like to thank the publishing house for considering me for this project and giving me this opportunity to interact with experts across the globe. This book has been an enriching experience for me.

Editor

Current Status of Reproductive Management in Buffalo in West Sulawesi Province, Indonesia

Muhammad Yusuf, Sudirman Baco, Muhammd Nasir Karim

Department of Animal Production, Faculty of Animal Science, Hasanuddin University,
Makassar, Indonesia

The aim of this study was to know the application of reproductive management of buffalos in West Sula-wesi Province. This study was conducted during a period from July to August 2012 in three different sub-districts; Mehalaan, Messawa, and Nosu. Purposive sampling was used to choose the location where the study was taking place with consideration that those sub-districts have different population of buffalos ranging from highest to smallest. Data collection both primary and secondary data was done through observation and interview to obtain both quantitative and qualitative data. The results of this study showed that mating system of buffalos in this area were two different methods; natural mating without any help and natural mating with the help of the farmers, while artificial insemination (AI) method for mating the buffalos did not apply yet. A total of 85% farmers/respondent did mating system for the buffalos with the help of the farmers, while the remaining 15% of the farmers did mating system without any help. This study also showed that most of the farmers had good knowledge about estrus (98.3%) and the remaining 1.7% farmers had poor knowledge about estrus. Most of them were capable to identify buffalos in estrus by observing their behavior.

Keywords: Buffalo; Reproductive Management; Mating System

Introduction

Buffalo is a triple purpose animal that provides milk, meat and mechanical power to mankind (Pasha & Hayat, 2012) and has been an integral part of livestock agriculture in Asia for over 5000 years in producing draft power, milk, meat and hides (Nanda & Nakao, 2003). Furthermore, they stated that the buffalo did not receive the attention of the policy makers and the researchers in accordance with its merits, which resulted in buffalo population decline in several eastern Asian countries. On the other hand, the pivotal role of buffalo in overall social development through their contributions has been well reviewed (Nanda & Nakao, 2003). In Indonesia the population of buffalo has also been declined for a few recent years. The population of buffalo in 2007 was 2,085,779 heads and decreased to 1,305,011 heads in 2011. Likewise, the population of buffalo in West Sulawesi Province decreased from 14,833 to 8112 heads during the same period. The causes of this problem were not fully understood. Most probably, the causes are multifactor that is including the reproductive management of this animal. Dobson and Kamonpatana (1986) stated that the reproductive performance of buffalo remained much lower than in cattle. Therefore, the aim of this study was to describe the current status of reproductive management of buffalo in West Sulawesi Province, Indonesia.

Materials and Methods

Animals and Management

This study was conducted between June and August 2012 in three different sub-districts in Mamasa Regency, West Sulawesi Province, Indonesia. The study enrolled 60 farmers with the total of buffalo were 158 heads. The buffalo were raised in a small holder system without any modern technology involved. The housing system for the animals was in the simple house during nighttime and free during daytime for grazing without any additional feeds such as concentrate, mineral, and feed additive. The buffalo are usually sent out in the morning time to the field for grazing and return back in the late afternoon. During grazing, some owners of buffalo sometimes observe their animals for estrous signs. Animal showed signs of estrus were naturally mate with the buffalo bull if available at the time of estrus. Otherwise, the owner will seek for the buffalo bull of the other farmer.

Data Collection

In order to know the reproductive management and the development of buffalo population, both primary and secondary data were used in the study. Primary data was collected with the help of a questionnaire to the 60 farmers that included the number of buffalo, raising management, and reproductive management. While secondary data was collected from related institution such as local government and livestock service.

Statistical Analyses

All data were tabulated using Excel program (Microsoft Excel, 2007). Buffalo population at different years was analyzed using simple linear regression. Chi-square was used to analyze the differences between natural mating system and mating system with the help of the farmers and the knowledge of the farmers for the signs of estrus.

Results and Discussion

Herd Size and Population

A total of 60 farmers were interviewed in the present study. The average buffalo owned by each farmer were only (±SD) 2.63 ± 1.77 heads, ranging from 1 to 10 heads. The median and mode were 2 and 1, respectively. This indicated that mostly farmers owned very small number of buffalo (1 to 2 heads) and very few farmers owned 5 or greater of buffalo. Basically, the role of buffalo in this area is for draft, however the farmers also using the buffalo as life saving. Nanda and Nakao (2003) noted that for more than 5000 years, buffalo have been used for draft, that are particularly suited to work on wet fields with a strong body, broad hooves, flexible pastern and fetlock joints. Furthermore, Pasha and Hayat (2012) stated that buffalo products and their contribution as a triple purpose animal that provide milk, meat and mechanical power to mankind. They also stated that among different products obtained from buffalo, milk, meat and hides are more important. However, in this study, the purpose of buffalo is mainly used for draft operations in agriculture.

Figure 1 shows the trend of buffalo population in Mamasa Regency during a period from 2007 to 2011. The data of this population was obtained from local government as a secondary data of this study. Buffalo population in this region linearly increased significantly (P = 0.0025) by year. On the primary data, we calculated that calving rate of buffalo was 30.9%. No attempt was made in the present study to calculate the number of buffalo sent out from this area.

Reproductive Management in Buffalo

In the present study, reproductive management was focused to ascertain the mating system applied, knowledge of the farmers regarding estrus, the use of any reproductive technology such as estrous induction or estrous synchronization, and future prospect for application of artificial insemination (AI). **Figure 2** shows that all mating in buffalo in this area were conducted naturally (natural mating system). However, this mating system was mainly guided by the farmers; approximately 85% farmers gave attention to their animals for mating after estrus was detected. The remaining 15% farmers did not have any special attention to their animals especially for mating. Such these farmers simply let their animals for mating as naturally. No effort was made by the farmers to the animals for mating.

In **Figure 3** shows that the proportion of farmers regarding their knowledge about estrus was much better than expected before the study was conducted. Approximately 98% farmers

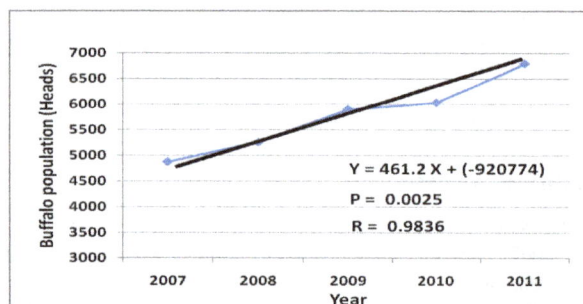

Figure 2.
Proportion of mating system in buffalo in Mamasa Regency.

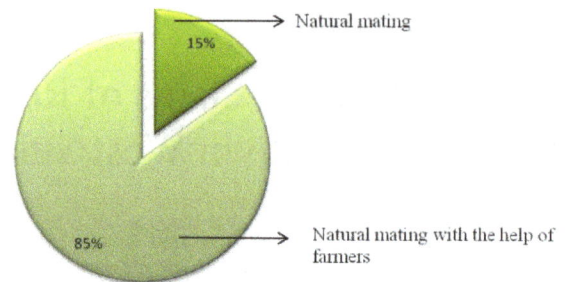

Figure 3.
Proportion the farmers based on their knowledge about estrus in buffalo.

have a good knowledge about estrus and the remaining 2% of the farmers do not have any understanding about estrus. **Table 1** shows the signs of estrus that farmers usually recognized. Mainly farmers recognized the animals in estrus if they are showing mucus. This sign is believed that the animals should be breed and after breeding the farmers also believed that the animal would become pregnant.

In the present study, the use of any reproductive technology such as estrous induction or estrous synchronization did not applied yet. No one respondent has been used this technology to improve their buffalo reproduction. Basically, the farmers knew that in case the buffalo have prolonged the time to become pregnant, so it means that the problem occurs in the animal. This is due to that some limitations to apply this technology such as farmers' knowledge about estrous induction or estrous synchronization and the cost for this treatment were the main reasons for application this technology. However, it is still a future prospect for application of artificial insemination (AI) in this area. The farmers would accept this technology in order to improve their buffalo reproduction. They would make their animals become pregnant and produce a lot of offspring during her life period. This indicates that the farmers in this area have potential to be improved from the nature of raising buffalo to more effectively and efficiently of raising buffalo. Therefore, provision of appropriate extension services for these farmers will improve the productive and reproductive performance of buffalo (Warriach et al., 2012).

Some problems that usually occur in buffalo reproduction include delayed puberty, seasonal breeding, long calving interval, and poor estrus detection (Pasha & Hayat, 2012). Furthermore, they stated that these problems hampered the reproductive efficiency in the female buffalo. Similarly, Terzano et al. (2012) stated that inherent reproductive problems (delayed puberty, higher age at first calving, long postpartum anestrus period, long inter calving period, silent heat coupled with poor expres-

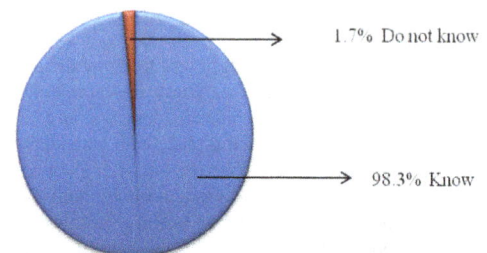

Figure 1.
Buffalo population in Mamasa Regency for last five years.

Table 1.
Knowledge of the farmers regarding the signs of estrus.

Sign of estrus	No. of farmer	Percentage (%)
Restlessness	25	18.8
Mucus	58	43.6
Standing estrus	10	7.5
Swollen	12	9.0

sion of estrous, seasonality in breeding and low conception rate) limit the productivity of buffalo. For the future use of AI in this area, the weakness of estrus symptoms and variability of estrus length would be expected as the limitation of this technology. This problem has been experienced and stated in the study of Pasha and Hayat (2012). As buffalo are polyestrous and are capable of breeding throughout the year (Perera, 2011), however, in many countries a seasonal pattern of breeding activity, and consequently calving is one of the limiting factors for buffalo reproduction.

The present study concluded that the reproductive management in buffalo was very poor; nonetheless, most of the farmers had good knowledge about estrus and were capable to identify buffalo in estrus by observing their behavior.

REFERENCES

Nanda, A. S., & Nakao, T. (2003). Role of buffalo in the socioeconomic development of rural Asia: Current status and future prospectus. *Journal of Animal Science, 74,* 443-455.

Pasha, T. N., & Hayat, Z. (2012). Present situation and future perspective of buffalo production in Asia. *The Journal of Animal and Plant Sciences, 22,* 250-256.

Perera, B. M. A. O. (2011). Reproductive cycles of buffalo. *Animal Reproduction Science, 124,* 194-199.

Terzano, G. M., Vittoria, L. B., & Borghese, A. (2012). Overview of reproductive endocrine aspects in buffalo. *Journal of Buffalo Science, 1,* 126-138.

Warriach, H. M., McGill, D., Bush, R. D., & Wynn, P. C. (2012). Production and reproduction performance of Nili-ravi buffaloes under field conditions of Pakistan. *The Journal of Animal and Plant Sciences, 22,* 121-124.

Current Status of Reproductive Management in Bali Cows in South Sulawesi Province, Indonesia

Sudirman Baco, Muhammad Yusuf, Basit Wello, Muhammd Hatta

Department of Animal Production, Faculty of Animal Science, Hasanuddin University, Makassar, Indonesia

The objective of this study was to elucidate the current status of reproductive management of Bali beef cows in south Sulawesi Province, Indonesia. This study was conducted in Bantaeng Regency, South Sulawesi Province, Indonesia. Purposive sampling was used to choose the location where the study was taking place with consideration that this region has potential place for Bali cows. Data collection both primary and secondary data was done through observation and interview to obtain both quantitative and qualitative data. The results of this study showed that reproductive management applied to Bali cows in this region was very low. However, knowledge of the farmers regarding reproductive management such as estrus and its signs were much better, although the time of insemination to the estrous cows sometimes was too late. Mating system of Bali cows in this area were two different methods; natural mating and artificial insemination (AI). To adopt AI technology, the farmers should follow the standard operation of this technology. This study also showed that most of the farmers had good knowledge about estrus synchronization; however, they are still having difficulty to obtain or to access this technology, especially the price of the hormones used for estrus synchronization.

Keywords: Bali Cows; Reproductive Management; Estrus; Mating System

Introduction

Reproductive efficiency has long been recognized as the most important aspect of commercial beef production (Dyer, 2009). For a cow-calf operation, good reproductive rates are critical to operational success and profitability (Parish, 2010). Furthermore he stated that cows that do not produce calves on at least annual basis use resources that could be better used to support more productive cattle. Therefore, closely monitor cattle reproductive efficiency to identify and address reproductive problems quickly. This means that it is necessary to manage the reproductive cows to achieve high level of profitability. However, to manage the reproductive rate of beef cattle, it is important to assess or measure the reproductive efficiency of the cattle as well as reproductive management that applied in the herds.

In South Sulawesi Province, Indonesia, Bali cattle are the type of beef cattle that are commonly raised by the farmers. This type of beef cattle has many advantages in rising according to the farmers' perception in this region. For example, this cattle has better reproductive performance, high tolerance to the heat stress, and having dual purpose for draught and for beef. These reasons make the farmers to maintain this type of cattle with them. However, since this type of beef cattle used for draught in the field, in which affecting their reproductive performance to produce many offspring during their lifetime, for a few last years their reproductive performance was decreased. This is probably caused by low level of reproductive management applied by the farmers. However, to our knowledge, there was no study to describe the reproductive management of Bali beef cows in this region. Therefore, this study tried to elucidate the current status of reproductive management of Bali beef cows in south Sulawesi Province, Indonesia.

Materials and Methods

Animals and Management

A total of 130 cows from 25 farmers in Bantaeng Regency, South Sulawesi Province, Indonesia were involved the present study. The Bali cows were raised in a small holder system without any modern technology involved. The housing system for the animals was in the simple house during nighttime and free during daytime for grazing without any additional feeds such as concentrate, mineral, and feed additive. The cows are usually sent out in the morning time to the field for grazing and return back in the late afternoon. During grazing, some owners of Bali cows sometimes observe their animals for estrous signs. Animal showed signs of estrus were naturally mated with the bull if available at the time of estrus or inseminated by inseminator.

Data Collection

For reproductive management and the development of Bali cattle population, both primary and secondary data were used in the study. Primary data was collected with the help of a questionnaire to the 25 farmers that included raising management and reproductive management. While secondary data was collected from related institution such as local government and livestock service. Clinical examination was implemented by the authors with the help of local technician and/or management staffs of the herd. All cows were subjected to trans-rectal palpation for pregnancy status and/or the genitalia to assess uterine conditions and ovarian structures. Trans-rectal palpation of the uterine was performed to determine the consistency of uterine including contraction, elasticity, tonicity, symmetry of uterine horns, and the presence of any fluid in the uterus (Gautam et al., 2010). The presence of any palpable ovarian structures, ovarian cysts was defined as one or more follicle-like structures >25 mm in diameter without a concurrent corpus luteum (CL). Ovaries without palpable structures (i.e. ovarian follicles > 10 mm in diameter and/or a functional CL) were considered inactive (Yusuf et al., 2010).

Results and Discussion

Reproductive Management of Bali Cows

In the present study, the farmers were interviewed regarding the simple reproductive management applied for their Bali cows, such as cow and heifer management, knowledge about puberty, estrus and its signs, reproductive disorders, mating system, estrous synchronization, and calving management. For cow and heifer management, mostly farmers did not pay special attention to how efficient their cattle will produce as many as calves during her lifetime. One main problem facing the Bali cows after calving was the duration of anestrus (Yusuf et al., 2012a). Prolonged postpartum anestrus in suckled beef cows is one of the main restrictions to obtain a calf every year (Miller & Ungerfeld, 2008). Short et al. (1990), Stated that prolonged suckling, nutritional deficiencies, climatic stress, parity, time of the year, and management practices were the main causes of prolonged calving intervals. However, in many cases, the farmers did not realize this condition. Therefore, effort to reduce this problem is necessary. The duration of anestrus in cattle was usually shortened when cows exposed to bulls (Rekwot et al., 2000; Landaeta-Hernández et al., 2004; Miller & Ungerfeld, 2008). Likewise, regular examination of postpartum cows in order to achieve shortening the duration between calving and first estrus seems to be one of solving the problems (Yusuf et al., 2012a).

In heifers, there was no selection of replacement beef heifers on genetic improvement and phenotype. Parish (2010) stated that for cow and heifer management, it should be considered indicators of reproductive performance when selecting replacements. In addition, it is important to have heifers cycling before the first breeding season to ensure the highest possible fertility at first service. The other thing that the farmers in this region did not pay attention was the age of heifers at first breeding and probability of heifers' age at first calving.

In the present study, knowledge of the farmers regarding estrus and its signs was quite good. They knew both primary and secondary estrus signs. When the signs of estrous occurred in their cows, especially standing estrus, they had a good effort to mate the cows with the bull or they inform to the inseminator to inseminate their cows as soon as possible. However, in many cases, it was often the time of insemination was too late. This indicated when the authors asking the inseminator especially uterine tone condition at the time of insemination was not in contraction. This suggests that although the knowledge of the farmers was much better about estrus signs but the time of insemination should also be improved.

For reproductive disorders, all farmers described no idea regarding this problem for their cows. Anestrus postpartum was the main problem in this region. Long duration of postpartum anestrus was the major reproductive disorders facing their cows

(Yusuf et al., 2012b). In addition, there was no treatment conducted for this problem. The authors tried to introduce reproductive management for anestrus cows with the use of hormonal treatment to shorten the interval from calving to first service. This was not accepted well since the farmers have difficulty to buy such kind of hormones. For instant, introduction of GnRH or progestin (CIDR) by authors for free to the farmers seemed to be successful. However when authors stopped this program, there was no more treatment conducted by the farmers for this reproductive problem.

In the present study, we have tried to assess the reproductive physiology of all the cows as shown in **Table 1** (Yusuf et al., 2012b).

Based on clinical examination of Bali cows as shown in **Table 1**, it indicated that in a population of these cattle, there were 37% cows become pregnant after repeating insemination or natural mating. The remaining 63% cows were not pregnant with various reproductive statuses. For these high non-pregnant cows required special attention to increase the number of cows to become pregnant. This means that it is necessary to improve reproductive management in Bali cows to achieve the highest level of pregnancy rate as well as to shorten the calving interval among the cows. Therefore, an effort to increase the reproductive efficiency in Bali cattle in this region is necessary in order to increase both population and profitability to the farmers.

Table 1.
Results of clinical examination of Bali cows.

Variable	No. of cows	Percentage (%)
No. of animal	111	100
Pregnant cow	41	37
Non-pregnant	70	63
Cyclic	15	14
Cyclic + nimpomania	1	1
Cyclic + metritis	1	1
Anestrus	52	47
Anestrus + metritis	1	1

Future Prospect to Improve Reproductive Management in Bali Cows

Since potential good reproductive performance of Bali cows in this region occurred in the past time, it is not difficult to improve this potential in the future. Improvement in reproductive management as well as nutrition especially micro-nutrient would be the way out to increase the population of Bali cattle through improvement of reproductive performance by conducting good reproductive management.

The present study concluded that the reproductive management in Bali cows was very low. Farmers' knowledge and limitation to the skills to manage the cows were the major factors for this problem. However, to improve the reproductive performance of Bali cows, it is necessary to improve the farmers' knowledge as well as skill in raising animals.

REFERENCES

Dyer, T. G. (2009). *Reproductive management of commercial beef cows* (pp. 1-7). Cooperative extension. The University of Georgia and Ft. Valley State College, the US Department of Agriculture and Counties of the State Cooperating.

Gautam, G., Nakao, T., Koike, K., Long, S. T., Yusuf, M., Ranasinghe, R. M. S. B. K., & Hayashi, A. (2010). Spontaneous recovery or persistence of postpartum endometritis and risk factors for its persistence in Holstein cows. *Theriogenology, 73,* 168-179.

Landaeta-Hernández, A. J., Giangreco, M., Meéndez, P., Bartolomé, J., Bennet, F., & Rae, D. O. (2004). Effect of biostimulation on uterine involution, early ovarian activity and first postpartum estrous cycle in beef cows. *Theriogenology, 61,* 1521-1532.

Miller, V., & Ungerfeld, R. (2008). Weekly bull exchange shortens postpartum anestrus in suckled beef cows. *Theriogenology, 69,* 913-917.

Parish, J. A. (2010). *Reproductive management of beef cattle herds* (pp. 1-5). Extension Service. Mississippi State University.

Rekwot, P. I., Ogwu, D., & Oyedipe, E. O. (2000). Influence of bull biostimulation, season and parity on resumption of ovarian activity of zebu (*Bos indicus*) cattle following parturition. *Animal Reproduction Science, 63,* 1-11.

Short, R. E., Bellows, R. A., Staigmiller, R. B., Berardinelli, J. G., & Custer, E. E. (1990). Physiological mechanisms controlling anestrus and infertility in postpartum beef cattle. *Journal of Animal Science 68,* 799-816.

Yusuf, M., Nakao, T., Ranasinghe, R. M. S. B. K., Gautam, G., Long, S. T., Yoshida, C., Koike, K., & Hayashi, A. (2010). Reproductive performance of repeat breeders in dairy herds. *Theriogenology, 73,* 1220-1229.

Yusuf, M., Rahim, L., Asja, M. A., & Wahyudi, A. (2012a). The incidence of repeat breeding in dairy cows under tropical condition. *Media Peternakan, 35,* 28-31.

Yusuf, M., Toleng, A. L., & Hasbi, N. S. (2012b). Reproductive disorders in Bali cows raised under small-holders (a preliminary study). In *Proceedings on Seminar National for Livestock Sustainability (in Indonesian language)* (pp. 9-14).

Ponderosa Pine Family Growth Comparisons in the Central Great Plains of Kansas

Wayne A. Geye[*], Keith D. Lynch
[1]Division of Forestry, Kansas State University, Manhattan, USA

Ponderosa pine (Pinus ponderosa Laws.) has been planted widely in the Great Plains. Recommendations based on a 1968 study were to use material from south central South Dakota and north central Nebraska. A second test to further delineate seed sources (provenance/families) in this region was established in 1986. This paper reports results for survival, height, diameter, and D2H measurements in Kansas at 15 years. Results identify a wide range of suitable families within the Great Plains region. A majority of the tested sources performed well especially those from central Nebraska. Those sources from eastern Montana and western Nebraska performed poorly where environmental or geographic conditions were the poorest, thus verifying the original recommendations.

Keywords: Ponderosa Pine, Pinus Ponderosa, Provenance, Seed Sources, Tree Selection, Growth Characteristics

Introduction

Ponderosa pine (*Pinus ponderosa* Laws) is an important component of the windbreak agroforestry system in the Great Plains. Its drought tolerance, dense crown form, and tall growth habit make ponderosa pine excellent for windbreaks, sight barriers, and ornamental plantings (Flint, 1983).It is one of the few tall trees that grow in the region and also provides full year-round protection to fields and farmsteads because of its ever-green nature (Schaefer & Baer, 1985).The natural range of ponderosa pine extends from British Columbia, Canada, southward into northern Mexico and from California eastward into the Great Plains, except for Kansas (Crichfield & Little, 1966). It has been widely planted in the plains region, but has shown inconsistent performance.

Western pine tip moth (*Rhyacioniabushnelli*) has caused widespread damage in the plains (Kopp et al., 1987), but outstanding performance of some individual trees in the plains plantations suggests that proper selection could improve tree quality.

Early studies determined that trees grown from seed collected from the northeastern range of ponderosa pine performed best in most of the provenance test plantations (Deneke & Read 1975; Baer & Collins 1979; Read, 1983; Schaefer & Baer, 1985, 1992; Van Haverbeke, 1986). Also, 6-year datafrom a Kansas plantation showed that early growth appeared to be clinally related to elevation of seed provenances (Deneke & Read, 1975). Therefore, plains nurseries have focused much of their ponderosa pine production on seed collections near Ainsworth and Valentine, Nebraska, and Rosebud, South Dakota. In addition, trees from Jordan, Montana, performed well in more than half of the early plantations (Read, 1983).

In 1986, a second cooperative ponderosa pine study was initiated by the GP-13 Technical Committee of the Great Plains Agricultural Council in cooperation with the North Central and Rocky Mountain Forest Experiment Stations. The intent of the study was to more intensively sample recommended provenances identified in the 1968 study. Collection origins are shown in Figure 1. Nine progeny tests were established in Saskatchewan Canada, Montana, North Dakota, South Dakota, Nebraska, Kansas, Oklahoma, Texas, and Minnesota. This paper reports data from the Kansas tests. This paper reports data from the South Dakota, Nebraska, and Kansas tests. No additional tree improvement studies have been initiated in the United States since this effort. Recently in Argentina (Meier et al., 2004) a genetic tree improvement effort was initiated with intention to established seed orchards in Patagonia. Tree improvement studies have attributed approximately 2% of the total variation to differences among geographic locations in the Southwestern United States (Yow et al.). One generation of tree improvement may lead to gains in yield of 1% - 15% or reduce rotations by 1 years - 20 years in ponderosa pine in the Inland Empire Tree Improvement Cooperative in the northern Rockies of the United States (Hamilton et al., 1994).

Materials and Methods

The tree plantation reported here used seedlings representing 75open-pollinated families from 13 geographic provenances

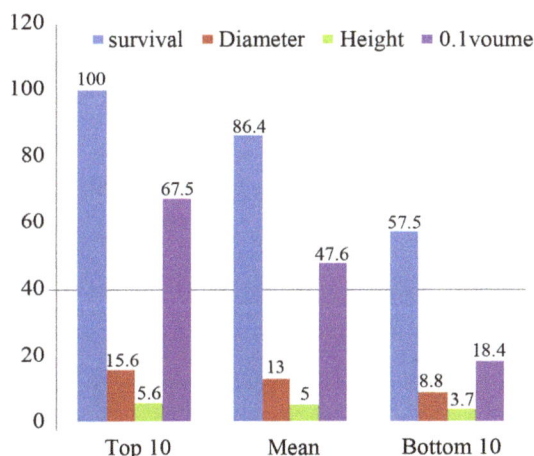

Figure 1.
Means of growth characteristics of the 10 ten and bottom 10 families in the kansas.

(Table 1) and were planted in an individual tree factorial plant-ing design with 8 replications at 3.7 m × 3.7 m (12 × 12 ft) in Kansas for total of 1167 trees. Spacing within each replication was represented by five trees in single-tree, noncontiguous plots. Two border rows surrounded the plantation. Weeds were con-trolled by cultivation for the first 3 years. The Kansas plantation was near Milford Lake, Kansas, on an alluvial sandy loam site. Superior height growth potential can be accurately identified at an early age (i.e., 5 to 15years; Lambeth, 1980; Read, 1983; Van Haverbeke, 1986; Schaefer & Baer, 1992). In this analysis, data were also analyzed separately for the plantation.

Analysis consisted of ANOVA using the GLM procedure of SAS (SAS Institute, 2003) for height, trunk diameter, and D^2H (a measure of trunk volume—volume index); Duncan's multi-ple range test for mean separation; andchi-square for survival. In addition, correlations were determined among height, di-ameter, D^2H, and latitude and longitude. Seventy five seed sources (families) were compared. Most of the sources were from north central Nebraska and southern South Dakota (Table 1).

Results and Discussion

Fifteen-year survival and growth of the top 25% were from four provenances 757, 990, 994, and 996; generally at the edge of its natural range. Tests of effects are shown in Table 2 for di-eter, height, and D^2H. Family performance differed signify-ntly (<1%) level across the plantation (Table 3).Means for the top 10, mean, and bottom 10 are shown in Figure 1. Replicates were significant at the <1% level and interactions were mixed. No winter dieback or diseases were detected. Similar results in a 28 plantation study in the Canada and the United States showednorth central Nebraska sources were best as reported

from a greater study (Read 1983).

Survival

Survival after 15 growing seasons was good for all of the 75 families compared. The mean survival of all families was 86.4% with a range of 38.1% to 100%. The top 10 families are as follows: 99010, 72005, 72109, 75713, 99001, 99010, 99201, 99404, 99506, 99613. All are from the north central part of Nebraska. Most families had 100% survival. The five poorest families were far western sources from Jordan, Montana and one from South Dakota at 48.1%.They were significantly dif-ferent (<1% level) from the other families. The other 74 sources were not. A similar test at age 10 among provenances showed 720 and 721 (central Nebraska) with 72 and 82 percent survival in the Northern Great Plains (Read 1983).

Diameter

The mean diameter was13.0 cm (6.0 in.)with a range from 17.2 cm - 5.3 cm (2.1 to 6.8 n). Mean diameters of the top 10 and bottom 10 families (Table 2) were 15.7 and 8.9 cm (6.2 and 3.4 in), respectively. Families from the 720 and 721 prove an-ces were among the best. A similar test at age 10 among prove-nces showed 720 and 721 (central Nebraska) were the best in the Northern Great Plains (Read 1983).The largest 59 families were not significantly different ranging from 17.2 to 14.1 cm (6.8 to 4.9 in). Five families (provenance 811) from Jordan, Montana, had significantly smaller diameters than the other families (where environmental factors were poor having very low precipitation and low annual temperatures), whereas di-ameters of the poorest 10 families were significantly smaller than those of the best (Table 3).Trees with the largest diameters had the tallest heights (r = 0.77096 at the 1% level).

Table 1.
Collection zones (provenances) of ponderosa pine seed sources.

Geographic origin(#)	Tree additions (families)(#)	Local area (town/state)	Annual precipita-tion((mm)(in))	Annual mean temp. C^0 (F^0)	Elevation m (ft)	Latitude (^{0}N)	Longitude (^{0}W)
720	02 - 11	Ainsworth NE	579 (22.8)	8.6 (47.6)	780 (2560)	42.59	100.00
721	01 - 11	Valentine NE	495 (19.5)	8.4 (47.2)	800 (2625)	42.88	100.55
757	01 - 40	Rosebud SD	955 (37.6)	8.5 (47.7)	850 (2789)	43.25	100.82
811	13,15 - 16,19,22	Jordan MT	327 (12.9)	6.9 (44.5)	808 (2625)	47.31	106.89
990	01 - 10	Springview NE	573 (22.7)	8.3 (47.0)	740 (2428)	42.82	99.75
991	01 - 05	Kilgore NE	516 (20.3)	8.3 (47.0)	800 (2625)	42.94	100.97
992	01 - 04	Drinkwater NE	495 (19.5)	8.3 (47.0)	859 (2800)	42.47	101.07
993	01 - 07	Nenzel NE	526 (20.3)	9.0(48.3)	950 (3117)	42.93	101.11
994	01 - 05	Bassett NE	635 (25.0)	9.0 (48.3)	710 (2329)	42.60	99.54
995	01 - 06	Snake River NE	495 (19.5)	8.3 (47.00)	866 (2840)	42.71	100.97
996	01 - 14	Sparks NE	495 (19.5)	8.3 (47.0)	800 (2625)	42.94	100.24

Table 2.
Test of source effects on 15-year-old ponderosa pine families.

Effect	DF	Diameter	Height	D^2H
		Pr > F		
Family	74	<0.0001	<0.0001	<0.0001
Replicate	5	0.0020	0.0436	0.1047
Family X replicate	345	0.02262	0.03992	0.2912

Table 3.
Ranking of the 10 top and bottom 10 families at 15 years in Kansas.

Survival		Diameter		Height		D²H - - volume index	
Family	(%)	Family	cm (in)	Family	m (ft)	Family	Volume
Top 10 families							
72004	100a	99010	17.2 (6.8a)	72109	6.0(19.6a)	72109	926a
72005	100a	72109	16.5 (6.5ab)	72103	5.8(19.1ab)	99010	906ab
72109	100a	99305	16.4 (6.5abc)	99101	5.8(18.9abc)	99305	815abc
75713	100a	96601	15.7 (6.2a-d)	72104	5.6(18.3a-d)	99006	698bcd
99001	100a	72004	15.2 (6.0a-f)	75710	5.5(18.2a-)	72004	687b-d
99010	100a	75710	15.2 (6.0a-f)	72004	5.5(18.2a-f)	75710	678b-f
99201	100a	75704	15.2 (6.0a-f)	99305	5.5(18.2a-f)	72104	676b-g
99404	100a	72104	15.0 (5.9a-f)	99404	5.5(18.1a-f)	72103	676c-g
99506	100a	99090	15.0 (5.9a-e)	99002	5.5(17.7a-h)	99008	647c-g
99613	100a	75707	15.0 (5.9a-e)	99405	5.4(17.5a-h)	72011	647fc-g
Bottom 10 families							
99102	71.4a-c	75719	10.7 (4.2o-s)	75719	4.3(14.0p-o)	75735	282p-x
81119	70.0a-d	75718	10.2 (4.0o-s)	75718	4.3(14.0p-o)	99501	240q-x
99006	68.8a-d	99501	9.9 (3.9p-s)	75735	4.2(13.7o-q)	99101	240q-x
99204	68.2a-d	99302	9.9 (3.9p-s)	99501	4.2(13.7n-q)	99302	235r-x
99502	63.6a-d	81113	9.9 (3.9p-s)	99302	4.1(13.5n-q)	75718	224t-x
81115	56.2b-d	99101	9.7 (3.8r-t)	81119	3.4(11.2r)	81113	187t-x
81116	50.0cd	81119	8.1 (3.2r-t)	81115	3.4(11.2r)	81115	180u-x
75718	50.0cd	81115	7.1 (2.8t-u)	81113	3.3(10.7r)	81119	155v-x
81122	38.9d	81116	6.9 (2.7u-v)	81116	3.1(10.3r)	81111	116w
81113	38.1d	81122	5.3 (2.1v)	81122	2.9((9.4r)	81112	80w
N	1167		1167		1167		1167
Mean	86.4		13.0 (6.0)		5.0 (16.5)		476
Sign.	< 1%		<1%		<1%		<1%
Range	38.1 - 100		5.3 - 17.2 (2.1 - 6.8)		2.9 - 6.0 (9.4 - 19.6)		80-906

Height

The tallest families were from central Nebraska. They were significantly different (<1% level) from the other families. Mean height of all sources was 5.0 m (16.5 ft) with a range of 2.9 to 6.0 m (9.4 to 19.6 ft). The top 10 sources (Table 3) differed by 0.6m (2.13 ft). Five sources from Jordan, Montana, (provenance 811) were significantly shorter than all of the remaining sources. The poorest 10families ranged in height from 2.9 to 4.3 m (9.4 to 14.0 ft) and were significantly shorter (1% level) than the best 71. There were no significant differences in height among the best 55 families.

A proceeding test to this one (Koop, 1987) using a much larger number of provenances found 720, 721 among the tallest seedlings at four years. A provenance test in North Dakota in 1968 and 1969 (Van Deusen, 1980) with many of the same provenances as in this study had similar results. At ages of 5 and 10 years, the best were 721, 757, 720. Many families from provenances 720 and 721 were among the tallest in our study. Ponderosa pine from this area should maintain height growth superiority whenplanted in the central Plains States. Evidently these sources contain genes endowing them with a broad range of site tolerances (Read, 1983). Sources from lower elevations

grew faster as reported by others (Mirov et al., 1952; Callaham & Hasel, 1961; Squillace & Silen, 1962; Hanover, 1963), but we did not observe this trend.

D²H (Volume Index)

The trees with the greatest size were the same as found for both diameter and height.

Volume index among sources, indicated by D²H values, was significantly different (1% level; Table 3). Mean D²H of all sources was 476 units with a range from 80 to 926. The best 51 sources differed significantly from the shortest24.Five sources from Jordan, Montana, were among the 10 sources with the smallest volume.

Correlation Analysis

Environmental factors did not influence tree growth as there was no strong relationship between annual precipitation and temperatures or elevation, latitude, and longitude. Diameter, height, and D²H values were highly significant; diameter and height at 0.77096, and 0.9345with D²H while height was 0.8056 with D²H. Read (1983) and Schafer & Baer (1985) have found a close relationship between juvenile-mature correlations.

Superior sources may be found at a fairly early age. Read (1983) noted that 2- and 3-year-old seedlings from the NE-SD areas among the tallest in his nursery study.

Summary and Conclusion

Fast-growing trees are desirable for establishment in the Great Plains. Ponderosa pine is often planted in homestead and field plantings in the Central and Northern Plains. Plains-wide studies conducted in the 1960s showed that sources from south central South Dakota and north central Nebraska were the best planting material. The present study was conducted to test a greater number of families to further refine selection areas for ponderosa pine sources. Materials from 75 individual trees were planted in the Central Plains sites for evaluation. Within this relatively small area, analyses indicated that ponderosa pine provenances affected growth of this 15-year-oldplantation. The best families came from central Nebraska. Environmental or geographic factors did not influence growth.

References

Baer, N, & Collins, P. (1979). Ten-year performance of a ponderosa pine provenance study in eastern South Dakota. South Dakota State University. Agricultural Experiment Station. TB-52, Brookings, 6.

Callaham, R. Z., & Hasel, A. A. (1961). *Pinus ponderosa* height growth of wind-pollinated progenies. *Silvae Genetica, 10,* 32-42.

Critchfield, W. B., & Little, Jr. E. (1966). Geographic distribution of the pines of the world .USDA. *Miscellaneous Publication, 99,* 97.

Flint, H. L. (1983). *Landscape Plants for eastern North America* (p. 677). New York: John Wiley & Sons.

Deneke, F. J, & Read, R. A. (1957). Early survival and growth of ponderosa pine provenances in East-Central Kansas. *USDA Forest Service, Research Note RM-297,* 4pp.

Hamiliton, D. A., & Rehfeldt, G. E. (1994). Using individual tree growth projection models to estimate stand-level gains attributable to genetically improved stock. *Forest Ecology and Management, 68,* 189-207.

Hanover, J. W. (1963). Geographic variation in ponderosa pine leader growth. *Forest Science, 9,* 86-95.

Lambeth, C. C. (1980). Juvenile-mature correlations in Pineaceae and implications or early selection. *Forest Science, 26,* 571-580.

Koop, R. F., Geyer, W. A., & Argent, R. M. (1987). Evaluation of ponderosa pine seed sources for the eastern Great Plains. *Journal Arboriculture, 13,* 139-144.

Read, R. (1983). Ten-year performance of ponderosa pine provenances in the Great Plains of North America. *USDA Forest Service Research Paper 250,* 17.

Meier Martinez, A., Mondino, V., & Gallo, L. (2004). Criteria for the selection of plus trees of ponderosa pine (*Pinus ponderosa*) and Oregon pine (*Pseudosugamenziesii)* for the establishment of seed orchards in the Andean Patagonia Region of Argentin. *Sgapya Forestal, 33,* 2-9.

Mirov, N. T., Dufield, J. W., & Liddicoet, A. R. (1952). Altitudinal races of Pinus ponderosa—A 12-year progress report. *Journal of Forestry, 50,* 825-831.

SAS Institute Inc. (2004). SAS 2002-2003 SAS/STAT User's guide. Version 9.1. Cary NC: SAS Institute Inc.

Schaffer, P. R., & Baer, W. (1985). Ponderosa pine provenance for windbreaks in eastern South Dakota.North. *Journal of Forestry, 2,* 105-107.

Schaffer, P. R., & Baer, N. W. (1992). Stability of ponderosa pine provenance: results after 21 years in eastern South Dakota. *Northern Journal of Forestry, 9,* 102-107.

Squillace, A. E., & Silen, R. (1962). Racial variation in ponderosa pine. *Forest Science Monograph, 2,* 27.

Van Deusen, J. L. (1980). Ponderosa pine provenances for the northern Great Plains. *USDA Forest Service, Research Paper RM-223,* 8.

Van Haverbeke, D. (1986). Genetic variation in ponderosa pine: A 15-year test of provenances in the Great Plains. *USDA Forest Service Rocky Mountain Forest Range Experiment Station Research Paper 80526,* 16.

Yow, T. H., Wagner, M. R., Wommack, D. E., & Tuskan, G. A. (1992). Influence of selection for volume growth on the genetic-variability of southwestern ponderosa pine. *Silvae Genetica, 41,* 326-333.

4

Community Structure, Diversity, Biomass and Net Production in a Rehabilitated Subtropical Forest in North India

Bajrang Singh[1], Karunakar Prasad Tripathi[2], Kripal Singh[1]

[1]National Botanical Research Institute, Council of Scientific and Industrial Research,
Rana Pratap Marg, Lucknow, India;
[2]Dolphin (PG) Institute of Biomedical and Natural Sciences, Manduwala, Dehradun, India.

Gangetic alluvial plain in north India constitutes significant proportions of barren sodic lands. A representative site, where afforestation was carried out during 1960s to rehabilitate the site under forest ecosystem, was selected to assess the restoration success. Three stands (S1, S2, and S3) were selected in a semi-natural subtropical forest at Banthra, Lucknow (26°45'N, 80°53'E) on the basis of different vegetation morphology and basal area gradient. Species composition and their growth forms were studied in overstory, understory and ground layer vegetation, in which dominants were assorted. Among the dominants few species were common in the three stands as also in different strata, which perhaps indicate their natural regeneration. Classification of individuals among the different size classes indicated "L" shape distribution in which most of the individuals remained confined in younger groups. Biomass increased from the stand S1 to S3 stand in overstory, and vise versa for understory. Stand S2 consisted of predominance of ground layer biomass over the other stands. Biomass allocation in different plant components differed significantly between the overstory and understory for aerial woody components (stem and branch). Annual litter fall did not differ significantly among the stands, where as fine root biomass (up to 45 cm soil depth) decreased from S1 to S3 stands. Rainy and summer seasons contributed to two-third proportion of total annual fine root production. The state of this rehabilitated forest when compared with the degraded and reference forest of the region indicated that structural complexity, biomass and production levels have been achieved to 70% of the reference forest site even after having a different species composition.

Keywords: Diversity, Community Structure, Concentration of Dominance, Biomass, Production, Litter, Fine Root

Introduction

Tropical forests are disappearing at an alarming rate of 13.5 million hectare per year globally Kobayashi (2004). In India about 20% of the geographical area is under forest in which tropical forests contribute nearly 83% of the forest area. Almost half of the forest area is classified as degraded forest with poor population density and species abundance. Deforestation and forest degradation are widely recognized as major threats to environmental stability, economic prosperity and social welfare and also to perform the statutory function of biodiversity conservation and ecosystem services. Often the forest management considers primarily commercially important monoculture species and rehabilitation of site for ecosystem/landscape management/species conservation or societal services assumes secondary importance. When the degraded and desolated lands do not turn out an economic yield, many sites are abandoned where natural succession proceeds and over a period of time several biotic communities colonize there and perform a variety of ecosystem function (Jha & Singh, 1991). The widespread degradation of alluvial soil in the Indo-Gangetic plains affected by varying degree of sodicity or salinity has received priority attention for afforestation during past few decades. It has been emphasized in the Indian forest policy to enhance the forest cover, biodiversity conservation and to provide the multiple goods and services to the ever-increasing human population of the country. Forest area is shrinking day by day and new forests are not being developed proportionally. It is estimated that about 53% of the total geographical area of the country is subjected to erosion and land degradation problems. Intensive af-forestation efforts are required to rehabilitate such sites under productive forest ecosystems. Both exotic and native species may be planted to rehabilitate degraded lands depending on site conditions (Singh et al., 2002; Datta & Agarwal, 2003; Singh et al., 2004; Singh & Singh, 2004; Shukla et al., 2011).

Natural succession remains arrested on sodic lands, and does not proceed automatically without some anthropogenic interventions. As a consequence, these sites do not have any significant vegetation growing on them except sporadic patches of some salt tolerant grasses. Creation of new forest on barren sodic land is therefore a critical task due to several soil constraints restricting the growth and development of plants. Mainly, an exotic *Prosopis juliflora* has been planted successfully (Bhojwaid & Timmer, 1998) which showed a fairly good adaptability to generate fuel wood quickly, but the drawback with this invasive alien species (exotic) is that *Prosopis julifora* does not accommodate the native species in their niche and overrides on the native species diversity. There is strong evidence that plantations can facilitate forest succession in their understories through modification of both physical and biological site conditions. Changes in light, temperature and moisture at the soil surface enable germination and growth of seeds transported to the site by wildlife and other vectors from adjacent forest remnants. It has been observed that at the stand level mixed species performed well for volume, basal area, biomass, and carbon sequestrations in comparison to pure monoculture stands (Piotto et al., 2003ab; Alice et al., 2004; Petit & Montazinimi, 2004, 2006). Economically viable and adoptable technology for afforestation of sodic land has been under experimentation for a long time (Sharma et al., 1992). Sandhu & Abrol (1981) studied the method of site preparation using augur holes and effect of establishment on *Eucalyptus tereticornis* and *Acacia nilotica* on sodic soil sites. Later on, these plantations suffered with the girdling, stress growth, and poor yield. Sodic soils contain exchangeable sodium in excess quantity which interferes with the growth of most crop plants and trees. The pH of such soils usually ranges from 8.5 to 10, which disturbs the ionic equilibrium of soil solution limiting to the growth of plants (Kelley, 1951). Such soils are generally poor in organic matter and nitrogen contents (Abrol & Bhumbla, 1971; Agrawal & Gupta, 1968; Khanduja et al., 1986; Pandey et al., 2011) and therefore its enhancement is vital for better growth and productivity (Shukla & Misra, 1993; Singh, 1996, 1998). Additionally the encrustation of calcium carbonate gravels and iron granules into a hard cemented bed in sub soil impedes not only root developments but water permeability too. Such characteristic properties resulting in root deformations, growth reductions and ultimately significantly lower yields do not support and pro- mote the extension of production forestry on sodic lands in commensuration with their inputs (Gupta & Abrol, 1990). However, (Abrol & Joshi, 1986) reported the economic viability of utilizing highly alkaline soil for the plantations of *Acacia nilotica*, *Eucalyptus tereticornis* but it could not gather enough momentum due to high initial investments and a large gestation period.

Afforestation with salt tolerant species was initiated from 1980 onwards (Sissay, 1986; Totey et al., 1987; Sharma, 1988). The afforestation trials those have been succeeded led to the identification and selection of tolerant species *in situ* on the basis of their growth performance (Khan & Yadav, 1962; Pandey, 1966; Srivastava, 1970; Yadav & Singh, 1970; Ahuja et al., 1979; Yadav 1980; Khoshoo 1987). Tolerance limit of several species was evaluated for sodic soils in pot culture experiments (Singh et al., 1994). Performance of trees at varying sodicity levels was also observed (Ashwathappa et al., 1986). Since the characteristics of salt affected soils vary greatly from place to place ranging from low to high ESP (exchangeable sodium percent) tolerant species were assorted accordingly (Yadav & Singh, 1970; Toth, 1981). Tolerance of *Dalbergia sissoo* in varying sodicity conditions was studied at different sites (Singh et al., 1990). Tree species are supposed to be more tolerant to adverse soil conditions, particularly *Prosopis* and *Acacias* (Garg & Jain, 1992). *Prosopis juliflora* can survive well on the calcarious soils which have on the average a maximum pH value of 9.5. *Acacia nilotica* was also found to be more resistant to soil salinity and sodicity (Yadav & Singh, 1970; Garg & Khanduja, 1979; Tomar & Yadav, 1980; Singh et al., 1986). Abrol (1986) suggested that, apart from identifying appropriate species and cultural practices, there is a need to evaluate the social and economic consequences of planting on saline land with fuel and forage species. A standardized silvicultural technology for afforestation on sodic lands was developed thereafter (Chaturvedi, 1985; Prasad & Sharma, 1990).

Earlier studies have been limited to the growth observations in height and diameter of the plants. Though some account is available for biomass and productivity of few trees grown on sodic soils (Chaturvedi, 1985; Dogra, 1989; Singh, 1991, 1998; Chaturvedi et al., 1991, Chaturvedi & Behl, 1996; Jain and Singh, 1998) that too pertains to juvenile stage of plant growth with a very little applicability to understand the community development process on a degraded land. Many previous trials failed to rehabilitate the sodic bare lands under tree cover due to lack of proven technology, proper financial support and

dedication (Yadav, 1975, 1980; Abrol, 1986). A man-made forest developed on sodic land at Banthra Research Station of the National Botanical Research Institute, Lucknow, India is the oldest successful endeavor of afforestation with multiple species. Very little information is available to our understanding in restoration of sodic land under forest ecosystem. Srivastava (1987) described the occurrence of species in this forest in which some were introduced, while others invaded naturally and colonized by the induced succession. Verma et al. (1982) reported that a mixed canopy cover was more effective in the reduction of pH than that of individual species. This is a good indication to diversify the monocultures with various indigenous species which also aid to counter the effects of epidemic and allelopathy. A generalized impact of soil reclamation in this forest was assessed by (Singh, 1996, 1998).

The tolerant tree and shrub species made the soil hospitable for less tolerant species. Thus a portion of land that was once totally barren and desolate is now recognized as a functional forest ecosystem with a top story of trees, middle story of small trees and shrubs; and a ground layer of herbs and seedlings of the perennials. Structure and composition of overstory trees determine the understorey vegetation diversity and their complexity (Barbier et al., 2008). Basic changes in demography, tree size and growth comprise classic descriptors of stand development (Brinkley, 2004). Efforts were made to observe the relationships between vegetation productivity and species diversity (Sculze & Mooney, 1993; Huston, 1997; Chapin et al., 2000; Loreau et al., 2003; Liang et al., 2007). Such forest inventories are useful to rehabilitate the other barren sites in an efficient way as well as to ensure a desired composition and structure of the developing forest which could be self sustainable even after rational extraction or mild disturbances from environmental perturbations.

This study was carried out in a 40-yr-old rehabilitated forest on sodic land to identify the development of community structure and productivity levels, their diversity, dominance and compatibility with natural forest of the region. An attempt was made to characterize the various species performance and their interrelations in the constitution and development of new "biotope".

Methods and Materials

Site

The forest was established on abandoned sodic soil during the 1960s at Banthra, Lucknow, situated in a subtropical, semiarid region of north India (26°45'N, 80°53'E). Geographically, this region is classified as Gangetic alluvial plains of the Uttar Pradesh state due to transported deposition of minerals from the Himalayan rocks by the Ganga River. A large tract of this region consists of cultivated land interspersed with barren sodic land measured about 1.3 million hectares (Figure 1). The prehistoric natural forests in this region were sparse and most of them were replaced by Sal (*Shorea robusta* Gaertn. f.) and Teak (*Tectona grandis* L. f.) forests in the middle of the nineteenth century. A few patches of those forests are still available in Katarniaghat Wildlife Sanctuary, in Behraich (lat 27°55'N, long 81°25'E) district of U.P. The natural dry tropical forests of Varanasi (lat 24°55'N, long 83°3'E) and Mirzapur (lat 24°55'N, long 82°32'E) were compared as degraded forest sites. Average annual rainfall at Lucknow ranged from 840 to 980 mm during the past 10 years, which is slightly less than that at the

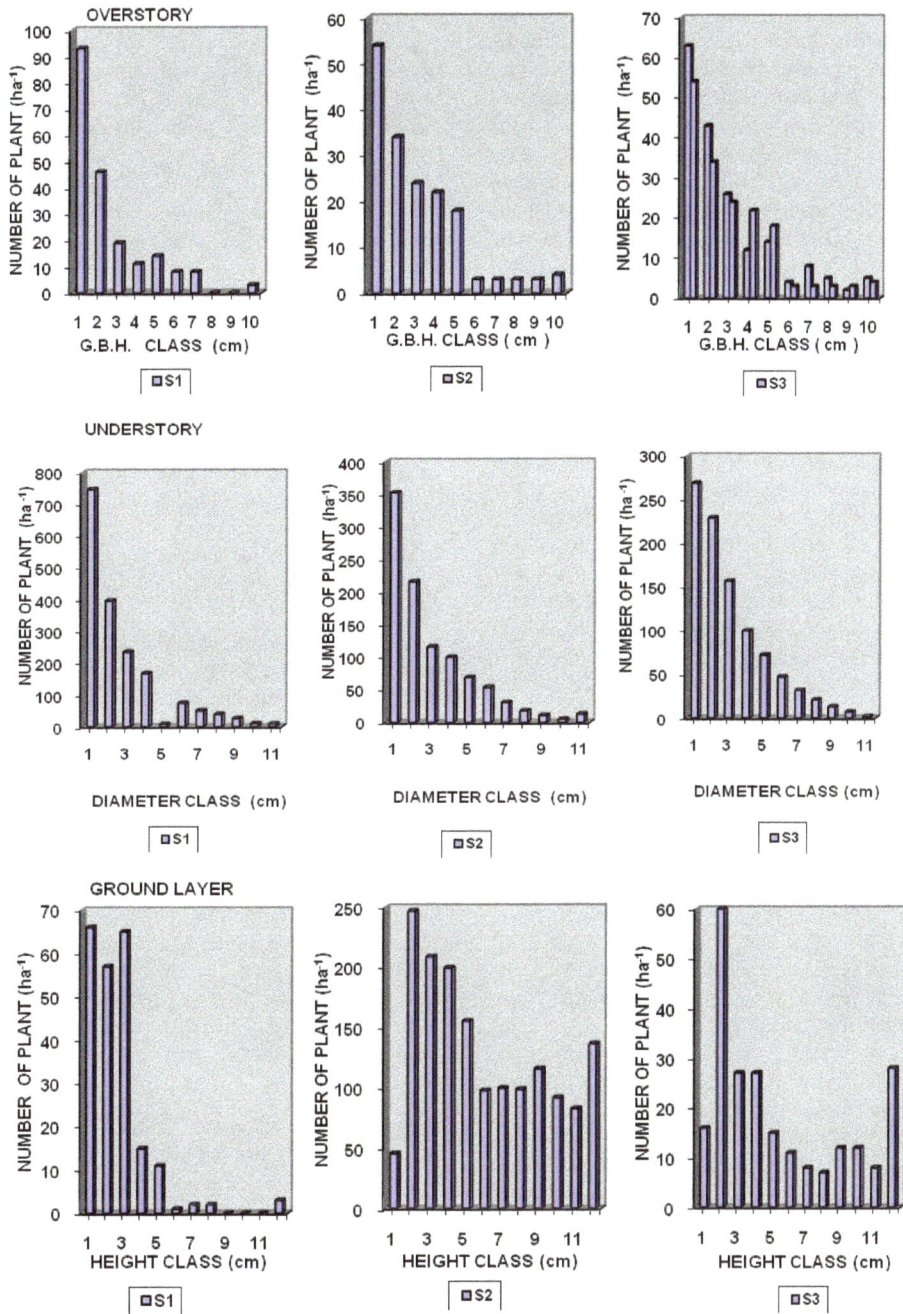

Figure 1.
Distribution of plant population in different size classes in three vegetation layers in three stands of 40 yr old man-made forest. Overstory classes Girth (cm): 1 =20 - 40, 2 = 40 - 60, 3 = 60 - 80, 4 = 80 - 100, 5 = 100 - 120, 6 = 120 - 140, 7 = 140 - 160, 8 = 160 - 180, 9 = 180 - 200, 10 ≥ 200. Understory classes Diameter (cm): 1 = 0 - 1, 2 = 1 - 2, 3 = 2 - 3, 4 = 3 - 4, 5 = 4 - 5, 6 = 5 - 6, 7 = 6 - 7, 8 = 7 - 8, 9 = 8 - 9, 10 = 9 - 10, 11≥ 10. Ground layer classes Height (cm): 1 = 0 - 10, 2 = 10 - 20, 3 = 20 - 30, 4 = 30 - 40, 5 = 40 - 50, 6 = 50 - 60, 7 = 60 - 70, 8 = 70 - 80, 9 = 80 - 100, 10 = 100 - 110, 11 ≥ 110.

reference site (1057 mm/year). More than 80% of the precipitation occurs in the monsoon season (July-September), and the remainder of the year is dry. Average minimum and maximum temperature differ significantly from winter (8°C, night; 20°C, day) to summer (27°C, night; 40°C, day), indicating a seasonally distinct climate. Average relative humidity was 63% during the last 10 years.

The site soil is an Inseptisol (Typic Natrustalf) with silty clay loam texture. A compact layer of indurated pan comprising $CaCO_3$ gravels and iron granules exists between 0.4 m and 0.8

m depths in these sodic soils. Structural degradation of the heavy (high bulk density) and impervious soil leads to crusting in winters and waterlogging condition during rainy seasons. During summer, efflorescence of $NaCO_3$ salt occurs as a powdery layer on the soil surface. Consequently, suspended particulate matter is quite high in the atmosphere during the day. The soil was characterized by a high pH (>10) and exchangeable sodium percentage (ESP) (>50) and low electrical conductivity (EC) (<1.0 dS/m) and organic carbon content (<0.2%) prior to planting (Garg, 1987). Carbonate and bicarbonate of Na and Ca were the dominant ions. When the content of soluble salts (EC) is low and exchangeable Na high, the physical condition of the soil is usually unfavorable for the tillage as well as establishment and growth of desired plants. As a consequence, only a few grasses, viz *Sporobolus* and *Desmostachia*, are found sporadically under natural conditions. Attempts were made to rehabilitate such barren land through afforestations as well as other land use systems over an area of about 50 ha acquired during the 1960s. The entire area was demarcated with barbed wire fencing and designated as Banthra Research Station (BRS). Some of the native tree species commonly occurring in tropical forests of north India (*Acacia nilotica, Albizia lebbeck, Albizia procera, Bauhinia variegata, Ficus bengalensis, F. rumphii, Syzygium heyneanum, Syzygium cumini, Terminalia arjuna, Derris indica*) were planted in plantation pits of 1 m^3 that had been filled with a mixture of soil, compost manure, and decomposed leaf litter in 2:1:1 proportion on a 5-ha site. They were also planted along the marked boundary of the BRS. The initial population density was around 1,000 trees/ha. A drainage channel 1.5 m deep and 2 m wide was developed around the plantation site to prevent water logging, which was intensively observed during the rainy season for a proper drainage of stagnating water. Mortality during the initial years was greater than 50%, and trees were replanted in consecutive years. Several species invaded and colonized this area over time through natural succession process due to changes in microrelief. Seed dispersal and natural regeneration of trees and many other species gradually extended to cover about 17 ha of total forest area in the 2010.

Vegetation Analysis

Three stands were selected in this forest according to gross morphology and a basal area gradient within the original 5-ha revegetated area. Sample plots, each 1 ha, were marked in all three stands denoted as S1, S2, and S3. Vegetation analysis was carried out on belt transects (10 m wide). The method and quadrat size were standardized using a species–area-curve relationship. Thirty-four quadrats of 10 × 10 m along three transects spaced 10 m apart were laid out contiguously in each stand. Plants were enumerated and measured for growth parameters. Ninety-five percent of the species were identified through the use of the National Botanical Research Institute's herbarium records. Girth of trees (>20 cm·gbh) was measured at 137 cm above the ground for overstory species, whereas diameter of young trees and shrubs occupying less than 10 cm dbh was measured 50 cm above the ground, using an electronic vernier caliper. These were classified as understory species. Height of small seedlings less than 50 cm and of herbaceous species were measured and placed in the ground layer community. Species structure (frequency, density, abundance, basal area/cover, importance value index (IVI), etc.) was determined from the field data (Misra, 1968). The cross-sectional area of the stem at measured levels is the basal area of all woody species. Leaf area cover (basal cover) for the ground layer was computed by specific leaf area ratio, which is defined as area per unit weight of the leaf (Misra, 1968). Species having greater than 10 IVI or >10% of the total basal area were considered dominant species in each stratum.

Species Diversity Index

Species diversity index (H) for overstory, understory, and ground layer vegetations were determined separately from the Shannon Wiener's information function (Shannon & Weaver, 1963).

$$H = -\sum (ni/N)\overline{log}_2 (ni/N) \qquad OR - \sum pilog_2 pi$$

where: ni = importance value for each species, N = total of importance values

pi = importance probability for each species = ni/N

Concentration of Dominance

Concentration of dominance (Cd) was measured by Simpson's index which is also known as index of dominance (Simpson's 1949).

Index of dominance (Cd) = $\Sigma (ni/N)^2$

where: ni = importance value for each species, N = total of importance value

Equitability or Evenness (e)

Equitability refers to the degree of relative dominance of each species in that area. Following Pielou (1966), equitability or eveness index was calculated as:

Equitability (e) = H/logs

Where: S = number of species and H = Shannon Wiener index

Productivity Assessment

All the stems were classified in 7 girth/diameter classes in overstory and understory vegetations and 7 height classes in ground layer. Overstory biomass was estimated by a common regression equation already developed from 13 tree species in a MAB project at BHU, Varanasi, because the permission for harvesting of green trees was not granted by the UP Forest Department in view of the felling restrictions on the trees planted under restoration programmes. Harvesting of sample plants of the understory could be possible and therefore three representatives from each size classes were sampled for stem, branch, and leaf and root biomass. Species of ground layer were sampled according to the height classes for stem, leaf and root components. Regression equations were developed with the help of data of sample plants for diameter (cm) on "x" axis and oven dry weight (g) of the particular component on "y". The form of regression was:

$$\ln Y = a + b \ln x$$

A software programme "SYSTAT 9.0 SPSS" was used for regressions. With the component biomass, total biomass of all species per unit area was computed as per their respective population density. The ground layer species biomass was computed with their equation where "x" variable was heights of plant in cm. Net productions were obtained as the differences in biomass during the two consecutive years (2007 to 2009. Biomass of ground layer was assumed to net production in view of their little contribution in total forest ecosystem productivity.

Litter

Litter fall was collected monthly in 1 m^2 trays during the year. Four trays were placed in each of the three stands. Components of litter were separated out as leaf, twig, flower-fruits and bark. Apart from, six quadrats of 1 m^2 were laid on underneath each stand to sample forest floor litter layers as L, F and H fractions.

Fine Root

Fine root biomass was extracted by wet sieving of ten soil cores of 100 cm^3 in each stand at two depths (0 cm - 15 cm and 15 cm - 30 cm) in five seasons. Similarly fine root production was estimated by establishment of root free in-growth cores in each season and extraction of root by wet sieving. These roots were classified into three diameter classes (<0.5, 0.5 mm - 1 mm and 1 mm - 2 mm). The live and dead roots were separated on the basis of gross morphology and degree of cohesion between cortex and periderm according to Vogt and Persson (1991). Both fractions were oven-dried at 80°C to constant weight.

Results

Community Structure and Diversity

There were only a few species which contributed more than 10% of the total basal area in each stand; therefore, dominants were assorted including >10 IVI (Table 1). On the basis of IVI and relative basal area, S1 stand constituted more dominant species in comparison to other two stands. Among the dominants few species were common in all the three stands such as *Syzygium cumini, Syzygium heyneanum, Streblus asper, Azadirachta indica, Albizia lebbeck* in overstory vegetations. In understory the common dominants among the stands were *Lantana camara, Streblus asper, Syzygium cumini*. Besides, some of the species were found common in all the three vegetation strata viz. *Leucaena leucocephala, Sterculia alata, Streblus asper, Syzygium cumini, Syzygium heyneanum*. But *Clerodendrum vescosum, Ichnocarpus frutescens*, and *Putranjiva roxburghii* remained confined to under stay & ground layer vegetation.

Species distribution pattern and their natural associations provide the clues for rehabilitation of barren sodic land under forest ecosystems. Basal area increased from S1 to S3 stand in overstory, whereas in understory it was decreased in same order. In ground layer, maximum basal cover was found in S2 stand. In overstory vegetation of S1 stand, dominant species covered 44.2% of the total basal area which indicates that other species have no less importance in organization of plant communities occupying rest of the 56% of the total basal area. In S2 stand other species contributed relatively less with the proportions of dominants contributing to 72% of the total basal area of overstory vegetation. S3 stand had almost similar value to that of S1 stand. In S1 stand, *Albizia lebbeck* and *Azadirachta indica* had 30% of the total basal area. In S2 stand *Albizia lebbeck, Syzyum heyneanum* and *Terminalia arjuna* hold about 51% of the total basal area. In S3 stand *Albizia lebbeck, Albizia procera, Ficus rumphii* consisted of about 45% of the total basal area. In understory vegetation, *Leucaena leucocephala* had greatest IVI in S1 stand. Dominant understory vegetation constituted 61%, 72% and 52% of the total understory basal area of the respective S1, S2 and S3 stands. In ground layer vegetation, S1 stand had more dominant species in comparison to other stands with greatest IVI in S2 stand. In S1 stand ground layer vegetation of the dominants occupied 87% of the total basal cover. S2 and S3 stands had 89% and 95% basal cover of dominants.

Population size of overstory and understory trees was largest in S1 stand, whereas S2 stand consisted of the greatest number of individuals of the ground layer (Table 2). However, population size on the basis of number of individuals per unit area does not contribute much in ecosystem function, as it does not give any additional weight to the size of individuals. Basal area is a composite function of the number and size of the individuals

Table 1.

Classification of species as a percent of the total species represented by their populations from unit to thousands in respective vegetation strata.

Strata	Unit	Tens	Hundreds	Thousands	Total
Overstory	37	38	23	2	100
Understory	36	36	23	5	100
Ground layer	25	37	25	13	100

Table 2.

Population size and plant diversity in a 45-yr old rehabilitated forest community developed on barren sodic land.

Parameter	Form	Stands			Mean ± SE
		S1	S2	S3	
	Overstory	610	517	535	554 ± 28
Population density (No. ha^{-1})	Understory	5554	2871	2759	3728 ± 913
	Ground layer	2190	15930	2320	6813 ± 4558
	Overstory	25.8	30.5	33.6	29.9 ± 1.9
Basal area (m^2·ha^{-1})	Understory	7.25	3.13	2.31	4.20 ± 1.2
	Ground layer	5.39	271.5	58.18	111.6 ± 66.4
	Overstory	35	27	28	30 ± 2
Species richness (number)	Understory	38	40	30	36 ± 1.6
	Ground layer	15	17	13	15 ± 0.47
	Overstory	1.29	1.0	1.09	1.1 ± 0.07
Equitability (e)	Understory	0.72	1.06	0.99	0.9 ± 0.08
	Ground layer	0.65	0.52	0.87	0.7 ± 0.08
	Overstory	3.99	3.25	3.64	3.6 ± 0.17
Shannon Wiener's index (H)	Understory	2.65	3.80	3.35	3.3 ± 0.27
	Ground layer	1.76	1.45	2.24	1.8 ± 0.19
	Overstory	0.11	0.18	0.13	0.14 ± 0.02
Concentration of dominance	Understory	0.40	0.14	0.15	0.23 ± 0.07
	Ground layer	0.45	0.46	0.29	0.40 ± 0.05

per unit area (1 ha) which shows the relative contribution of the species in structure and function of the forest ecosystems. On the basis of basal area, three stands contributed in different proportions for each vegetation types. The predominance of overstory basal area in S3 stand and understory in S1 stand and a relatively large basal cover of ground layer vegetation in S2 stand differentiates these stands each other in the community structure and its organization during developmental process. However, these sites differences do not appear to be significant in species richness. It is because of the fact that various tree species planted initially are now representing presently the overstory. Whereas, understory and ground layer mainly consist of the progeny of same species along with the few natural invaders.

Measurement of biodiversity in a specific area (local scale) on the basis of richness does not provide a complete understanding about the individuals of the species as it suffers from the lack of evenness or equitability (e). Richness index decreased from overstory to understory by 50% followed by the ground layer. It ranged from maximum 2.47 (S1 stand) in overstory, to minimum 0.43 (S2 stand) in ground layer vegetation. Equitability (e) was greatest in S1, S2 and S3 stands, for overstory, understory and ground layer vegetations respectively. Both indices decreased from overstory to ground layer on average of stands. The species were more evenly distributed in ground layer showing lowest in comparison to the overstory vegetation, so it was decreased from overstory to ground layer vegetation. Shannon Wiener's index (H) is one of the most popular measures of general species diversity in a forest. This index decreased from overstory to ground layer in accordance with the richness index. In this forest, Shannon Wiener's diversity index ranged from 3.99 (S1 stand) in overstory, to least 1.45 (S2 stand) in ground layer vegetation (Table 2). Although

S2 stand had greatest ground flora even then their diversity index was lowest for the ground flora which indicates that the ground flora consisted of only few abundant species such as *Barleria prionitis*, *Blepharis maderaspatensis* and *Clerodendrum vescosum*. Concentration of dominance (c) = Σ pi^2 indicates that the dominance was more concentrated with fewer species in the ground layer in comparison to overstory vegetation where dominance was shared by multiple species. It showed an opposite pattern with the diversity index among the three vegetation types.

Entire plant population of the three vegetation strata were classified over a range of size classes according to girth (overstory), diameter (understory) and height for ground layer (Figure 2). In general, these stands appear to be still immature and ecosystem has not yet reached equilibrium. Their configuration and structure indicate a "L" shape positively skewed asymmetrical distribution of plant populations with increasing size of girth and diameter for both overstory and understory vegetation, respectively. Population of the understory vegetation was nearly half in S2 and S3 stands in comparison to S1 stand, whereas S2 stand predominated in ground layer vegetation. Population of ground layer vegetation was almost evenly distributed across the height class in S2 and S3 stands, whereas in S1 stand most of the population was confined to first few groups only. It was observed that the number of species decreased with increasing plant size in each of the three vegetation forms. In overstory, maximum number of species occurred in initial girth class (20 cm - 40 cm) in all the three stands. Similarly in understory species richness was greatest in initial diameter class. Such types of species composition depicted that species diversity reduced among older individuals with the growth and developments of plants. The pattern of species area relations was almost similar for each of the three stands.

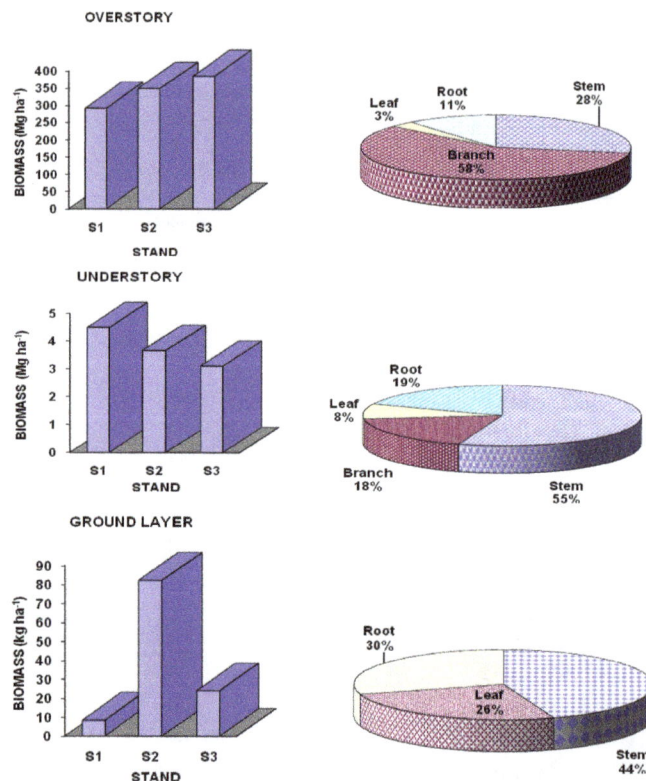

Figure 2.
Biomass of three vegetation layers along with proportional contribution in different components in three stands of 40 year old man made forest.

Stand Biomass

Overstory biomass increased from 292.8 Mg·ha^{-1} (S1 stand) to 386.2 Mg·ha^{-1} (S3 stand) corresponding to increase in 30% basal area from S1 stand to S3 stand (Figure 3). The contribution of plant components in total biomass was 28% stem, 58% branches, 3% leaf and 11%root. Typically branches cover more biomass than boles in most of the tropical forests. The contribution of root in this forest was relatively low in comparison to an average root proportions of about 20% of total biomass found in most of the natural forests. It may be due to the soil compactness and presence of a calcic compact (calcium carbonate gravels) layer in subsoil. As a consequence roots do not spread and proliferate freely. The understory biomass slightly decreased from 4.5 Mg·ha^{-1}, (S1 stand) to 3.12 Mg·ha^{-1} (S3 stand) corresponding to decrease in basal area (Figure 3). The percent contribution of the total biomass in different components of understory vegetation showed about 55% in stem, 18% in branches, 8% in leaf, and 19% roots. Biomass distribution pattern among the components in understory vegetation changed to that of overstory vegetation. Besides, pattern of stand biomass also reversed among stands showing a decrease from S1 to S3 stand. Total biomass of ground layer vegetation decreased from 82.3 kg·ha^{-1} (S2 stand) to 8.4 kg·ha^{-1} (S1 stand) corresponding to decrease in basal cover from 271.49 m^2·ha^{-1} (S2 stand) to 5.39 m^2·ha^{-1} (S1 stand). The percent contribution of different components of ground layer vegetation in stem, leaf and root was 44%, 26% and 30% respectively. The root/shoot ratio increased considerably from overstory (0.12) to understory (0.23) and ground layer (0.42) vegetations.

Net Production

Mean annual increment in biomass (MAI) of overstory vegetation had little difference within stands, but current annual increment (CAI) and net primary production (NPP) were greatest in S2 stand (Figure 3). Thus, a stand superior in total biomass (S3) could not succeed as well in net production although with minor reduction, which may be due to the fact that S3 stand would have been optimized maximum production during 45 yr. However, at the same time MAI, CAI and NPP of understory vegetation were in the order of S1 > S2 > S3 stand in accordance with their biomass. Overstory production contributed to 95% of total forest production in a year. Understory and ground layer contributed to 4% and 1% of the total forest production, respectively.

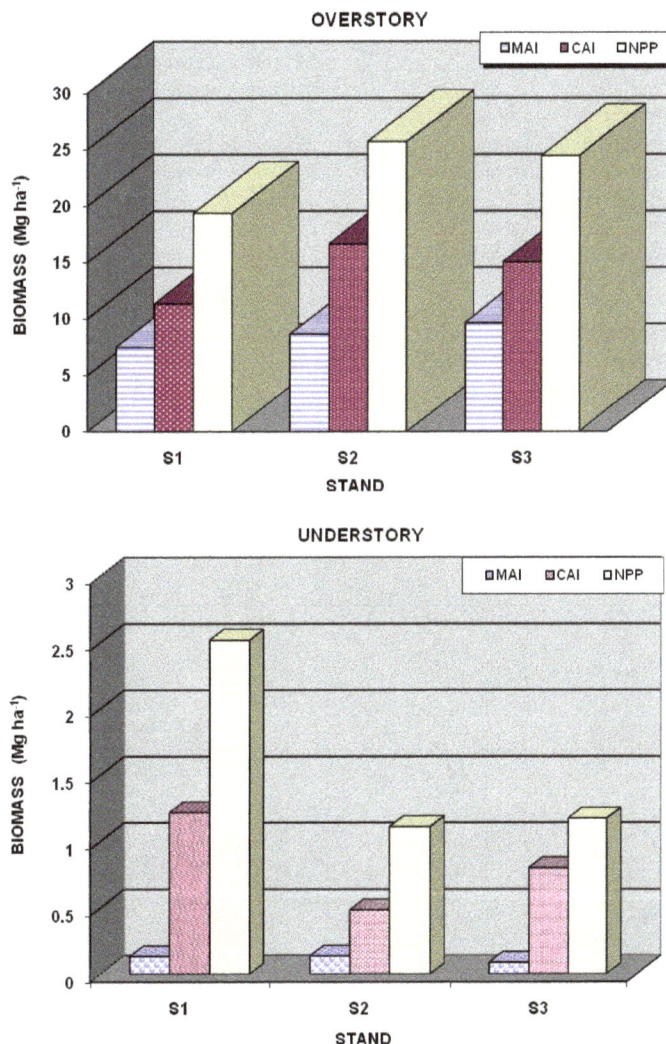

Figure 3.
Annual increment in dry matter of overstory and understory in three forest stands. MAI = mean annual increment, CAI = current annual increment, NPP = net primary production.

Litter and Fine Root

Litter Fall

Litter fall is an important flux to transfer the organic matter and nutrients from trees to the soil which maintains soil sustainability in general and reclamation of sodic soils in this particular case. About 8 $Mg \cdot ha^{-1} \cdot yr^{-1}$ - 9 $Mg \cdot ha^{-1} \cdot yr^{-1}$ litter fall occurred among the stands of the 40 yr-old forest under study (Table 3). Maximum input of litter to the soil was recorded in S2 stand and minimum in S3 stand although these were not significantly different. In all stands, litter was separated in four components in varying proportions. On stand mean basis leaf, twig, flower-fruits and bark contributed to 68%, 20%, 9% and 3% of the total litter fall, respectively. Earlier, a relatively high annual litter fall of 11.2 $Mg \cdot ha^{-1} \cdot yr^{-1}$ was reported in the same forest (Singh, 1996). Thus annual litter fall decreased slightly with further maturity of forest. Since the forest constituted both deciduous and evergreen species, litter fall occurred throughout the year with its maximum quantity in March due to more deciduous species. Seasonal variation in litter fall indicated a relatively high proportion in summer and spring seasons, each about 25% of the annual flux. Generally the wind velocity increases in these seasons which promote the abscission of mature and old leaves and dead twigs and branches. Least amount of litter fall was observed in autumn season (12% of annual).

Forest Floor Litter

Although litter fall and decomposition are instantaneous processes but entire litter fall during one year does not decompose completely during the next year and a substantial amount remains to be decomposed in subsequent years. The residue accumulated on the forest floor was estimated as 6.3 $Mg \cdot ha^{-1}$ under different stage of decay forming ecto-organic layers (L, F and H) on mineral soil. The magnitude of forest floor was relatively large in S3 stand corresponding to their standing biomass and small in S2 stand (Table 3). The distribution of organic matter in these layers varied in different proportions such as 38% (L), 29% (F) and 33% (H) of the total forest floor litter on the basis of seasonal mean. This litter serves as 'nutrient bank', checks evaporation and conserves soil moisture to optimize the microbiological activity in the underlying mineral soil.

Fine Root Biomass

Fine roots play a crucial role in water and minerals uptake and moisture retention in the soil. On account of their ephemeral nature, fine roots may also contribute to soil organic matter and carbon sequestration. Besides, fine roots act well in reclamation of sodic soil by reducing the pH, improving the soil structure, porosity and water permeability of heavy impervious sodic soils. Fine root biomass in different season (rainy, autumn, winter, spring and summer) was extracted with respect to soil depth, intermittently at 15 cm intervals up to 45 cm. Total fine root biomass (live and dead) was 532 $g \cdot m^{-2}$ up to 0 cm - 45 cm depth in which live and dead proportions were classified as 93% and 7% respectively (Table 4). Fine root biomass decreased with depth up to 45 cm. Total fine root biomass varied significantly ($P < 05$) between the stands with maximum in S1 stand and minimum in S3 stand. About 47% of total fine root biomass occurred in superficial 0 cm - 15 cm depth. Fine root biomass decreased significantly from rainy to summer season. Total fine and small roots were classified into 5 diameter classes from < 1 mm to 25 mm. Biomass of fine roots increased with size class in all the three stands accounting maximum in 10 mm - 25 mm size class ranging from 27% (S3 stand) to 42% (S1 stand) and minimum in the 1 mm - 2 mm class with 7% (S1 stand) to 12% (S2 stand).

Table 3.
Annual litter fall in a rehabilitated forest on sodic land at Banthra, Lucknow (Mg·ha⁻¹).

Components	Stands			
	S1	S2	S3	Mean ± SE
Leaf	6.18	5.43	5.57	5.73 ± 0.230
Twig	1.49	2.02	1.71	1.74 ± 0.154
Flower and Fruits	5.70	1.23	4.70	7.57 ± 0.238
Bark	1.00	1.30	4.60	2.30 ± 0.115
Total	8.34	8.81	8.21	8.45 ± 0.182
Forest floor litter	6.00	5.40	7.50	6.30 ± 0.28

Table 4.
Fine root biomass in a rehabilitated forest on sodic land (g·m⁻²).

Depth (cm)	State	Stands			
		S1	S2	S3	Mean ± SE
0 - 15	Live	273 ± 88.67	195 ± 51.5	210 ± 71.39	226 ± 23.89
	Dead	17 ± 2.63	33 ± 17.14	17 ± 8.15	22 ± 5.33
15 - 30	Live	283 ± 134.7	223 ± 84.73	140 ± 55.14	215 ± 41.45
	Dead	5 ± 1.53	16 ± 9.57	4 ± 1.24	8 ± 3.84
30 - 45	Live	93 ± 37.3	47 ± 13.14	30 ± 11.22	57 ± 18.8
	Dead	6 ± 3.5	2 ± 0.36	3 ± 0.95	4 ± 1.2
Total	Live	649	465	380	498 ± 79.38
(0 - 45)	Dead	26	51	24	34 ± 8.68
	Total	675	516	404	532 ± 79.22

Fine Root Production

Fine root production was less than half (233 g·m^{-2}·yr^{-1}) to that of their biomass up to a depth of 30 cm (Table 5). About two third i.e. 67% of the annual fine root production was measured during rainy and summer season. Corresponding to biomass, fine root production was also less in lower strata of 15-30 cm depth (101 g·m^{-2}) in comparison to surface soil of 15 cm (132 g·m^{-2}). It appears that about 70% of the total annual production of fine roots undergoes to mortality. However, in the periodic in-growth core extractions, fractions of dead roots were found to only 20% of the total annual production during different seasons, but cumulative mortality during the year would be many more times. Fine root production differed significantly between the stands and it decreased from S3 to S1 stand.

Discussions

Rehabilitation of barren sodic land under forest almost performs the same ecological functions to that of their natural allies despite of in different species composition plant community structure and species abundance might be different, yet the degraded state was renewed to an extent where life support systems could be operated by the several tropical biotas. However, the process was enough difficult to create a healthy ecosystem on account of many soil constraints commonly found in sodic soils which inhibit the establishment and growth of plants (Gupta & Abrol, 1990; Sumner, 1993; Naidu & Rengasami, 1993; Garg, 1998), nevertheless it was successful effort to accommodate a wide range of species in the new biotope. Knowledge of the species adaptation to such sites and process involved in succession can be utilized for rehabilitating the other similar sites in general and sodic ones in particular. This study, however, does not indicate temporal changes, yet it examines the diversity of a rehabilitated forest from several angles which exerts a strong influence to restore stability and resilience of rehabilitated ecosystem. Forest developed in such a way under protection forestry programs is important to determine the rate of restoration, structural and ecological diversities, establishment of steady state in biogeochemical cycle and patterns towards the climax communities. Any major disturban-

ces may destabilize the building of niche (composition, aggregation and organization) which may extend the restoration process.

Community Structure

A semi natural forest developed on sodic soil constituted 74 species belonging to 35 families. These species were classified in overstory (44), understory (19), ground layer (8) and climber (3). Several species of overstory are also found in understory and ground layer vegetation which are supposed to be offspring of the parents at different growth stage. Several parameters of this rehabilitated forest have been compared with a degraded and reference forest of the region (Table 6). Plant population density decreased from degraded, rehabilitated to reference forest, respectively for overstory species but understory population in our rehabilitated forest was exceedingly high due to biotic protection. The average basal area (30 m^2·ha^{-1}) of the forest studied lies in between the degraded (Jha & Singh, 1990, Singh & Singh, 1991) and reference (Tripathi & Singh, 2009) forests. It appears that the carrying capacity for supporting the tree stock on sodic soils differs with species ranging from 12 to 38 m^{2}ha^{-1} basal area in *Acacia nilotica* and *Eucalyptus camaldulensis* plantations at same age (Singh et al., 2000). These values compare fairly well with 17 m^2·ha^{-1} - 40 and 20 m^2·ha^{-1} - 75 m^2·ha^{-1} for dry and wet forests of the world, respectively (Murphy & Lugo, 1986b). The basal area of our rehabilitated forest is also well comparable to that of deciduous and evergreen forest (16 m^2·ha^{-1} - 33 m^2·ha^{-1}) of Brazil (Haase 1999), beech forest 28 m^2·ha^{-1} of Japan (Nagaike et al., 1999). Basal area of hardwood forest in USA was relatively high in the range of 40 m^2·ha^{-1} - 45.5 m^2·ha^{-1} (Gilliam et al., 1995) which might be expected with better silvicultural management on a good soil type.

The species richness of overstory vegetation was relatively high from understory and ground layer. Number of species in overstory can also be compared well with rainforest (43), moist forest (45), temperate forest (45) in one hectare plot area (Brockway, 1998). The classification of species as a percent of the total species in respective vegetation strata according to their population in numerals revealed that there were only a few species representing their populations in thousands and a high percent of total species had their individuals either in units or tens as under (Table 1).

Table 5.
Fine root production in a 45-yr old rehabilitated forest on sodic soil (g·m^{-2}).

Depth (cm)	State	Rainy (%)	Autumn (%)	Winter (%)	Spring (%)	Summer (%)	Total annual
0 - 15	Live	35(33)	11(10)	8(8)	10(9)	43(40)	107
	Dead	7(28)	5(20)	2(8)	3(12)	8(32)	25
15 - 30	Live	31(38)	13(16)	6(8)	6(8)	24(30)	80
	Dead	3(14)	10(47)	2(10)	2(10)	4(19)	21
Total (0 - 30)	Live	66(35)	24(13)	14(7)	16(9)	67(36)	187
	Dead	10(22)	15(32)	4(9)	5(11)	12(26)	46
Grand total		76(33)	39(17)	18(7)	21(9)	79(34)	233

Table 6.
Comparative evaluation of vegetation structure and productivity in degraded, rehabilitated and reference forest.

Parameters	Vegetation strata	Degraded forest*	Rehabilitated forest**	Reference forest***
Population density (No/ha)	Overstory	1055 ± 68.7	554 ± 28	49 ± 45
	Understory	343 ± 121	3728 ± 913	245 ± 116
	Ground layer	-	6813 ± 456	4015 ± 730
Basal area (m²·ha⁻¹)	Overstory	16.53 ± 0.82	29.9 ± 1.9	35.22 ± 6.54
	Understory	1.34 ± 0.47	4.2 ± 1.2	14.00 ± 8.20
	Ground layer	-	112 ± 66.4	274 ± 61
Species richness	Overstory	9 ± 0.24	44 ± 3.1	16 ± 1.76
	Understory	8 ± 0.28	18 ± 1.6	25 ± 2.5
	Ground layer	-	12 ± 0.72	21±5.78
Richness index	Overstory	0.39 ± 0.11	2 ± 0.1	1.74 ± 0.29
	Understory	0.67 ± 0.09	1 ± 0.1	1.50 ± 0.57
	Ground layer	-	0.8 ± 0.1	0.91 ± 0.21
Equitability	Overstory	0.72 ± 0.11	1.1 ± 0.07	0.93 ± 0.03
	Understory	1.06 ± 0.05	0.9 ± 0.08	0.77 ± 0.02
	Ground layer	-	0.7 ± 0.08	0.73 ± 0.06
Shanna Wiener's Index	Overstory	1.07 ± 0.32	3.6 ± 0.17	1.84 ± 0.14
	Understory	1.98 ± 0.09	3.3 ± 0.27	2.27 ± 0.17
	Ground layer	-	1.8 ± 0.19	2.23 ± 0.18
Concentrate of dominance	Overstory	0.58 ± 0.107	0.14 ± 0.02	0.18 ± 0.02
	Understory	0.31 ± 0.04	0.23 ± 0.07	0.15 ± 0.02
	Ground layer	-	0.40 ± 0.05	0.186 ± 0.01
Biomass (Mg·ha⁻¹)	Overstory	89.87 ± 8.62	343±27	480 ± 37
	Understory	4.28 ± 0.34	3.8 ± 0.4	70 ± 5
	Ground layer	0.82 ± 0.15	0.04 ± 0.02	5 ± 0.2
Net production (Mg·ha⁻¹·yr⁻¹)	Overstory	12 ± 1.2	23±2.1	35 ± 2.5
	Understory	0.9 ± 0.15	1.2 ± 0.15	15 ± 1.6
	Ground layer	1.1 ± 0.17	0.05 ± 0.01	7 ± 1.1

*After Singh and Misra (1979) and Singh and Singh (1991). **Present study. ***Tripathi and Singh (2009).

About 28 species were found naturally regenerating in this forest which was about 64% of the total species listed in the forest. All these species may be considered to be well adopted in sodic soils viz. *Aegle marmelos, Alangium salvifolium, Albizia lebbeck, Azadirachta indica, Bauhinia variegata, Cassia siamea, Cassia fistula, Cordia dichotoma, Callistemon lanceolatus, Dalbergia sissoo, Dryopteris embryopteris, Ficus glomerata, Holoptelea integrifolia, Leucaena leucocephala, Mangifera indica, Phoenix sylvestris, Pithecellobium dulce, Derris indica, Putranjiva roxburghii, Streblus asper, Sterculia alata, Syzygium cumini, Syzygium heyneanum, Tamarindus indica, Thevetia peruviana, Terminalia arjuna, Ziziphus nummularia* etc.

Importance value index (IVI) of this forest in general ranged from 10 to 77 including 10 to 48 for overstory species. These values match with the range of 11 to 52 for sub-tropical tree species of a wet hill forest, India (Rao et al., 1990). The IVI of tree species of a protected forest in Orissa (India) ranged from 12 to 55 (Verma et al., 1997), and trees of dry tropical forest of Vindhyan region constituted 3 to 32 IVI (Singh & Singh, 1991). Thus, our estimate for a rehabilitated forest compares fairly well with the natural forests in India. Shannon Wiener's index (H) of general diversity obtained from 1.8 (ground layer) to 3.6 (Overstory vegetation) indicated that the variability of trees was apparently higher in comparison to ground flora. Since many tree species were planted in this forest, tree diversity index exceeded from dry tropical forests of India, whereas ground layer diversity was almost similar to that of other native forests (Jha, 1990; Singh & Singh, 1991). Shannon Wiener's diversity index of this forest was relatively low from tropical rainforests (3.8 to 4.8) of Silent Valley, India, (Singh et al., 1984). A more generalized relation of the species diversity is derived when stands were pooled together for certain correlations. For instance, the Shannon Wiener's index increased positively with the increase in IVI and the correlation was highly significant (P < 01). The Shannon Wiener's index was negatively correlated with the concentration of dominance, and redundancy, whereas; it had a direct relation with equitability and richness index.

Successional patterns on plant species diversity during rehabilitation of barren land in India are not known and thus the species recruitment/replacement rate is yet to be understood with temporal scale from the initial establishments. The way through which the succession approaches to attain equilibrium, alike to that of natural forests of this region and is stabilized might be interesting to understand for creating a new biotope of our own choice. Manipulation at time to time may be made to divert the ecological processes in the best interest of the entire organism associated with the forests. However, most of the natural forests are disturbed to various degrees on account of a high population pressure for timber, industrial pulp and fuelwood which affect the species diversity significantly. For instance a dry tropical forest of Vindhyan region in India consisted of lower species diversity and basal area in comparison to our study due to several biotic disturbances (Singh and Singh, 1991). However, if a forest is not disturbed during the development of dominant species, then also the species richness is reduced (Odum, 1960). Therefore moderate disturbance may be in favor of high species richness. The studies made elsewhere on species diversity with succession reported the conflicting patterns. McCormick (1968) and (Nicholson & Monk, 1974) found that diversity increased with succession, while Shafi and Yarantan (1973) reported decline in diversity with age. A few noted the highest diversity in the early stages of succession (Habeck, 1968; Long, 1977; Peet, 1978), whereas, several others have depicted a polynomial increase followed by a decrease during succession (Margalef, 1963, 1968; Loucks, 1970; Auclair & Goff, 1971; Schoonmaker & Mackee, 1988). In some cases diversity may show multiple peaks during succession as found by (Halpern & Spies, 1995). Therefore, no generalized trend is maintained and such variations if examined along the site quality gradients would be useful in modeling of diversity patterns.

The different theories of the community organization stated that the diversity is a structural concept which relates to stability, maturity, productivity and evolutionary time, predation pressure and spatial heterogeneity (Hill, 1973). Species diversity (richness) and dominance (Simpson index) were inversely related to each other in agreement with Zobel et al. (1976). Most of these studies suggest that major diversity changes occur during early forest formation and time of species saturation during succession varies greatly in different forests (Nicholson & Monk, 1974). The trends of equitability with succession has not been yet resolved as the high diversity of undisturbed tropical forest implies high equitability levels for mature tropi- cal forest communities (Janzen, 1970; Shafi & Yaranton, 1973), whereas other data indicate a strong decrease in equitability with forest age in USA (Loucks, 1970; Auclair & Goff, 1971; Nicholson & Scott, 1972). It might be possible that a high scale catastrophic perturbation would have reduced the equitability in mature forest vegetation. In general diversity indices (richness, Shannon Wiener's and equitability) tend to stabilize from ground layer to overstory vegetation. Nicholson & Monk (1974) expressed a basic change in the strategy of the plant communities in initial forest formation from a low to high equitability. The young communities (0 yr - 20 yr.) are characterized by plentiful resources and growing space with low equitability, whereas older communities, highly competitive in space and

resource, exhibited a high equitability. Thus rapid increase and stabilization of plant equitability early in succession is viewed as a necessary adjustment to resource scarcity. These findings entail that the community structure and species diversity are related to several environmental factors which lead to specific changes at various scale (region, landscape, biome).

Productivity

Standing biomass and net primary production (NPP) are the aggregate response of the plant species in a particular set of environmental conditions. Climatic and edaphic factors across the region along with the species intrinsic potential determine the limits of ecosystems productivity. Forest productivity differs considerably with environmental conditions from arid to humid climates (8 Mg·ha^{-1}·yr^{-1} - 40 Mg·ha^{-1}·yr^{-1}), standing biomass and net primary production are generally found to be closely associated with each other. Forest established on degraded land tends to acquire the climate based yield in a particular habitat.

Biomass

Total biomass of the three stands varied about 12% - 14 % from their mean in this rehabilitated forest. Our mean value (347 Mg·ha^{-1} ± 27 Mg·ha^{-1}) is supposed to be far better from the other dry tropical forests (degraded) of India, however it is yet to achieve the status of a reference forest of this region (Table 6). Murphy & Lugo (1986b) have reported the global pattern of biomass of dry tropical forests (78 Mg·ha^{-1} - 320 Mg·ha^{-1}), in which the biomass of this rehabilitated forest was relatively higher, but it was lower to that of wet tropical forest (269 to 1186 Mg·ha^{-1}) The total biomass of our rehabilitated forest was less than half of the humid sal forest of India (Singh & Singh, 1989). It was also below by 24% and 37% from the tropical rain forest of Sarawak and Karnataka, India (Proctor et al., 1983; Rai & Proctor, 1986). In a particular habitat biomass depends much on the composition of the tree species in a forest because species to species variations, even on the sodic soil sites have been found to be quite high ranging from 202 (*Acacia nilotica*) to 405 Mg·ha^{-1} (Eucalyptus) on the same site at same age (Singh et al., 2000). The average biomass of our rehabilitated forest was less (43%) than the mixed dry deciduous forest at Haryana India (Gupta & Bhardwaj, 1993). Such variations may be considered as the proportional responses of the site quality interactions to various species in order to perform the forest ecosystem functioning as efficiently as it can optimize the climate based yield.

Biomass Allocation

Above ground biomass in this study was low in comparison to some tropical forests including wet forests. Biomass in above ground plant parts was 88% of the total (stem 28%, branch 57% and leaf 3%), indicating that allocation in root was below the average of 20% observed in most of the natural forests. This might be expected due to soil compactness and presence of "Kankar pan" (a stony layer of CaCO$_3$ gravels) in sub soil. As a consequence, root could not spread and proliferate freely. In this rehabilitated semi-natural forest the mean root biomass contributed about 12.4% of total forest biomass in which coarse root contributed 11% and fine roots contributed to 1.4% of total forest biomass. The contribution of roots in total forest biomass was comparatively low from dry tropical forests supporting 12% - 18% of total biomass (Singh & Misra, 1979; Murphy & Lugo, 1986a; Singh & Singh, 1991). The contribution of coarse root in present forest was comparable to lower range of value (8% - 50%) reported for a number of dry forests (Murphy & Lugo 1986b). It was also slightly less from the average contribution of 16% in 33 moist and wet tropical forests cited by Brown & Lugo (1982).

In global pattern, below ground biomass was 10 Mg·ha^{-1} - 45 Mg·ha^{-1} for dry tropical forest and 11-135 Mg ha^{-1} for tropical wet forest (Murphy & Lugo, 1986b). However, tropical rain forest in India consisted of a relatively low root biomass of 14 Mg·ha^{-1} - 20 Mg·ha^{-1} (Rai & Proctor, 1986). The root/shoot ratio in our forest (0.124) was slightly less than that of global pattern (0.181) of dry tropical forest (Murphy & Lugo, 1986b). This forest had a very less root/shoot value in comparison to other subtropical forests ranging from 0.39 - 0.42 (Jordan 1971a; Vyas et al. 1977). The understory biomass in our rehabilitated forest (3.76 Mg·ha^{-1}) contributed a very little proportion of 1.2% of total biomass of the forest, which is far less from dry tropical forests of India, in which about 13 % - 40% of total biomass was shared by understory (Singh & Singh, 1991). However, (Singh & Misra, 1979) reported only 2% - 7% of total biomass in understory of dry tropical forests on a different site in same geographical region.

Net production of this rehabilitated forest was estimated to 25 Mg·ha^{-1}·yr^{-1}, which appears to be better than dry tropical forests of India, but it is less than the reference forest (Singh & Misra, 1979; Singh & Singh, 1991; Tripathi & Singh, 2009). Net production of this forest compares fairly well at higher level in the global pattern of 13 Mg·ha^{-1} - 28 Mg·ha^{-1} and 8 Mg·ha^{-1} - 21 Mg·ha^{-1}·yr^{-1} for wet and dry tropical forests, respectively (Murphy & Lugo, 1986a, b). The net production of present forest was also within the range (10.3 to 28.6 Mg·ha^{-1}·yr^{-1}) of montane rain forest of Puerto Rico and tropical rain forest of Thailand (Kira et al., 1967; Jordan, 1971b). Managed plantations may generate a relatively high yield from the rehabilitated forests as observed in case of *Eucalyptus saligna* and *Albizia falcataria* species at north east cost of Island Hawai (Binkley & Ryan, 1998). Therefore the net production of tropical forest plantations is considered to be one of the most productive eco- systems in the world, showing 40 Mg·ha^{-1}·yr^{-1} of above ground net primary production (Lugo et al., 1998; Evans, 1992; Binkley et al., 1997). The enhanced productivity has been found to be closely associated with the environmental factors and generic potential.

Fine Root Biomass and Production

Fine roots perform some principal physiological functions in absorption and conduction of solute, nutrient uptake, transpiretion, water retention in soils and on the death and decay, contribute to soil organic matter. Fine roots also act well in reclamation of sodic soils by reducing the pH, improving the soil structure and water permeability of heavy impervious sodic soils. In general fine root biomass and production in afforested sodic soils was relatively less in comparison to many other forests of India and abroad. This may be due to the high soil compactness, low rainfall, high pH, and poor water permeability in sodic soils, which adversely affected the fine root development in hostile conditions. Total fine root biomass (live + dead) measured as 532 gm^{-2} up to 0 cm - 45 cm depth, was comparatively low in the present study from the of tropical deciduous forest up to 30 cm depth (Singh & Misra, 1979). The variation in fine root biomass among the different stands was 14% (S1) to 20% (S3) from their mean value .Fine root biomass was relatively high from the three plantation forests on same site (Singh et al., 2000). In an earlier study, fine root bio-

mass (live + dead) was estimated as 222 g·m^{-2} up to 30 cm depth in the same forest under the canopy of few species (Singh 1998). Thus various estimates of fine root biomass on the same site differ significantly from each other corresponding to species and soil depth considered in their studies. Fine root biomass of this forest compares fairly well with that of dry tropical forest (2.9 to 5.3 Mg·ha^{-1}) in India (Singh & Singh, 1991). The contribution of fine roots to total dry matter turnover in the forest including litter was about 59% which lies within the range 20% - 77% reported for a variety of forests (Vogt et al., 1986). However in some dry tropical forests, fine roots contribution was relatively low about 40% of the total dry matter turnover (Singh & Singh, 1991).

Biomass of fine root was estimated as low as 0.5 Mg·ha^{-1} (Gower, 1987) and as high as 39.5 Mg·ha^{-1} from the rain forest (Cavelier, 1992). The fine root biomass and production both depend on environmental condition and community structure. A high precipitation zone (10372 mm·a^{-1}) showed significant seasonal difference (Khewtan & Ramakrishnan, 1993). Fine root biomass of a hard wood forest estimated as 4.71 Mg·ha^{-1} by Fahey & Hughes (1994) compared well with this forest, but it was quite less in comparison to a semi-deciduous rain forest at Panama (9.45 Mg·ha^{-1}) observed by Cavelier (1992). In some Indian tropical evergreen forests including eucalypt plantations fine root biomass varied from 0.32 to 3.65 Mg ha^{-1} respectively (Bargali et al., 1992; Vasalakshi, 1994), and our estimates compared fairly well with some natural forests of the arid and semiarid regions of India cited by Singh (1996). Fine root biomass of our forest decreased with soil depth and about 47% of total fine root biomass is accommodated in superficial layer (0-15 cm). Decrease in fine root with soil depth is observed in many forests occupying most of the proportion's (43% of the total) in floor horizon alone (Fahey & Hughes, 1994). Simmons (1993) measured a much high value of fine root biomass (3.9 Mg·ha^{-1}) in the thicker forest floor (10 cm) of a mature northern hardwood forest. Fine root mortality (necromass) have been observed as 10% - 15% of total fine roots in various forests (Singh, 1998; Singh et al., 2000) and our estimate belongs to lowest end of this range.

Seasonal fluctuation in fine root biomass varied from 42% (summer) to 62% (rainy) of the yearly mean in our forest. Maximum fine roots were extracted in rainy season and minimum in summer season. However, Singh (1998) observed minimum fine root biomass in winter season. Seasonal variation has been observed to 22% of the yearly mean in small roots in the tropical dry deciduous forests in India (Singh & Singh 1981). The same for fine roots varied from 12 % - 17% from their seasonal means in two dry tropical evergreen forests (Vasalakshi 1994). Fluctuations of 50% or more have commonly been found in many other forests (Grier et al., 1981; McClaugherty et al., 1982; Persson, 1978, 1979, 1980). However, several oak and pine forests did not show any marked seasonal variations and had almost stable biomass pools (Keyes and Grier 1981; Aber et al. 1985). Comparing fine root biomass between forests is difficult because of differences in measuring techniques, soil depth and root diameter classes considered. In most Indian forests of dry zone it has varied from 32 to 340 gm^{-2} for <2 mm diameter to a depth of 30 cm (Parthasarthy, 1988; Vasalakshi, 1994). In our forest, fine roots of < 2 mm diameter were categorized to 20% of total fine roots, which was comparatively less than other Indian dry zone forests. Different soils, vegetation intensity and climatic factors constitute variable quantities of fine roots, nevertheless our forest compared fairly well with some natural forests of the arid or semi arid zone of India (Singh 1998).

Fine root production was measured as 2.33 $Mg \cdot ha^{-1} \cdot yr^{-1}$ in this forest, which was nearly half to that of fine root biomass to a depth of 30 cm. Live fine root production ranged from 2.5 $Mg \cdot ha^{-1} \cdot yr^{-1}$ in chir pine forest to 3.61 $Mg \cdot ha^{-1} \cdot yr^{-1}$ in banj oak evergreen forest (Usman & Rawat, 1999). According to Satoo (1971) the fine root production in evergreen broad leaved forest ranged between 3.7 to 5.3 $Mg \cdot ha^{-1} \cdot yr^{-1}$. The fine root production in this study lies within the range of 1.4 to 11.5 $Mg \cdot ha^{-1} \cdot yr^{-1}$ reported for various tropical and temperate forests (Keys & Grier, 1981; Fogel, 1985; Santantonio & Hermann, 1985; Santantonio & Santantonio, 1987; Adhikari, 1992; Garkoti, 1992; Fahey & Hughes, 1994). In Indian forests, fine root production ranged from 0.5 $Mg \cdot ha^{-1} \cdot yr^{-1}$ of tropical dry deciduous forest to 3.2 $Mg \cdot ha^{-1} \cdot yr^{-1}$ in humid tropical forest (Singh & Singh, 1981; Khewtan & Ramakrishnan, 1993). Fine root production in tropi- cal evergreen forest in India ranged from 1 to 1.17 $Mg \cdot ha^{-1} \cdot yr^{-1}$ Vasalakshi (1994). Fine root production differed with seasons and maximum production occurred during rainy season and summer season of about 67% of the annual production. Fine root production of <2 mm diameter occurred about 90% of total size classes studied which decreased with increasing root size in this forest.

Litter

Litter fall is an important flux of nutrient cycle to maintain the soil sustainability for perpetual production. Besides, it also contributes significantly in reclamation of sodic soil. Litter constitutes several parts i.e. leaves, twigs, bark, dead branches, flower, fruits, seed etc in which leaf litter generally contributes more than 50% of the total litter. Littoral and swamp forest and tropical moist deciduous forest had the highest total as well as leaf litter fall, while tropical dry deciduous forest had lowest total and leaf litter fall. While comparing in four contrasting forest types in India, Singh et al. (1992) also reported that dry deciduous forest had lower litter fall than other types. The relatively low net primary productivity of dry deciduous forests might be one of the reasons. The litter quantity also depends on the population density, age and species of the forest. Since most of these forests are degraded ones, the litter fall was relatively low. Annual litter fall of 8.5 $Mg \cdot ha^{-1} \cdot yr^{-1}$ was estimated in the rehabilitated forest which was greater than the litter fall (5.8 $Mg \cdot ha^{-1} \cdot yr^{-1}$) of dry tropical forest in India (Singh & Misra, 1979; Singh & Singh, 1991). Mean litter fall of 6.4 $Mg \cdot ha^{-1} \cdot yr^{-1}$ was reported for tropical montane forests (Vitousek & Sanford, 1986). A review of 44 published studies of Indian forests documented total and leaf litter fall in the range of 4.3 to 8.5 $Mg \cdot ha^{-1} \cdot yr^{-1}$ and 3.4 to 6.9 $Mg \cdot ha^{-1} \cdot yr^{-1}$ respectively (Dadhwal et al., 1997). Vogt et al. (1986) compiled a range of 2.44 to 9.44 $Mg \ ha^{-1} yr^{-1}$ for various forest types of the world. The range of litter for a variety of tropical and dry forests of the world was reported to be 0.8 $Mg \cdot ha^{-1} \cdot yr^{-1}$ - 15.3 $Mg \cdot ha^{-1} \cdot yr^{-1}$ (Vitousek, 1984; Murphy & Lugo, 1986b; Dantas & Phillipson, 1989). So our data lie in the middle of the said range. Leaf litter in the present study contributed 68% of total litter fall which is comparable to that of (75%) tropical dry deciduous forests of India (Singh & Misra, 1979). In our forest, of the total annual litter fall, summer season contributed maximum (25%) and autumn minimum (12%). Corresponding figures in deciduous forest may reach to 42% of the total annual in summer and 18% in rainy season (Singh & Misra, 1979).

The data generated here with on plant community students, species diversity, biomass, net production, litter fall and fine roots indicated that the restoration of barren sodic land in a new

forest ecosystem has acquired most of the characteristics properties of natural forests of this region, even after differing in species composition. This case study suggests the adoption of the most successful species for the restoration of other identical sites in amore efficient way. If we could control the fire, livestock grazing and invasive alien species, organization of new plant communities, succeeds with the little anthropogenic efforts. Mixed plantation with native species, particularly leguminous, has become more successful. It would be better to introduce the medicinal herbs, found in natural forests to develop amore useful and compatible ground layer in the rehabilitated forest.

Acknowledgements

Authors are grateful to the Director, National Botanical Research Institute, Lucknow for extending the necessary support to carryout the said work at an institutional site.

References

Aber, J. D., Melillo, J. M., Nadelhoffer, K. J., McClaugherty, C. A., & Pastor, J. (1985). Fine root turnover in forest ecosystems in relation to quantity and form of nitrogen availability: A comparison of two methods. *Oecologia, 66,* 317-321.

Abrol, I. P. (1986). Fuel and forage production from salt affected wasteland in India. *Reclamation Revegetation Research, 5,* 65-74.

Abrol, I. P., & Bhumbla, D. R. (1971). Saline and alkali soils in India—their occurrence and management. Rome: World Soil Resources, FAO Report.

Abrol, I. P., & Joshi, P. K. (1986). Economic viability of reclamation of alkali lands with special reference to agriculture and forestry. In H. R. Yadav (Ed.), Wasteland diagnosis and treatment (pp. 149-168). New Delhi: Concept Publishing Co.

Adhikari, B. S. (1992). Biomass productivity and nutrient cycling of Kharsu Oak silver fir forests in central Himalaya. Ph.D. Thesis, Naintial: Kumaun University.

Agrawal, R. R., & Gupta, R. N. (1968). Saline alkali soils in India. New Delhi: ICAR Technical Bulletin (Agriculture Series).

Ahuja, P. S., Singh, .K. N., & Rana, R. S. (1979). Evaluation of forages and other plants for utilization of salt affected soils. Karnal: Trees Annual Report CSSIR.

Alice, F., Montagnini, F., & Montero, M. (2004). Productividad en plantacionespuras y mixtas de especies forestales nativas en La Estación Biológica La Selva, Sarapiquí. *Agronomy Costarricense, 28,* 61-71.

Auclair, A. N., & Goff, F. G. (1971). Diversity relations of upland forests in western Great Lakes area. *American Naturalist, 105,* 499-528.

Barbier, S., Gosselin, F., & Balandier, P. (2008). Influence of tree species on understory vegetation diversity and mechanisms involved—A critical review for temperate and boreal forests. *Forest Ecology & Management, 254,* 1-15.

Bargali, S. S., Singh, S. P., & Singh, R. P. (1992). Structure and function of an age series of eucalypts plantations in Central Himalaya. I. Dry matter dynamics. *Annals of Botany, 69,* 405-411.

Bhojvaid, P. P., & Timmer, V. R. (1998). Soil dynamics in an age sequence of Prospis juliflora planted for sodic soil restoration in India. *Forest Ecology & Management, 106,* 181-193.

Binkley, D., O'Connel, A. M., & Sankaran, K. V. (1997). Stand growth: pattern and controls. In E. K. S. Nambiar and A. Brown (Eds.), *Management of soil, water nutrients in tropical plantation forests* (pp. 719-442). Canberra: ACIAR Monograph 43.

Binkley, D., & Ryan, M. G. (1998). Net primary production and nutrient cycling in replicated stands of *Eucalyptus saligna* and *Albizia facaltaria. Forest Ecology & Management, 112,* 79-85.

Brinkley, D., White, S. C., & Gosz, J. R. (2004). Tree biomass and net increment in an old aspen forest in New Mexico. *Forest Ecology &*

Management, 203, 407-410.

Brockway, D. G. (1998). Forest plant diversity at local and landscape scales in the cascade mountains of south-western Washington. *Forest Ecology & Management, 109*, 323-341.

Brown, S., & Lugo, A. E. (1982). The storage and production of organic matter in tropical forests and their role in the global carbon cycle. *Biotropica, 14*, 161-187.

Cavelier, J. (1992). Fine root biomass and soil properties in a semideciduous and lower montane rain forest in Panama. *Plant and Soil, 142*, 187-201.

Chapin, F. S. III, Vitousek. P. M., & Cleve, K. V. (1986). The nature of nutrient limitation in plant communities. *The American Naturalist, 127*, 48-58.

Chapin, F. S., Zavaleta, E. S., Eviner, V. T., Naylor, R. L., Vitousek, P. M., Lavourel, S., Reynolds, H. L., Hooper, D. U., Sala, O. E., Hobbie, S. E., Mack, M. C., & Diaz, S. (2000). Consequences of changing biotic diversity. *Nature, 405*, 234-242.

Chaturvedi, A. N. (1985). Firewood farming on degraded lands in the gangetic plain. Uttar Pradesh: Forest Bulletin.

Chaturvedi, A. N., & Behl, H. M. (1996). Biomass production trials on sodic site. *The Indian Forester, 122*, 439-455.

Chaturvedi, A. N., Bhatia, S., & Behl, H. M. (1991). Biomass assessment for shrubs. *The Indian Forester, 117*, 1032-1035.

Dadhwal, V. K., Shukla, N., & Vora, A. B. (1997). Forest litter fall in India: A review and an estimate of litter fall carbon flux. *The Indian Forester, 123*, 45-52.

Dantas, M., & Phillipson, J. (1989). Litter fall and litter nutrient content in primary and secondary Amazonian terra firm rain forest. *Journal Tropical Ecology, 5*, 27-36.

Dogra, P. D. (1989). Fuel wood production potential in India through short rotation intensive culture (SRIC) system applied to tree growing for fuel wood on reduced land area or on wasteland sites. *The Indian Forester, 115*, 11-16.

Dutta, R. K., & Agrawal, M. (2003). Restoration of opencast coal mine spoil by planting exotic tree species: A case study in dry tropical region. *Ecological Engineering, 21*, 143-151.

Evans, J. (1992). Plantation forestry in the tropics (2nd ed., p. 403). Oxford: Oxford Science.

Fahey, T. J., & Hughes, J. W. (1994). Fine root dynamics in a northern hardwood forest ecosystem, Hubbard Brook Experimental Forest, N.H. *Ecology, 82*, 533-548.

Fogel, R. (1985). Roots as primary producers in below ground ecosystems. In A. H. Fitter et al., (Eds.), *Ecological interaction in soil plants, microbes and animals* (pp. 23-26). Special Bulletin Number 4 of the British Ecological Society. Oxford: Blackwell Scientific Publication.

Garg, V. K. (1998). Interaction of tree crops with a sodic soil environment: Potential for rehabilitation of degraded environments. *Land Degradation Development, 9*, 81-93.

Garg, V. K., & Jain, R. K. (1992). Influence of fuel wood trees on sodic soils. *Canadian Journal of Forest Research, 22*, 729-735.

Garg, V. K., & Khanduja, S. D. (1979). Mineral composition of leaves of some forest trees grown on alkali soils. *The Indian Forester, 105*, 741-745.

Garkoti, S. C. (1992). High altitude forests of central Himalaya: productivity and nutrient cycling. Ph.D. Thesis, Nainital: Kumaun University.

Gilliam, F. S., Turrill, N. L., & Adams, M. B. (1995). Herbaceous-layer and overstory species in clear-cut and mature central applachian hardwood forests. *Ecological Applications, 5*, 947-955.

Grier, C. C., Vogt, K. A., Keys, M. R., & Edmonds, R. L. (1981). Biomass distribution, above and below ground production in young and mature *Abies amabilis* zone ecosystem of the Washington cascades. *Canadian Journal Forest Research, 11*, 155-167.

Gupta, R. K., & Abrol, I. P. (1990). Reclamation and management of alkli soils. *Indian Journal of Agricultural Sciences, 60*, 1-60.

Gupta, S. R., & Bhardwaj, B. B. (1993). Biomass and productivity of forest ecosystems of Shivalik hills. *Journal of Tree Sciences, 12*, 57-63.

Haase, R. (1999). Litterfall and nutrient return in seasonally flooded and non-flooded forest of the Pantanal, Mato Grosso, Brazil. *Forest Ecology & Management, 117*, 129-147.

Habeck, J. R. (1968). Forest succession in the Glacier Peak Cedar-hemlock forests. *Ecology, 49*, 872-880.

Halpern, C. B., & Spies, T. A. (1995). Plant species diversity in natural and managed forests of the pacific northwest. *Ecological Applications, 5*, 913-934.

Hara, T. (1988). Dynamics of size structure in plant populations. *Trends in Ecology & Evolution, 3*, 129-133.

Hill, M. O. (1973). Diversity and Evenness: A unifying notation and its consequences. *Ecology, 54*, 427-432.

Hurlbert, S. H. (1971). The non concept of species diversifies: a critique and alternative parameters. *Ecology, 52*, 577-586.

Huston, M. A. (1997). Hidden treatments in ecological experiments: re-evaluating the ecosystem function of biodiversity. *Oecologia, 110*, 449-460.

Jain, R. K., & Singh, B. (1998). Biomass production and soil amelioration in a high density *Terminalia arjuna* plantation in sodic soils. *Biomass & Bioenergy*, 1-6.

Janzen, D. H. (1970). Herbivores and the number of tree species in tropical forests. *American Naturalist, 104*, 501-528.

Jha, C. S. (1990). Landuse and vegetation analysis of a dry tropical forest region. Ph.D. thesis, Varanasi: Banaras Hindu University.

Jha, C. S., & Singh, J. S. (1990). Composition and dynamics of dry tropical forest in relation to soil texture. *Journal of Vegetation Science, 1*, 609-614.

Jordan, C. F. (1971a). Productivity of tropical forest and its relation to a world pattern of energy storage. *Journal of Ecology, 59*, 127-142.

Jordan, C. F. (1971b). A world pattern in plant energetic. *American Scientist, 59*, 425-433.

Kelley, W. P. (1951). Alkali soils: their formation, properties and reclamation. New York: Reinhold Publishing Corporation.

Keyes, M. R., & Grier, C. C. (1981). Above and below ground net production in 40 year old Douglas fir stands on high and low productivity sites. *Canadian Journal of Forest Research, 11*, 599-605.

Khan, W. M. A., & Yadav, J. S. P. (1962). Characteristics and afforestation problems of saline alkali soils. *The Indian Forester, 88*, 259-271.

Khanduja, S. D., Chandra, V., Srivastava, G. S., Jain, R. K., Misra, P. N., & Garg, V. K. (1986). Utilization of alkali soils in the plans of northern India—A case study (p. 54). New Delhi: Commonwealth Science Council.

Khiewtan, R. S., & Ramakrishnan, P. S. (1993). Litter and fine root dynamics of a relict sacred grove forest at Cherapunji in northeastern India. *Forest Ecology & Management, 60*, 327-344.

Khoshoo, T. N. (1987). Ecodevelopment of alkaline land. A case study (pp. 141). New Delhi: Publication and Information Directorate, CSIR, on behalf of Director NBRI, Lucknow.

Kira, T., Ogawa, .H, Yoda, K., & Ogino, K. (1967). Comparative ecological studies on three main types of forest vegetation in Thailand. Dry matter production with special reference to the Khaochong rain forest (4th ed.). Kyoto: Nature and Life in South East Asia.

Kobayashi, S. (2004). Landscape rehabilitation of degraded tropical forest ecosystems. Case study of the CIFROR/Japan project in Indonesia and Peru. *Forest Ecology and Management, 201*, 13-22.

Kohyama, T., Hara, T. (1989). Frequency distribution of tree growth in natural forest stands. *Annals of Botany, 64*, 47-57.

Liang, J., Buongiorno, J., Monserud, R. A., Kruger, E. L., & Zhou, M. (2007). Effect of diversity of tree species and size on forest basal area growth, recruitment and mortality. *Forest Ecology and Management, 243*, 116-127.

Long, J. N. (1977). Trends in plant species diversity associated with development in a series of Pseudotsuga menziesiil, Gaultheria shallon stands. *Northwest Science, 51,* 119-130.

Loreau, M., Mouquet, N., & Gonzalez, A. (2003). Biodiversity as spatial insurancein heterogeneous landscapes. *Proceedings of National Academy of Sciences, 100,* 12765-12770.

Loucks, O. L. (1970). Evolution of diversity, efficiency, and community stability. *American Zoology, 10,* 17-25.

Lugo, A. E., Clark, J. R., & Child, R. D. (1998). Ecological development in the humid tropics: guidelines for planners. Morilton, Arkansas: Winrock International Institute of Agriculture Development.

Margalef, R. (1963). On certain unifying principles in ecology. *American Naturalist, 97,* 357-374.

Margalef, R. (1968). Perspectives in ecological theory. Chicago: University of Chicago Press.

McClaugherty, C. A, Aber, J. D., & Melillo, J. M. (1982). The role of fine roots in the organic matter and nitrogen budgets of two forested ecosystems. *Ecology, 63,* 1481-1490.

Misra, R. (1968). Ecology work book. Calcutta: Oxford and IBH Publishing Company.

Murphy, P. G., & Lugo, A. E. (1986a). Ecology of tropical dry forests. *Annual Review of Ecology, Evolution and Systematics, 17,* 67-88.

Murphy, P. G., & Lugo, A. E. (1986b). Structure and biomass of a subtropical dry forest in Puerto Rico. *Biotropica, 18,* 89-96.

Nagaike, T., Kamitani, T., & Nagashizuka, T. (1999). The effect of shelter wood logging on the diversity of plant species in a beech (*Fagus crenata*) forest in Japan. *Forest Ecology and Management, 118,* 161-171.

Naidu, R., & Rengasamy, P. (1993). Ion interactions and constraints to plant nutrition in Australian sodic soils. *Australian Journal of Soil Research, 31,* 801-819.

Nicholson, S. A., & Monk, C. D. (1974). Plant species diversity in old field succession on the Georgia Piedmont. *Ecology, 55,* 1075-1085.

Nicholson, S. A., & Scott, J. T. (1972). Successional trends in plant communities in the Lake George drainage basin. *Bulletin of Ecological Society of America, 53,* 17.

Odum, E. P. (1960). Organic production and turnover in old-field succession. *Ecology, 41,* 34-49.

Odum, E. P. (1969). The strategy of ecosystem development. *Science, 164,* 262-270.

Pandey, G. C. (1966). A short not on introduction of tree species on wild "Usar" lands of Uttar Pradesh (p. 377). *Proceedings of XI Silva Conference,* Dehradun.

Pandey, V. C., Singh, K., Singh, B., Singh, R. P., (2011). New approaches to enhance eco-restoration efficiency of degraded sodic Lands: critical research needs and future prospects. *Ecological Restoration, 29,* 322-325.

Parthsarthy, N. (1988). Seasonal dynamics of fine roots in a tropical forest in south India. *Journal of Indian Botanical Society, 66,* 338-345.

Patten, B. C. (1962). Species Diversity in net phytoplankton of Raritan Bay. *Journal of Marine Research, 20,* 57-75.

Peet, R. K. (1978). Forest vegetation of the Colorado Front Range: Patterns of species diversity. *Vegetation, 37,* 65-78.

Persson, H. (1979). Fine root production, mortality and decomposition in forest ecosystems. *Vegetation, 41,* 101-109.

Persson, H. (1980). Structural properties of the field and bottom layers at Ivantjarnsheden. In T. Person (Ed.), *Structure and function of northern coniferous forest—An ecosystem study* (pp. 153-163). Stockholm: Swedish Natural Science Resarch Council.

Persson, H. A. (1978). Root dynamics in a young Scots pine stand in central Sweden. *Oikos, 30,* 508-519.

Petit, B., & Montagnini, F. (2004). Growth equations and rotation ages of ten native tree species in mixed and pure plantations in the humid Neotropics. *Forest Ecology & Management, 199,* 243-257.

Pielou, E. C. (1966). The measurement of diversity in different types of biological collections. *Journal of Theoratical Biology, 13,* 131-144.

Piotto, D., Montagnini F., Ugalde, L., & Kanninen, M. (2003a). Growth and effects of thinning of mixed and pure plantations with native trees in humid tropical Costa Rica. *Forest Ecology and Management, 177,* 427-439.

Piotto, D., Montagnini, F., Ugalde, L., & Kanninen, M. (2003b). Performance of forest plantations in small and medium-sized farms in the Atlantic lowlands of Costa Rica. *Forest Ecology and Management, 175,* 195-204.

Prasad, K. G., & Sharma, S. D. (1990). Status report on afforestation of salt affected soils in India. Proc. National Seminar on Technology for Afforestation of Wastelands (pp. 41). Dehradun, Uttar Pradesh: Forest Research Institute.

Proctor, J., Anderson, J. M., Chai, P., & Vallack, H. W. (1983). Ecological studies in four contrasting lowland rain forests in Gunung Mulu National Park, Sarawak. I. forest environments structure and floristics. *Journal of Ecology, 71,* 237-260.

Rai, S. N., & Proctor, J. (1986). Ecological studies on four rainforests in Karnataka, India. I. Environment, structure, floristic and biomass. *Journal of Ecology, 74,* 439-454.

Rao, P., Barik, S. K., Pandey, H. N., & Tripathi, R. S. (1990). Community composition and tree population structure in a sub-tropical broad-leaved forest along a disturbance gradient. *Vegetation, 88,* 151-162.

Sandhu, S. S., & Abrol, I. B. (1981). Growth responses of Eucalyptus tereticornis and Acacia nilotica to selected cultural treatments in a highly sodic soil. *Indian Journal of Agriculture Science, 51,* 437-443.

Santantonio, D, & Hermann R. K. (1985). Standing crop production and turnover of fine roots on dry moderate and wet sites of mature Douglas fir in Western Oregon. *Annals Forestry Science, 42,* 113-142.

Santantonio, D., & Santantonio, E. (1987). Effect of thining on production and mortality of fine roots in a Pinus radiata plantation on the fertile site in New Zealand. *Canadian Journal of Forest Research, 17,* 919-928.

Satoo, T. (1971). Primary production relations of coniferous forests in Japan, In P. Duvigneaud (Ed.), *Productivity of forest ecosystem* (pp. 191-205). Paris: Unesco.

Schoonmaker, P., & McKee, A. (1988). Species composition and diversity during secondary succession of coniferous forests in the Western Cascade Mountains of Oregon. *Forest Science, 34,* 960-979.

Schulze, E. D., & Mooney, H. A. (1993). Biodiversity and Ecosystem Function. Berlin: Springer-Verlag.

Shafi, M. I., & Yaranton, G. A. (1973). Diversity floristic richness and species evenness during a secondary (post fire) succession. *Ecology, 54,* 897-902.

Shannon, C. E., & Weaver, W. (1963). The mathematical theory of communication. Urbana: University of Illinois Press.

Sharma, S. D., Prasad, K. G., Rai, L., & Malik, N. (1992). Development of technology for afforestation of sodic solis. I-leguminous species. *The Indian Forester, 108,* 547-559.

Sharma, S. K. (1988). Recent advances in afforestation of salt affected soils in India. *Advances in Forest Research, 2,* 17-31.

Shukla, A. K., & Misra, P. N. (1993). Improvement of sodic soil under tree cover. *The Indian Forester, 119,* 43-52.

Shukla, S. K., Singh, K., Gautam, N. N., Singh, B. (2011). Biomass productivity and nutrient availability of *Cynodon dactylon* (L.) Pers. growing on soils of different sodicity stress. *Biomass & Bioenergy, 35,* 3440-3447.

Simmons, J. (1993). Lime effects on carbon and nitrogen dynamics of a northern hardwood forest floor. Ph.D. Thesis, Ithaca, New York: Cornell University.

Simpson, E. H. (1949). Measurement of diversity. *Nature, 163,* 688.

Singh, A. N., Raghubanshi, A. S., & Singh, J. S. (2004). Impact of native tree plantations on mine spoil in a dry tropical environment. *Forest Ecology and Management, 187,* 49-60.

Singh, A. N., Raghubanshi, A. S., & Singh, J. S. (2002). Plantation as a tool for mine spoil restoration. *Current Science, 82,* 1436-1441.

Singh, B. (1996). Influence of forest litter on reclamation of semiarid sodic soils. *Arid Soil Research Rehabilitation, 10,* 201-211.

Singh, B. (1997). Deforestation consequences and rehabilitation of sodic wastelands in the gangetic plains. In S.K. Gupta and K. Kumar (Eds.), *Environmental Degradation* (pp. 119-134). Varanasi: Tara Book Agency.

Singh, B. (1998). Biomass potentials of some firewood shrubs of North India: Short communication. *Biomass, 16*, 199-203.

Singh, B. (1998). Biomass production and nutrient dynamics in three clones of Populus deltoids planted on Indo-Gangetic plains. *Plant and Soil, 203*, 15-26.

Singh, B., Garg, V. K., Singh, P. K., & Tripathi, K. P. (2004). Diversity and Productivity effect on the Amelioration of Afforested Sodic Soils. *The Indian Forester, 130*, 14-26.

Singh, B., Tripathi, K. P., Jain, R. K., & Behl, H. M. (2000). Fine root biomass and tree species effects on potential N mineralization in afforested sodic soils. *Plant and Soil, 219*, 81-89.

Singh, J. S., Rawat, Y. S., & Chaturvedi, O. P. (1984). Replacement of oak forest with pine in the Himalaya affects the nitrogen cycle. *Nature, 11*, 54-60.

Singh, K., Yadav, J. S. P., & Sharma, S. K. (1990). Performance of shisham (*Dalbergia sissoo*) in salt affected soils. *The Indian Forester, 116*, 154-162.

Singh, K., Yadav, J. S. P., & Singh, V. (1992). Tolerance of trees to soil sodicity. *Journal of Indian Society of Soil Science, 40*, 173-179.

Singh, K.P., & Misra, R. (1979). Structure and function of natural, modified and silvicultural ecosystem in eastern Uttar Pradesh. *Final Technical Report (1975-1978).* Banaras Hindu University, Varanasi: MAB Research Project.

Singh, K. P., & Singh, R. P. (1981). Seasonal variations in biomass and energy of small roots in tropical dry deciduous forest, Varanasi, India. *Oikos, 37*, 88-92.

Singh, L., & Singh, J. S. (1991). Species structure, dry matter dynamics and carbon flux of a dry tropical forest in India. *Annals of Botany, 68*, 263-273.

Singh, O., Sharma, D. C., Negi, M. S., & Singh, R. (1994). Leaf litter production and decomposition in *Dalbergia sissoo* and *Bombax ceiba* plantations in U.P. *The Indian Forester, 120*, 682-688.

Singh, P. (1991). Management strategies for wasteland development. In R. K. Singh, A. K. Saxena and I. S. Singh, (Eds.), *Agroforestry— Present status and scope for future development in farming systems* (pp. 133-142). Kumarganj, Faizabad: Narendra Dev University of Agriculture and Technology.

Singh, S. P., & Singh, J. S. (1989). Ecology of central Himalayan forests with special emphasis on sal forest ecosystem. In J. S. Singh and B. Gopal (Eds.), *Perspectives in Ecology, Professor S.C. Pandeya Commemoration Volume* (pp. 193-232). New Delhi: M/S Jagmander Book Agency.

Sissay, B. (1986). Salt wasteland in Ethiopia: Potential for production of forage and fuel. *Reclamation Revegetation Research, 5*, 59-64.

Srivastava, G. S. (1970). Behaviour and adoptability of some ornamental plants to "usar" soils. *Indian Journal of Ornamental Horticulture, 11*, 1-3.

Srivastava, G. S. (1987). A man-made forest on sodic soils. In T.N.

Khoshoo (Ed.), Ecodevelopment of alkaline land, Banthra—A case study (pp. 124-128). Lucknow: NBRI, CSIR.

Sumner, M. E. (1993). Sodic soils: New perspectives. *Australian Journal of Soil Research, 31*, 683-750.

Tomar, O. S., & Yadav, J. S. P. (1980). Effect of saline irrigation water of varying EC, SAR and RSC levels on germination and seedling growth of some forest species. *Indian Journal of Forestry, 3*, 306-314.

Totey, N. G., Kulkarni, R., Showmik, A. K., Khatri, P. K., Dahia, V. K., & Prasad, A. (1987). Afforestation of salt affected wasteland I— Screening of forest tree species of Madhya Pradesh for salt tolerance. *The Indian Forester, 113*, 805-815.

Toth, B. (1981). Afforestation on salt affected soils in Hungary. *Agrokemia Es Talajtan Tom, 30*, 205-212.

Tripathi, K. P., & Singh, B. (2009). Species diversity and vegetation structure across various strata in natural and plantation forests in Katarniaghat Wildlife Sanctuary, North India. *Tropical Ecology, 50*, 191-200.

Usman, S., Singh, S. P., & Rawat, Y. S. (1999). Fine root productivity and turnover in two evergreen Central Himalayan forests. *Annals of Botany, 84*, 87-94.

Vasalakshi, N. (1994). Fine root dynamics in two tropical dry evergreen forest of Southern Indian. *Journal Bioscience, 19*, 103-116.

Verma, R. K., Totey, N. G., & Gupta, B. N. (1997). Analysis of forest vegetation in the permanent preservation plot of Tamna in Orrisa. *The Indian Forester, 123*, 1007-1016.

Verma, S. C., Jain, R. K., Rao, M. V., Misra, P. N., & Murty, A. S. (1982). Influence of canopy on soil composition of man-made forest in alkali soil of Banthra, Lucknow. *The Indian Forester, 108*, 431-437.

Vitousek, P. M. (1984). Litterfall, nutrient cycling and nutrient limitation in tropical forests. *Ecology, 65*, 285-298.

Vitousek, P. M., & Sanford, R. L. (1986). Nutrient cycling in moist tropical forests. *Annual Review of Ecology and Systematics, 17*, 137-167.

Vogt, K. A., Grier, C. C., & Vogt, D. J. (1986). Production, turnover and nutrient dynamics of above and belowground detritus of world forests. *Advance Ecological Restoration, 15*, 303-377.

Vyas, L. N., Garg, R. K., & Vyas, L. N. (1977). Stand structure and aboveground biomass in dry deciduous forests of Aravalli Hills at Udaipur (Rajasthan), India. *Biologia (Bratislava), 32*, 265-270.

Yadav, J. S. P. (1975). Improvement of saline alkali soil through biological methods. *The Indian Forester, 101*, 385-395.

Yadav, J. S. P. (1980). Salt-affected soils and their afforestation. *The Indian Forester, 106*, 259-272.

Yadav, J. S. P., & Singh, K. (1970). Tolerance of certain forest species to varying degree of salinity and alkalinity. *The Indian Forester, 96*, 587-599.

Zobel, D. B., McKee, A., Hawk, G. M., & Dyrness, C. T. (1976). Relationships of environment to the composition, structure and diversity of forest communities of the central western cascades of Oregon. *Ecological Monographs, 46*, 135-156.

5

Forest-Climate Politics in Bangladesh's Media Discourse in Comparison to Global Media Discourse

Md. Nazmus Sadath[1,2], Max Krott[1], Carsten Schusser[1]

[1]Chair of Forest and Nature Conservation Policy, Georg August University Goettingen, Goettingen, Germany
[2]Forestry and Wood Technology Discipline, Khulna University, Khulna, Bangladesh

Forest and climate issues are prominent within the policies and media in Bangladesh, as well as on the global level. In this study, media discourses from 1989 to 2010 from the "International Herald Tribune" and "The Daily Ittefaq" of Bangladesh are analyzed. Quantitative content analysis classifies 16 frames of the forest and climate issue and 17 political actors. Substantial differences between the forest and climate discourses of the national and international media have been discovered. The national print media reports that the forest is in a crisis due to climate change, whereas the international print media describes the forest as a solution opportunity to climate change. The hypothesis that the international media drives the national media discourse is rejected. The national media forest and climate discourse in Bangladesh began five years earlier than in the international media, and the different framing of the forest and climate issues can be explained by the influence of strong actors on both the national and international level. Journalists and politicians are the strongest influences in the national print media (The Daily Ittefaq) and primarily frame the discussion around the adverse impact of climate change on the forest in Bangladesh, a country that faces potentially severe effects from climate change. By stressing that climate change has caused a forest crisis, the national media brings attention to a threat that they are not responsible for. Scientists, Non-Governmental Organizations and international organizations are the major voices in the international print media (International Herald Tribune). They shape the global forest and climate media discourse around the wider scope of forests' role in climate change. International scientists and NGOs present themselves as problem solvers of climate change by framing the discussion around the mitigating role of the forests. These strategic arguments explain the differences in media discourse.

Keywords: Climate Change; Forest-Climate Discourse; Political Actor; Media Framing

Introduction: The Dynamic Media Discourse about Climate Change and the Forest

In the last century, deforestation, desertification and forest degradation have been major environmental problems that have diminished sustainable forest use and caused a significant loss of biodiversity, including the extinction of both flora and faunal species. This loss of biodiversity is the sixth largest species loss in the earth's history (Leakey & Lewin, 1995; Pimm & Brooks, 2000). Additionally, global warming is magnifying these environmental threats for the forest. Now that it has become a global issue in political discourse, forest issues are no longer a concern of individual nations. Despite several global initiatives such as FSC, IPF and UNFF, the forest sector did not reach its goal of a united global forest regime or policy. However, in 2008, the forestry sector did find a niche within the already existent climate change regime, and a significant step was made for establishing a global forest regime within the climate change regime (Levin et al., 2008). This new global forest and climate regime (Levin et al., 2008) has contributed to the emergence of a forest and climate discourse in the scientific and political arenas at both the global and national levels.

There has been rigorous discussion at the international level on creating global initiative to tackle forest climate issue but there has also long been a national perspective to the forest and climate issue, even though the problem exists beyond the national state boundary, in part because there are countries such as Bangladesh that will be affected by this global problem more than others. Bangladesh is located on a delta and faces various adverse impacts from climate change, including on the currently stressed forest sector. A major portion of forested land in Bangladesh is situated within the coastal region. The world's largest single mangrove area, identified as "the Sundarbans", is a biodiversity hot spot with one of the richest gene pools in the world, but it is directly under threat from sea level rise under various scenarios of climate change (Ali, 1999). Climate change's impact on agriculture and the hydrological system will also pose an indirect anthropogenic threat to the existing natural forest of Bangladesh (Huq et al., 1999). Because of the issue's immense scope, it involves a diverse group of stakeholders at different political levels both nationally and globally. These diverse stakeholders utilize the media as a platform to spread their viewpoints with the purpose of exerting influence on environmental politics and public opinion; simultaneously, the media assists in the aggregation of interests within the political process, providing a channel for communication and facilitating the revision of shared goals and policies (Curran, 2002). To some extent, the media performs these functions for the public sphere because it provides more open access to various actors (Kleinschmit, 2012). The mass media also plays a crucial role in mediating the process of public deliberation and in the diffu-sion concerns and opinions (Hardt, 2004). In today's modern complex democratic society, the media is one of the most im-portant sources of information for individuals other than elec-tions and opinion polls; thus, the media reflects the public opinion (Kleinschmit & Krott, 2008). The media has an impor-tant place in both the national and international political level, as it provides the place where different actors build on their arguments with the interest of legitimizing their policies or decisions. By keeping in mind the accelerating political issue of climate change and forests in Bangladesh and the active politi-cal role of the media in forest policy, the study investigates how the media has reported on climate change and the forest issue since its first mention in the late 80s and early 90s. The com-parison between the national and international levels of media is of special interest. Finally, if we find differences between the media, we will look for explanations. Given this background, the following section will organize the research question into three hypotheses guiding the empirical media analysis.

Theoretical Framework and Hypotheses

Media Discourse and the Framing of the Forest and Climate Issue

Discourse is a social construction of reality; that is, discourse produces a specific picture of the issue of forest and climate change (Fairclough, 1995). Keller (1997) approaches discourse as a specific content that is a thematically institutionalized form of text production, comprising public discussions on certain political and/or environmental issues delivered through the me-dia where the conversation and exchange of opinion between relevant actors has occurred. In this study, the forest and cli-mate media discourse is understood as the communication about topics and actors present in the print media that are rele-vant to both the forest and climate change.

In the media, the specific topic of climate change and its ef-fects on forests can be viewed from many different perspectives. This phenomenon is addressed by the theory of "framing". According to Chong and Druckman (2007: p. 104), "the major principle of framing theory is that an issue can be construed as having implications for multiple values or considerations. Fra-ming refers to the process by which people develop a particular conceptualization of an issue or re-orient their thinking about an issue". Therefore, to make framing work, one must select specific "aspects of a perceived reality and make them more salient in a communication text, in such a way as to promote a particular problem definition, casual interpretation, moral eva-luation, and/or treatment recommendation for the item de-scribed" (Entman, 1993). Framing highlights particular pieces of information about the subject, thereby assigning more impor-tance to those pieces than to others (Prittwiz, 1990; Entman, 1993).

The framing of the topic by different actors in the media on the national and international political levels may be different, as the problem's definition, perception and interest for a single state that is heavily impacted by climate change might be dif-ferent from an international perspective (Takahashi, 2008). Although existing literature was not found regarding this exact phenomenon, Kingdon (2003) mentioned in his policy agenda setting theory that public opinion on a particular issue may differ with location (Kingdon, 2003). In addition, the audience of the global media is different from that of the national media,

and every media outlet is very careful to satisfy their audience or readership; thus, the framing of similar issues (Gerhards, 1994; Boykoff & Boykoff, 2004) may be different in the print media at the international level when compared to the national print media. Therefore, this study's first hypothesis as follows: "*The media actively frames forest and climate change issues. Therefore, how the issue is presented depends on the media; hence, the framing of national media is different than that of international media.*"

National and International Media

This study's objective is to analyse the difference in media discourse at both the international and national levels; therefore, it is imperative to explain the globalization of the media. Al-though the media is highly recognized as a driving force of globalization (Kleinschmit & Krott, 2008), the literature on the international or global media lacks a common definition. Many scholars refer to the international media within a technical con-text, e.g., new communications technologies or multi-national media industries (Held et al., 1999). In regard to the topics on which international media reports, the definitions become more abstract, referring to the scope and composition of the audience (McQuail, 2010). We follow the definition of Reese (2010), stating that trans- or international media are those who can ob-tain news from transnational boundary sources and can address a wider audience beyond national boundaries, such as the "In-ternational Herald Tribune" or the "Financial Times" (Reese, 2010). In contrast, national media such as "The Daily Ittefaq" of Bangladesh is characterized by content with respect to lan-guage and substance, which provides the public sphere with a national perspective on Bangladesh (Rahman, 2010). In recent years, he international and national media have shifted their focus from more local and regional environmental issues to more global issues, such as global warming, ozone layer deple-tion and the extinction of species (Mazur & Lee, 1993). Be-cause of this shift in environmental reporting, it is important to question whether there is any link between the international and national media. It is clear that international print media such as the "International Herald Tribune" have more resources to co-ver global events and, as a result, have a larger news pool (Wu, 1998). In contrast, national media, such as "The Daily Ittefaq" of Bangladesh, lack the resources to cover global environmen-tal issues first hand and tend to depend on international sources for their stories. Therefore, there is a greater chance that na-tional print media will follow international print media in re-porting global environmental issues such as forest and climate politics (Mazur & Lee, 1993). Based on this globalization of news, this study's second hypothesis as follows: "*International media claims to address important global issues first. Therefore, the forest and climate change issue is mentioned in the interna-tional media prior to the national media of Bangladesh.*"

Explaining Framing by Political Actors

Framing is part of the communicative strategy of political actors. It is influenced by the strengths of the actors, which is not only limited to their status and resources but also depends on how much value they bring to forest and climate issues. By combining these three factors, the standing of a certain actor or a group of actors is determined in the media discourse, i.e., the strength of a certain actor or group of actors having a voice in the media in comparison to others (Feindt & Kleinschmit,

2011). In this study, we used speakers as political actors. The frequency of the appearance of a certain actor in the media as a speaker on a certain topic or field is a good indicator of media standing. The higher standing of certain actors in forest and climate media discourse provides that speaker with more opportunities to shape and/or frame the discourse in accordance with his or her interests or viewpoint (Sadath et al., 2012). In depicting certain events, the speaker stresses some aspects of a situation and downplays other aspects (e.g., Schäfer, 2008), which in turn has implications for how the speaker might benefit from specific frames (Sadath et al., 2012). Based on this strategic framing theory, this study's last hypothesis is as follows: "*Topics and framing within the media are influenced by the actors that speak in the media. Therefore, strong actors and their interests can explain the content and the timing of frames in both the national and international media.*"

Methodology

Two reputable daily newspapers, "The Daily Ittefaq" and "International Herald Tribune", were selected to represent the national and international print media, respectively, for the analysis. The Daily Ittefaq was selected because of the newspaper's popularity in Bangladesh and its ability to reach Bangladeshi political elites and decision-makers (Sadath et al., 2012). The print media selection for international media is critical, as the definition of global media is not very clear among scholars (Park, 2009). Following UNESCO (1997) and Sparks (1998), The Wall Street Journal, the Financial Times and the International Herald Tribune are considered to be reputable international publications (UNESCO, 1997; Sparks, 1998). The first two print media do not focus on environmental issues, but rather focus on financial and economic reporting. The International Herald Tribune publishes articles with a broader range of issues and subjects, including the environment, politics, culture and sports. In addition, two thirds of the readers of the International Herald Tribune are non-American, and the newspaper is sold in more than 160 countries and territories (IHT, 2012). Therefore, the "International Herald Tribune" was selected to represent the print media discourse at an international level (Sadath et al., 2012).

Relevant articles of the "International Herald Tribune" were identified using the LexisNexis database, and the relevant articles from the national newspaper "The Daily Ittefaq" were collected manually using the national library archive of Bangladesh. The search was limited to the years between 1989 and 2010 because since 1989, the climate change issue had gained momentum in international and national environmental discussions and subsequently international negotiation had begun. The database search for relevant articles was performed using the keywords "Climate change" and "Forest". The resulting articles were then screened to obtain the relevant articles with the screening criterion containing at least one paragraph within the article that linked the forest to climate change. As a result, a number of articles were identified and in this sample: the "International Herald Tribune" yielded 90 articles with 149 statements, and the "The Daily Ittefaq" of Bangladesh yielded 49 articles with 52 statements.

Quantitative-qualitative content analysis was then performed on these newspaper articles to determine their general impression regarding forest issues connected to climate change and how these topics are framed. Content analysis is the method

used for elevating social reality, utilizing both a manifest and non-manifest context. According to Krippendorff (1980), content analysis is the research technique for making valid, explicative inferences from data about their context. In addition, content analysis is the appropriate method for objectively identifying significant text within a large volume of newspaper text (Neuman, 2006). A coding system was developed to interpret the data. This coding system used two units of analysis: the article and the statement. A statement refers to what has been spoken by a certain actor or speaker on the selected issue. Every speaker may have been coded in more than one place in the same article, but all of his or her discussions was coded together as one statement. The selected articles were first coded according to date, media sources, news factors, events and newspaper sections. Then, at the statement level, each speaker was coded according to their typology: politicians, civil administration, forest and environment administration, scientists, journalists, forest enterprises, forest and non-forest non-governmental organizations (NGO), interstate organizations (such as the UN and EU), the World Bank, local individuals and experts. This speaker classification is based on the various types of involved stakeholders or actors in the forest and climate change issues (Park, 2009; Real, 2009). The next step was to determine the frames from each statement made by the individual speakers. Framing is the process of highlighting certain aspects of the forest and climate problem in terms of its causal interpretation (i.e., diagnostic frame), graveness and urgency of the issue (i.e., motivational frame) and/or solution suggestions (i.e., prognostic frame) (Feindt & Kleinschmit, 2011; Park, 2009; Entman, 1993; Bendford & Snow, 2000; Semetko & Valkenburg, 2000). Based on these framing elements, this study formulates 24 frame categories covering the aspects of how the forest was related to climate change. For example, in using the forest as a carbon sink, the frames are the forest's use of CO_2, climate change and forest productivity, climate change and biodiversity, sea level rise and forest cover and climate change in relation to wildlife conflicts. These frames are drawn from the literature review on forest and climate change issues at the global level (Real, 2009; Etkin & Ho, 2007) and from the Bangladesh climate change assessment report (MoEF, 2008). Then, the statements made by each actor/speaker were categorized according to the ways they related the forest with climate change. Several roles of the forest were framed for mitigating global warming or climate change by explaining how several forest issues such as deforestation and illegal logging are contributing to climate change and how climate change will affect the forest and biodiversity (Park, 2009; Kleinschmit et al., 2009).

Results: The Two Worlds of National and International Media Discourse

Framing of the Forest within the Climate Change Media Discourse

The analysis of framing the forest issues within climate change discourse using the different speakers indicated differences between the two media. The frequencies of specific frames are measured by the percentage of the total number of statements about forest and climate frames in each of the media (i.e., the "International Herald Tribune" (n = 149) and the "The Daily Ittefaq" (n = 52)). In total, 16 frames were found from the 24 pre-fixed frames. This reduction indicates that media does not report on every aspect of the forest and climate issues;

rather, they are selective in choosing the frames that pass through their selection filter. The 16 frames found are not equally represented in both media, but both international and national print media each stressed some of these frames. The national print media stressed on more motivational frames (i.e., the graveness and urgency of the issue), whereas the international print media stressed on prognostic frames (i.e., problem solving suggestions). In the national print media, motivational frames are dominant (see **Figure 1**), which depict the adverse impact of climate change on the forest, e.g., the loss of forest land due to sea level rise (34.62%), the forest's difficulty in coping against climate change (17.31%) and climate change's effect on biodiversity (3.85%). In addition, prognostic frames, which talk about the forest's role in adapting to climate change, are also present in the national media but in a smaller percentage, as the forest's positive role as a carbon sink (11.54%) and the role of afforestation in the adaptation against climate change (13.46%). Another share of frames (13.46%) are diagnostic frames that depict deforestation as a contributor to climate change, with 5.77% of the frames found in the national print media agreeing that REDD is the right instrument to handle the situation (see **Figure 1**). In the national print media, the frames are more inclined to support the crisis argument that the forests of Bangladesh are facing an imminent threat from the global environmental problem of climate change. This kind of framing in the national media may have resulted from the tendency to shift responsibility outside the national boundary in order to obtain more international assistance in solving the problem.

The media frames within the international media highlighted the forest's role in mitigating climate change, where the forest as a carbon sink was the most dominant frame with 21.52%; the use of the forest for mitigating and adapting CO_2 (8.86% and 11.39%, respectively) were also mentioned. While the potential threat of climate change to forests is present in a smaller percentage, with 0% of sea level rise as a threat to forest cover, the forest's difficulty to adapt against climate change represents 7.59%, and climate change's effect on biodiversity represents 3.80%. However, 7.59% of frames in the international media did conclude that REDD is the right instrument for reducing emissions from deforestation.

In international print media, framing mostly highlights the potential role of the forest in mitigating climate change problems and downplays the potential threat of climate change to the forest and its biodiversity. On the contrary, in national media, the framing of the forest and climate issue is characterized as the traditional forest in crisis. For example, Ataur Rahman, an environmental activist, used the forest in crisis argument in his framing of the forest and climate issue: "13% of Bangladesh coast with the Sundarbans may go under water during the next 50 years" (The Daily Ittefaq, 10.11.2007). The empirical findings support the first hypothesis, which states that the framing of the forest within climate change discourse is quite different in international print media when compared to the national print media of Bangladesh.

Early Forest and Climate Discourse in the National Media

The relative importance of the forest within the climate discourse is reflected in the growing attention it has received in the national and international print media between 1989 and 2010 (**Figure 2**). A total of 49 newspaper articles dealing with the forest and climate change issue were found in the national newspaper, The Daily Ittefaq. In the international media, i.e., the International Herald Tribune, 90 articles resulted from the collection. The reporting on the forest and climate issue in the national media was first observed in 1989 and 1990. Sporadic reporting was found until it increased beginning in 2006 and peaking in 2010. However, in the International Herald Tribune, reporting on forest and climate issues were first observed in 1995, reaching a strong peak in 2007 (23) with significant reporting in 2009 (14) and in 2010 (15). This analysis revealed that forest issues were not present in the climate change discussion prior to 1999-2000 in international media. However, a small amount of sporadic reporting was found in national media between 1989 and 2000. The trend shows that after 2007, forest issues are more prominent in the climate discussion in the national print media.

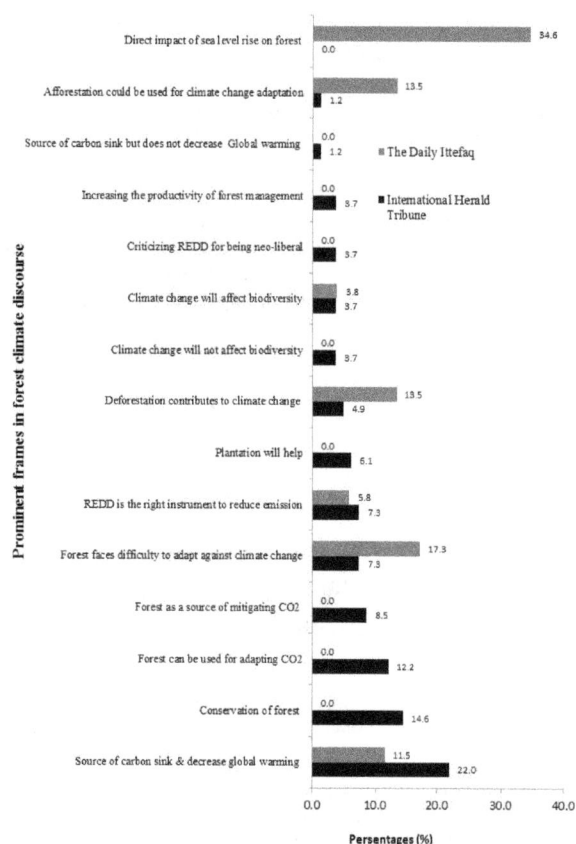

Figure 1.
Forest frames in the climate change media discourse (by share of all statements about frames in %).

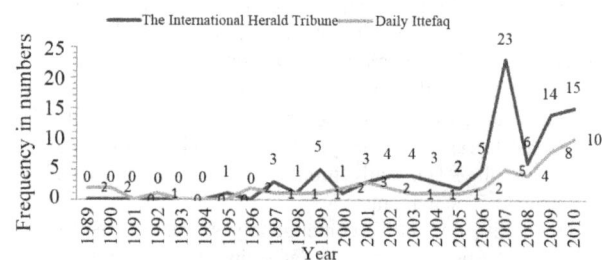

Figure 2.
Frequency of articles addressing climate change and forests.

The comparison of the frequency of articles proves that the attention drawn from the national and international media to the forest issues within the climate discussion shows a difference in timing (i.e., the national media linked the forest with global warming or climate change issues before the international media). Five articles were found in the national print media during the first 5 years in The Daily Ittefaq, which primarily expresses the grim danger global warming could bring to the forests of Bangladesh. For instance, the environment minister of Bangladesh tells the media that "Shundarban will be lost to the sea in the course of time due to sea-level rise" (The Daily Ittefaq, 13. 03.1989). Similarly, an environment staff reporter of The Daily Ittefaq reported that "lowlands of Bangladesh are under a serious threat including the world's largest mangrove forest" (The Daily Ittefaq, 18.07.1990). An identical statement from the environment minister Mr. Abdullah Al Noman was published: "low-lying coastal land of Bangladesh will be flooded by the year 2050 if the sea-level rose to 32 cm due to global warming and this affects the forest in terms of biodiversity and production" (The Daily Ittefaq, 28.02.1996). However, the forest as a topic gained more momentum in the climate change discourses after 2006 in both print media where the national media has a consistent growing trend, but in the international media, forests have been mentioned inconsistently in climate change topics; therefore, the empirical findings reject the second hypothesis as the Bangladeshi print media reported before the global media on new global issues, such as the relationship between climate change and forests.

There are differences in the content and in the timing of national and international media discourse. Our next hypothesis provides an explanation by referring to the strategic framing by political actors in the media.

Strategic Framing by Political Actors in Media Discourse

In the national media, journalists (46.2%) are the dominant speakers, followed by politicians (28.8%) and scientists (11.5%). Single individuals or communities (5.8%), NGOs (1.9%), development consultants (1.9%), interstate organizations (1.9%) and administrations (1.9%) represent the rest of the speakers in the national media. However, scientists (34.2%) have dominant standing in the international media, followed by the forest NGOs and other NGOs (22.1%), journalists (12.7%) and interstate organizations (8.7%). Politicians (6%) and the World Bank (5.4%) are the other two prominent speakers who shape the international forest and climate discourse (see **Figure 3**). It is evident that the national print media discourse is mostly shaped by the politicians and journalists; thus, the discourse is primarily a political one. In the national print media, the discourse is driven by the actors, who need to legitimize their policy decisions or need to create a public opinion in favour of their interests. This supports the argument of Bendford and Snow (2000) that framing processes are strategic, deliberative and goal oriented, which means that actors frame a certain issue within an broader discourse to pursue their interests (Bendford & Snow, 2000; Somorin et al., 2011). In contrast, the international media is dominated by scientists and NGOs. In this media, the minimal presence of politicians is evident. The differential standing of actors in both print media leads to different framings of the forest issue.

Politicians and journalists are dominant in the national print

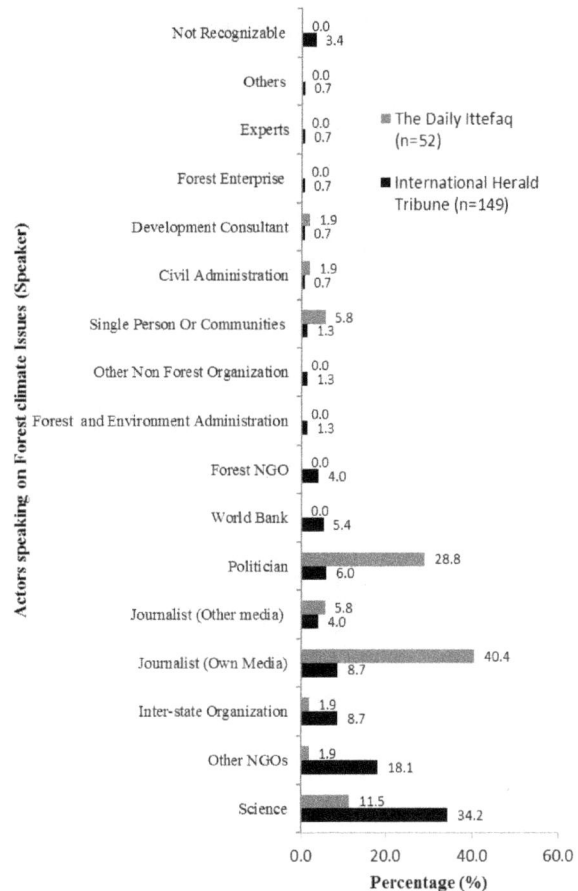

Figure 3.
Speaker in both media.

media, using the forest in crisis frame and stressing the adverse impact of climate change on forests, which is beyond their control, as it was in the case of forest dieback in Germany during the 1980s (Glück, 1986; Krott, 2005; Holzberger, 1995). This kind of crisis framing complies with the concept of the "paradox of disaster" in the media, which supports their policy standpoint and interest to scale up the potential danger to the forest from climate change. Such a framing helps the politician's attempts to mobilize public opinion and voters in favour of their policy interests by providing information in a particular order (Kleinschmit & Krott, 2008; Jacoby, 2000). In addition, it also legitimizes the argument that the degrading forest situation of Bangladesh needs foreign help and financing for its conservation (MoEF, 2008).

Scientists and NGOs focus the global perspective on the forest as a potential mitigator, representing the solution of global warming in the international print media. However, REDD and afforestation issues have recently been given importance by strong actors in national media, which indicates the motivation for the change of national political actors towards the current forest and climate issues that can bring international finance to the national sector; thus, they are now consistent with the global forest and climate discourse. That shift in focus supports Hajer's concept of discourse coalition, i.e., when political actors shift from one discourse coalition to another and adopt a different framing strategy based on the same issues because their interests have shifted or the new framing is more suitable for their cause (Hajer, 1995). For instance, the prime minister of Bangladesh, ShiekhHasina showed the government's intention to participate in the REDD process: "Bangladesh is now ready to participate in REDD process under CDM" (The Daily Ittefaq, 01.06.2010). Similarly, Mr.SalemulHuq, a development consultant, stated that "Afforestation is a solution for global warming and could also provide opportunity to Bangladesh". These two examples along with other similar statements made by actors in the national media during last 3 years indicate a shift in the framing of forest and climate politics from crisis argumentation to the forest as a solution to climate change. The empirical findings show that the powerful political actors are able not only to frame but also re-frame issues and discourses (Arts et al., 2010). Hence, the third hypothesis of the study is supported by the empirical findings. The topics and framing within the media is influenced by actors who speak in the media. Therefore, strong actors and their interests can explain the content and timing in the national and international media.

Conclusion: The Different Worlds of Media Discourse

The analysis revealed substantial differences between the forest and climate discourse of the national and international media. The first hypothesis is supported, which states that there are differences in framing. In national print media, the dominant frames exist around the forest in crisis due to climate change arguments (i.e., motivational framing), while in international print media, the dominant frames describe the forest as a solution to climate change (i.e., prognostic framing).

The second hypothesis is rejected, which states that international media are driving the national media discourse. In the national media of Bangladesh, the forest/climate discourse began five years earlier than in the international media. This initial national media discourse does not mean that international political discourse could not have been influential in Bangladesh. This may have occurred through political discourse but was not used directly on the media level.

The differences in framing can be explained by the influence of strong actors at the national and international levels. Journalists and politicians are the strongest speakers in the national print media (The Daily Ittefaq), and they framed the discussion primarily around the adverse impacts of climate change on the forests in Bangladesh, a country that will face severe impacts of climate change. The potential loss of the forest due to a rise in sea level is a major discussion in the national forest and climate discourse. Scientists, NGOs and international organizations are the major speakers in the international print media (International Herald Tribune). They shape the global forest and climate media discourse around the larger scope of the forest within climate change. International scientists and NGOs present themselves as problem solvers in the climate change issue by framing the mitigating role of forests. This solves problems solely on the global level but not for Bangladesh, where the danger of increasing sea level is a primary concern. The scientists and NGOs thus adopt the role of global helpers, therefore establishing the third hypothesis, which is that the framing of forest and climate politics is influenced by the speakers who follow their strategic interests.

In addition, at the end of the study period, i.e. 2008 and 2009, the powerful actors of Bangladesh (politicians and administration) emphasized the potential use of forest plantations and REDD+ for mitigation. This indicates that the politicians of Bangladesh sense opportunities to obtain financial resources from these global initiatives. This also supports the third hypothesis, which is that the strong actors are able to re-frame the media discourse according to the shift of their interests over time.

The analyses show that the media discourse exists on multiple levels, is diverse and offers different support for specific policies. However, support by the media for either crisis policy or mitigation policy has not yet induced policy change. In addition to the media discourse, many other factors influence policy making. Our media analysis allows for further research on whether the media discourse reflects and supports different forest and climate policies on the national and international levels.

Acknowledgements

I would like to thank Dr. Daniela Kleinschmit of SLU, Sweden, and Mi Sun Park for their help in preparation of the coding book for the content analysis. I am also grateful to the TIFF and SUFONAMA Class of 2011 from Georg-August University and the Students of FWT Discipline of Khulna University Bangladesh for coding the sample article. Finally, I would like to thank DAAD for financing my PhD project.

REFERENCES

Ali, A. (1999). Climate change impacts and adaptation assessment in Bangladesh. *Climate Research, 12,* 109-116.

Arts, B., Appelstrand, M., Kleinschmit, D., Pülzl, H., Visseren-Hamakers, I. et al. (2010). Discourses, actors and instruments in international forest governance. In: J. Raynor, A. Buck, & P. Katila (Eds.), *Embracing complexity: Meeting the challenges of international forest governance.* IUFRO.

Bendford, R. D., & Snow, D. A. (2000). Framing processes and social movements: An overview and assessment. *Annual Review of Sociology, 26,* 611-639.

Boykoff, M. T., & Boykoff, J. M. (2004). Balance as bias: Global warming and the US prestige press. *Global Environmental Change, 14,* 125-136.

Chong, D., & Druckman, J. N. (2007). Framing theory. *Annual Review of Political Science, 10,* 103-126.

Curran, J. (2002). *Media and power.* London/New York: Routledge.

Entman, R. M. (1993). Framing: Toward clarification of a fractured paradigm. *Journal of Communication, 43,* 51-58.

Etkin, D., & Ho, E. (2007). Climate change: Perceptions and discourses of risk. *Journal of Risk Research, 10,* 623-641.

Fairclough, N. (1995). *Critical discourse analysis: A critical study of language.* London: Longman.

Feindt, P. H., & Kleinschmit, D. (2011). The BSC crisis in German newspapers: Reframing responsibility. *Science as Culture, 20,* 183-208.

Gerhards, J. (1994) Politische öffentlichkeit. Ein system und akteurstheoretischer bestimmungsversuch. In F. Neidhardt (Ed.), *Öffentlichkeit, öffentliche meinung, soziale bewegungen, kölner zeitschrift für soziologie und sozialpsychologie, sonderheft 34/1994* (pp. 77-105). Opladen: Westdeutscher Verlag.

Glück, P. (1986). Seminar "Waldsterben und öffentlichkeitsarbeit". *Allgemeine Fosrtzeitung, 97,* 359-364.

Hajer, M. A. (1995). *The politics of environmental discourse.* Oxford: Oxford University Press.

Hardt, F. (2004). *Mapping the world: New perspectives in the humanities and social sciences.* Tübingen: Francke Verlag.

Held, D., McGrew, A. et al. (1999). *Global transformations: Politics, economics and culture*. Stanford: Stanford University Press.

Holzberger, R. (1995). *Das sogenannte waldsterben. Zur Karrier eines klisches: Das thema wald im journalistischen diskurs.* Bergatreute: Eppe .

Huq, S., Karim, Z., Asaduzzaman, M., & Mahtab, F. (1999). *Vulnerability and adaptation to climate change for Bangladesh*. Dordrecht: Kluwer Academic Publishers.

IHT (2012). *A short history on the international herald tribune.* URL. http://www.ihtinfo.com/media/59755/iht2064_short_history_2012-5.pdf

Jacoby, W. (2000). *Imitation and politics; redesigning modern Germany.* Ithaca, NY: Cornell University Press.

Keller, R. (1997). *Diskursanalyse.* In R. Hitzler, & A. Honer (Eds.), *Sozialwissenschaftliche hermeneutik,* Opladen: Leske Und Budrich.

Kingdon, J. W. (2003). *Agendas, alternatives and public policies.* New York: Addison-Wesley Educational Publishers Inc.

Kleinschmit, D. (2012). Confronting the demands of a deliberative public sphere with media constraints. *Forest Polica and Economic, 16,* 71-80.

Kleinchmit, D., & Krott, M. (2008). The media in forestry: Government, governance and social visibility. In T. Sikor (Ed.), *Public and private in natural resource governance: A false dichotomy?* London: Earthscan.

Kleinschmit, D., Sadath, N., Park, M. S., & Real, A. (2009). Mthods for media analysis on forest climate policies: A thematic coading book. Working Paper. Göttingen: Forest and Nature Conservation Policy Institue, Georg August University Göttingen.

Krippendorff, K. (1980). *Content analysis an introduction to its Methodology.* London: Sage.

Krott, M. (2005). *Forest policy analysi.* New York: Springer.

Leakey, R., & Lewin, R. (1995). *The sixth extinction: Biodiversity and its survival.* London: Weidenfeld and Nicolson.

Levin, K., McDermott, C., & Cashore, B. (2008). The climate regime as globel forest governance: Can reduced emissions from deforestation and forest degradation (REDD) initiatives pass a "dual effectiveness" test? *International Forest Review, 10,* 538-549.

Mazur, A., & Lee, J. (1993). Sounding the global alarm: Environmental issues in the US national news. *Social Studies of Science, 23,* 681-720.

mcQuail, D. (1994). *Mass communication theory: An introduction.* London, Thousand Oaks, New Delhi: SAGE.

McQuail, D. (2010). Global mass communication. In D. McQuail (Ed.), *McQuail's mass communication theory* (6th ed.) (pp. 247-270). Los Angeles, London, New Delhi, Singapore, Washington DC: SAGE.

MoEF (2008). *Bangladesh climate change strategy and action plan*

2008. Dhaka: Ministry of Environment and Forests, Government of the People's Republic of Bangladesh.

Neuman, W. L. (2006). *Social research methods qualitative and quantitavie approach* (6th ed.). Upper Saddle River: Pearson.

Park, M. S. (2009). *Media discourses in forest communication: The issue of forest conservation in the Korean and global media.* Göttingen: Cluvillier Verlag.

Pimm, S. L., & Brooks, T. M. (2000). *The sixth extinction: How large, howsoon, and where? Nature and human society: The quest for a sustainable world.* Washington DC: National Academy Press.

Prittwiz, V. (1990). *Das Katastrophen-Paradox, Elemente einer Theorie der Umvelt politik.* Opladen: Leske Und Budrich.

Rahman, M. (2010). Climate change coverage on the mass media of Bangladesh. *Global Media Journal pakistan edition, 3.*

Real, A. (2009). *Discourses and distortions: Dimensions of global and national forest science communication.* Göttingen: Faculty of Forest Science, Georg August University Göttingen.

Reese, S. D. (2001). *Framing public life. Perspectives on media and our understanding of the social world.* Mahwah.

Reese, S. D. (2010). Journalism and globalization. *Sociology Compass, 4,* 344-353.

Sadath, M. N., Kleinschmit, D., & Giessen, L. (2012). Framing the tiger: Incongruent media discourses at different political levels as an obstacle for biodiversity governance. *Forest Policy and Economics* (in Review Process).

Schäfer, M. (2008). Medialisierung der wissenschaft? Empirische untersuchung eines wissenschaftssoziologischen konzept. *Zeitschrift für Soziologie, 37,* 2006-225.

Semetko, H., & Valkenburg, P. M. (2000). Framing European politics: Acontent analysis of press and television news. *Journal of Communication,* 93-109.

Somorin, O. A., Brown, H. C., Visseren-Hamakers, I. J., Sonwa, D. J., Arts, B., & Nkemd, J. (2011). The Congo Basin forests in a changing climate: Policy discourses on adaptation and mitigation (REDD+). *Global Environmental Change, 22.*

Sparks, C. (1998). Is there a global public sphere? In D. K. Thussu (Ed.), *Electronic empires: Global media and resistance* (pp. 108-124). London: Arnold.

Takahashi, B. (2008). Framing climate change: A comparitive analysis of a US and a Canadian newspaper. *International Journal of Sustainability Communication,* 152-170.

UNESCO (1997). *The media and the challenge of the new tecnology.* World Communication Report, Paris.

Wu, H. D. (1998). Investigating the determinants of international news flow: A media analysis. *International Communication Gazette, 60,* 493-506.

6

The Concept of the Ergonomic Spectrum

Yozo Yamada[1], Efi Yuliati Yovi[2], Dianne Staal Wästerlund[3], John J. Garland[4], Janusz M. Sowa[5]

[1]Graduate School of Bio-Agricultural Sciences, Nagoya University, Nagoya, Japan
[2]Faculty of Forestry, Bogor Agricultural University, Bogor, Indonesia
[3]Department of Forest Resource Management, Swedish University of Agricultural Sciences, Umeå, Sweden
[4]Garland & Associates, Waldport, USA
[5]Department of Forest and Wood Utilization, University of Agriculture in Krakow, Krakow, Poland

Forestry conditions differ among regions and nations. Moreover, labor costs, forestry mechanization, and environmental impacts are also different. These factors directly or indirectly influence the ergonomic state of nations. The ergonomic state of a nation can be described in terms of ergonomic factors such as labor productivity, work accidents, physiological burden, and stress. Labor productivity and work accidents can be defined as income or condition factors, and physiological burden and stress as outcome or result factors. Thus, the value of outcome factors must be examined in relationship to income factors. On the ergonomic spectrum, each factor can be conceived as a continuum from a negative to a positive ergonomic status. All factors can be set in a line, and the present state of each nation is indicated by a profile formed by the assembled factors. The locations of nations along the two-dimensional coordinates of the world standard can be realized by an ergonomic spectrum. Moreover, future directions for improvement can be obtained by reference to the three-dimensional coordinates, which include the axis of time.

Keywords: Ergonomic Spectrum; Economic Situation; Forestry Condition; Working Condition; Work Accident

Introduction

Globally, forest engineering has been shifting from manual to mechanized work. Consequently, ergonomic issues have changed from physical or physiological burdens of forestry workers to psycho-physiological stressors of machine operators. However, many developing countries continue to rely on manual work for most of their forestry operations. Manual work is a classic ergonomic issue, and it remains significant in developing countries. Moreover, some countries, such as Japan, are located midway between manual and mechanized forestry. Forestry mechanization has spread in Japan except in the area of felling operations due to the steep and complicated terrain. This situation cannot be improved unless an innovative technique is developed to overcome the adverse natural conditions.

Manual felling operations with a chainsaw have been recognized globally as one of the most dangerous jobs in forestry, and many studies have been conducted from a safety perspective. For example, Hammond studied critical safety behaviors of differently skilled workers using a helmet camera in the USA (Hammond et al., 2011). Bentley noted the potential for injury among inexperienced fellers in New Zealand and claimed that felling safety is dependent upon an appropriate assessment of hazards and good judgment with respect to decisions regarding felling (Bentley et al., 2005). Felling operations in Japan have been the most dangerous forestry work during this half century (Oka et al., 2011), and many severe accidents have occurred during manual felling operations in Sweden among private forest owners (Lindroos & Burstrom, 2010).

Wearing protective devices to reduce work accidents during manual felling operations is one of the most effective direct safety countermeasures. Kashima claimed that wearing protective trousers while operating a chainsaw results in a 60% decreases in lower body injuries (Kashima & Uemura, 2008). However, many of the protective devices have been developed in cool and dry climate countries and are not suitable for use in tropical or wet temperate regions. Workers in those regions hate to wear protective devices, even helmets and gloves, because they are uncomfortable. Wästerlund researched heat stress in forestry and argued for a standard research method to determine protective clothing comfort (Wästerlund, 1998). Holland evaluated the ventilation capacity of various helmets for forest harvesting (Holland et al., 2002).

Ergonomic issues vary in every nation, as each has different forestry conditions. Although Feyer compared fatal occupational injuries among US, Australian, and New Zealand workers (Feyer et al., 2001), very few cooperative studies have involved multiple nations. Thus, it is difficult to describe the present ergonomic state of each nation using a simple world standard.

This report offers the ergonomic spectrum as a new idea to objectively identify current forestry ergonomic conditions in each nation and to identify the actual and more preferred future direction to improve each nation's ergonomic conditions. At first, we introduce the concept of ergonomic spectrum, and then explain how to select indices for using the ergonomic spectrum, and finally show two experimental examples.

Concept of Ergonomic Spectrum

Everyone wants to eliminate forestry accidents. If this occurs thorough forestry mechanization, forestry work accidents will decrease dramatically, including in felling operations. Unfortunately, it is quite difficult to realize this scenario because of the forestry conditions in each nation. Work accidents are influenced significantly by forestry conditions, which are influenced by national and regional indigenous characteristics such as geography, topography, climate, vegetation, population, wood demand, wood trade, and economic and social conditions. Moreover, labor costs, forestry mechanization, labor productivity, and environmental impacts also differ among nations. Thus, working conditions directly or indirectly influence the ergonomic state of each nation.

The ergonomic state of a nation can be estimated through ergonomic factors such as work accidents, physiological burden, and stress. These factors indicate the current ergonomic situation based on forestry activities in each nation, and improvement in these factors would result in safer conditions, lighter work and stress loads, and more comfortable working conditions. If forestry and working conditions are defined as income or conditions, then ergonomic factors can be defined as outcome or results. The value of outcome factors must be examined relation to the income factors.

However, it is unrealistic to comparatively evaluate factors that differ in substances, units, and scales. Thus, the spectrum concept is used to allow a clear comparison among different ergonomic situations in the world.

With respect to a spectrum, the emotional spectrum and the recreational opportunity spectrum are well known to include some vectors. The emotional spectrum (ES) can evaluate an emotional state of human from the electroencephalogram, and is composed of four element emotion vectors; anger, sadness, joy, and relaxation. The recreation opportunity spectrum (ROS) is an applied example. The ROS is a combination of physical, biological, social, and managerial conditions that give recreational value to a place (Clark & Stankey, 1979).

The basic concept of the ergonomic spectrum is similar to that of the ES and the ROS. Whereas the ROS separates nations into seven classes from the least to the most remote (British Columbia Resources Inventory Committee, 1998), the ergonomic spectrum is more complicated and not as easily divided into patterns.

As shown in **Figure 1**, each factor is represented along a continuum from a negative to a positive ergonomic state, with a negative state shown in orange, and a positive one in green. All factors are set in a line, and the present state of each nation is indicated by a cross-section of the assembled factors. It appears like a spectrum; thus, it is easy to recognize where a nation is located compared to the world standard using the two-dimensional coordinates. Moreover, the directions needed for improvement can be shown in a three-dimensional coordinates that include the axis of time.

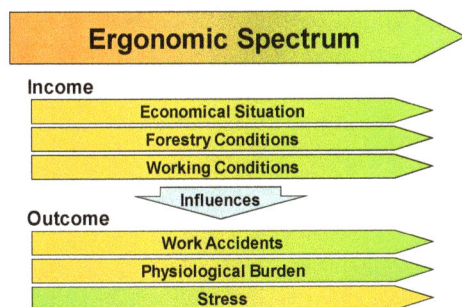

Figure 1.
Concept of the ergonomic spectrum.

Selection of Ergonomic Spectrum Indices

Indices for the ergonomic spectrum were defined as the economic situation, forestry conditions, working conditions, and work accidents. The first three factors are income factors, and the last is an outcome factor. Physiological burden and stress could not be included because these types of data are difficult to gather from official reports. Although researchers sometimes report these data, the studies are often case studies and usually do not represent average burdens of all workers in a country. Furthermore, other factors are applicable to the ergonomic spectrum, but data on these could not be obtained for every country; thus, they were eliminated from this study.

Economical Situation

National economic and social situations have great impact on forestry as an industry. The economic situation of a nation can be described by many factors such as nominal or real gross domestic product (GDP), nominal or real gross domestic index, the domestic population, consumer price index, and Engel's coefficient. In this study, the nominal GDP per person, as an index of the national economic situation, and Engel's Coefficient, as an index of rich and poor, were selected for use in the ergonomic spectrum.

Forestry Conditions

Forestry conditions are characterized by climate zone, average slope gradient, forest coverage rate, artificial forest rate, planted species, wood demand, wood self-support ratio, forest road density, and forestry mechanization rate. Data on some of these factors are difficult to obtain because no description is available in official reports. In this study, the forest coverage rate, as an index of potential wood resources, and the self-support ratio of timber, as an index of logging activities, were selected.

Working Conditions

Working conditions can have a crucial impact on work accidents, physiological burden, stress, and decision making during forestry work. Working conditions include wage, number of forestry workers, the age index, management area, and labor productivity. In this study, the average wage, average management area, and average productivity were selected as representative factors. The average management area per person was calculated by dividing the forest area by the number of forestry workers. Although the actual number of workers cannot be compared against figures for other countries, management area can be estimated by the number of workers. The average productivity per year per person, which is calculated by dividing felling volume by the number of forestry workers, also indicates the degree of mechanization.

Work Accidents

Work accidents are external factors or objective variables explained by the economic situation, forestry circumstances, and working conditions. In this study, the work accident rate and the fatal accident rate were selected as factors. These are expressed as the rate of workers injured or killed per 1 million felled trees. An effort must be made to decrease these objective factors through every possible means. Unfortunately, Indonesia has not surveyed the number of workers injured and killed in their forestry operations.

Experimental Examples

Data of nine factors for the ergonomic spectrum were obtained from official reports from five countries, Sweden, USA, Poland, Indonesia, and Japan, to compare the ergonomic situations among them. Moreover, data were gathered at five different times in Japan, namely 1970, 1980, 1990, 2000, and 2010, to observe changes in the ergonomic situation.

Comparison of Ergonomic Spectra among the Five Nations

Figure 2 shows a comparison of the ergonomic situations among the five countries. The right-hand side of the graph is the desirable situation. Sweden's profile shows values mostly well to the right side, with little variation, whereas values for the other countries range from values toward the left-hand side to values approaching the right-hand side. The five countries show quite different profiles. The USA maintains a positive status in terms of economic situation, number of work accidents, and management area but earns low or moderate scores for forest coverage rate, the self-support ratio, wages, and productivity. Poland maintains low or medium status for the economic situation, forestry conditions, and working conditions but scores high for work accidents. Indonesia maintains low scores for their economic situation and working conditions but gets the highest marks for self-support ratio. Japan fluctuates wildly among factors. Japan gets high marks for nominal GDP, forest coverage rate, and wages. However, the high wages are due to appreciation of the yen against the US dollar. In contrast, Japan gets low scores for the self-support ratio, management area, and productivity, which are significant problems in the Japanese forestry industry. Moreover, the work and fatal accident rates are worse than those in the other countries.

The forest coverage rate is fixed in each country, and the nominal GDP and Engel's coefficient cannot be improved by the forestry sector. The self-support ratio and wage could be changed if forestry realizes a revitalization. Management area and productivity can be improved by multi-functional mechanization, but it is not easy to promote multi-functional machines in Japan and Indonesia. Without high mechanization, the accident rates can be reduced only by wearing protective de-

vices and safety training for forestry workers.

Changes in the Japanese Ergonomic Spectrum

Figure 3 shows the changes in the Japanese ergonomic spectrum over four decades. In 1970, values for all factors except forest coverage rate were low. In 1980, nominal GDP, Engel's coefficient, and wage increased gradually. Notably, the fatal accident rate decreased largely, but the work accident rate was the worst. In 1990, the former three factors continued to increase gradually, and the work accident rate improved dramatically. Since 2000, the former three factors have continued to increase, and the work accident rate has improved slightly. The fatal accident rate has not changed since 1980.

Although wages increased dramatically due to the high yen, the actual wage was not so high and has actually decreased since 2000. The nominal GDP and Engel's coefficient have increased steadily, and productivity and management area have increased slightly. Improving those factors in Japan will not occur smoothly because of the difficulties of multi-functional mechanization.

The work accident rate improved largely from 1980 to 1990. However, improvement has slowed in the past two decades. Moreover, the fatal accident rate has become slightly worse since 2000. Manual felling remains common in Japanese forestry. The reduction in work injuries and fatal accidents is the most significant and urgent issue. In the future, Japan should struggle to improve work conditions, fatal accident rates, productivity, and the self-support ratio.

Conclusion

We can appreciate the ergonomic situation of Japan in relation the world using the ergonomic spectrum. Moreover, we can identify the direction for improvement in our forestry ergonomic situation based on the changes in the ergonomic situation through the past several decades. In other words, we can understand easily where we are now and which directions we should improve.

However, the following problems must be solved for the ergonomic spectrum to become a useful tool. The first problem is factor selection. Which factors should we choose for the ergonomic spectrum? We must discuss whether the factors are sufficient to show the actual ergonomic situation. Moreover, we must examine whether a factor can be corrected even if it is

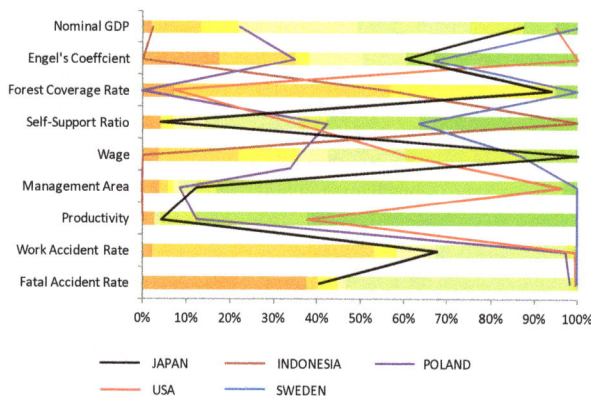

Figure 2.
Comparison of the ergonomic spectra among the five nations.

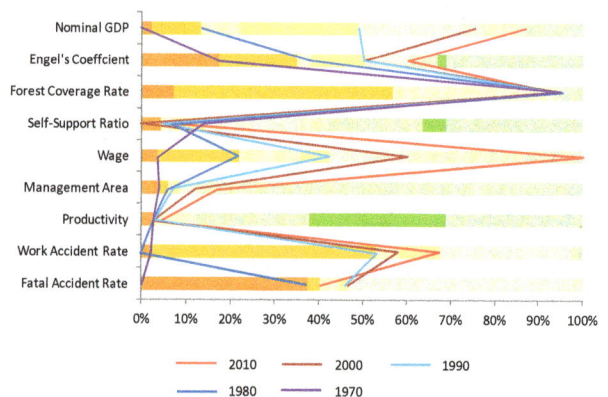

Figure 3.
Changes in the Japanese ergonomic spectrum.

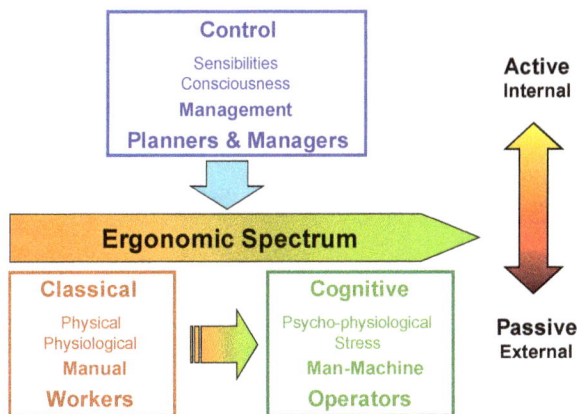

Figure 4.
Ergonomic strategy based on the ergonomic spectrum.

significant. Some countries may not have official reports or technical studies on some factors such as work accidents, physiological burden, and stress. Although it is sometimes suitable for some factors to be described qualitatively, whether those qualitative factors should become part of the ergonomic spectrum must be considered.

The second problem is differences among operations. Felling is the most dangerous manual operation, as it requires the use of a chainsaw. In Japan, delimbing and bucking operations have become safer due to more widespread use of processors. We can select suitable operations according to what we want to know. In this study, all forestry operations and all data were considered.

The third problem is classification of the ergonomic spectrum. In this study, the five countries displayed characteristically different spectra. However, it was not clarified whether those spectra represented each situation. Thus, more data must be collected from more countries if we are to describe more groups using the spectrum.

The fourth problem is the need to consider which elements of the ergonomic spectrum should be targeted for improvement. It may be true that the ultimate goal is the same throughout the world. Surely, no one wants accidents in the work place. However, in each nation, some factors can be improved and others are indigenous to the country and cannot be improved. For example, Japan cannot reach the same ergonomic situation as Sweden because its climate and terrain are different. Thus, we must find another goal that is suitable and can be accomplished. The direction of improvement is different for different countries. Studying changes in the ergonomic situation will help to identify appropriate directions for change.

The ergonomic spectrum is intended mainly for forestry workers and operators to improve their working conditions. Although improvements are traditionally basic and essential issues in ergonomics, they cannot be accomplished without the help of management, local officials, owners, planners, supervisors, and leaders. In this sense, improvements can be defined as external and passive from the point of view of workers and operators.

We must think of extending future ergonomics to forest management control spheres (**Figure 4**). Control ergonomics, which is based on the ergonomic spectrum, aims to enhance the sensibilities and consciousness of management and fosters internal change in each manager in working not only with staff but also with workers and operators. It is also concerned with classical ergonomics and cognitive ergonomics and considers management philosophy, design, decision making, social responsibility, evaluation, improvement, and education and training for workers and operators. In this sense, control ergonomics can be defined as active ergonomics.

REFERENCES

Bentley, T. A., Parker, R. J., & Ashby, L. (2005). Understanding felling safety in the New Zealand forest industry. *Applied Ergonomics, 36,* 165-175.

British Columbia Resources Inventory Committee (1998). Recreation opportunity spectrum inventory: Procedures and standard manual Ver.3.0. *Resources Inventory Committee, 39.*

Clark, R. N., & Stankey, G. H. (1974). *The recreation opportunity spectrum: A framework for planning, management, and research.* USDA Forest Service Research Paper PNW-98.

Feyer, A. M., Willamson, A. M., Stout, N., Driscoll, T., Usher, H., & Langley, J. D. (2001). Comparison of work related fatal injuries in the United States, Australia, and New Zealand: Method and overall findings. *Injury Prevention, 7,* 22-28.

Hammond, T. R., Rischitelli, G., & Wimer, J. A. (2011). Defining critical safety behaviors in a point-of-view video observation study of tree fallers at work. *International Journal of Occupational and Environmental Health, 17,* 301-306.

Holland, E. J., Laing, R. M., Lemmon, T. L., & Niven, B. E. (2002). Helmet design to facilitate thermoneutrality during forest harvesting. *Ergonomics, 45,* 699-716.

Kashima, J., & Uemura, T. (2012). Report on injured parts by saw-chain in chain-saw operations and protective effect of protective clothes. *Journal of the Ceramic Society of Japan, 22,* 275-278.

Lindroos, O., & Burstrom, L. (2010). Accident rates and types among self-employed private forest owners. *Accident Analysis & Prevention, 42,* 1729-1735.

Musha, T., Terasaki, Y., Haque, H. A., & Ivanitsky, G. A. (1997). Feature extaction from EEGs associated with emotions. *Artificial Life and Robotics, 1,* 15-19.

Oka, M., Nakazawa, M., Sasaki, T., Yoshida, C., Uemura, T., Kashima, J., & Kato, T. (2011). Studies on the forestry death disaster of introduction of high performance forestry machines in 10th year. *Journal of the Ceramic Society of Japan, 26,* 27-34.

Wästerlund, D. S. (1998) A review of heat stress research with application to forestry. *Applied Ergonomics, 29,* 179-183.

7

Growth Characteristics, Biomass and Chlorophyll Fluorescence Variation of Garhwal Himalaya's Fodder and Fuel Wood Tree Species at the Nursery Stage

Azamal Husen

Department of Biology, Faculty of Natural and Computational Sciences, University of Gondar, Gondar, Ethiopia

Fodder and fuel wood deficiency in the Himalayan region is well recognized. Rural inhabitants are exploiting these forest resources for their livelihood for generations which leads to severe deforestation. The aim of this study was to identify the fast growing fodder and fuel wood tree species of Garhwal Himalayas at nursery stage with wider relevance and great potential for extensive afforestation programmes. Seed of *Bauhinia purpurea* L., *Bauhinia retusa* Roxb., *Bauhinia variegate* L., *Celtis australis* L., *Ficus nemoralis* Wall., *Ficus roxburghii* Wall., *Grewia optiva* Drummond, *Leucaena leucocephala* (Lam.) de Wit, *Melia azedarach* L., *Ougeinia oojeinensis* (Roxb.) Hochr., *Quercus leucotrichophora* A. Camus, *Terminalia alata* Heyne ex Roth. and *Toona ciliate* M. Roem. were collected from the superior trees and seedlings were raised. After one year and one month of establishment at the nursery, the growth characteristics, biomass and chlorophyll fluorescence (dark-adopted F_v/F_m) of each species were also recorded. *G. optiva* had shown the highest growth in terms of height, basal diameter increment and number of branches, while production of leaves was more on *O. oojeinensis*. Biomass and chlorophyll fluorescence (maximum quantum yield or photochemical efficiency of PSII) was found highest in *Q. leucotrichophora* which indicates photosynthetically this species was most active among the studied fodder and fuel wood tree species. The information in this communication could be utilized for developing various conservation and sustainable strategies in the Garhwal Himalayas to mitigate the scarcity of fodder and fuel wood production.

Keywords: Scarcity; Fodder and Fuel Wood Species; Screening; Growth; Biomass; Chlorophyll Fluorescence

Introduction

The rural population of the Himalayan region have been exploiting forest resources for their livelihood for generations. The excessive and uncontrolled use of fodder and fuel wood has ended up with severe deforestation. In Garhwal Himalaya, about 77.4% of the total human population is rural (Anonymous, 1991). Several hilly regions of Garhwal Himalayas, the rural inhabitants, mostly women's are use to spend about 60 percent of their daytime for collection of fodder and fuel wood, and for this purpose they often travel 5 - 10 km distances per day. The foliage of tree forms is the key source of green fodder during dry months as the major areas is rain fed and no green fodder is grown in agricultural fields (Husen & Nautiyal, 2004). Moreover, for rural inhabitants due to poor connectivity with the urban areas, the alternative sources of fuel wood are not easily accessible consequently making the population to totally depend on wood resources (Bhatt et al., 2004); and it is evident that most of this demand is met from the adjoining forests almost uninterrupted (Shah, 1982; Khoshoo, 1987), and biomass (i.e. fuelwood and fodder) extraction is the major reason for such depletions (Singh, 1998). Most of the fodder and fuel wood tree species of Garhwal Himalayas are under stress because of unplanned or unscientific lopping (Husen & Nautiyal, 2004). It has been remembered that the improvement and pres-

ervation of life in the third world will largely depend on the presences of forest and also availability of fuel wood (Perschel, 1991). Therefore, to meet this critical requirement of fodder and fuel wood tree species, and to reduce the pressure on natural forests stand, there is a strong need to screen the fast growing fodder and fuel wood tree species in hilly regions of Garhwal Himalayas.

Growth characteristics and photosynthetic efficiency or chlorophyll fluorescence have been used to select high-quality seedlings or clones for a particular environment (Husen et al., 2004a; Husen, 2009). Chlorophyll fluorescence is a useful physiological test to detect perturbation of leaf metabolism and growth of seedling because it is a noninvasive, nondestructive and rapid procedure (Barbagallo et al., 2003). The chlorophyll fluorescence rises rapidly from a ground state (F_o) when all electron acceptors are fully oxidized, to a maximum fluorescence level (F_m) when electron acceptors are highly reduced and unable to accept and transfer electrons. Chlorophyll fluorescence then decreased slowly to a steady state (F_s) as photo-chemistry and CO_2 assimilation increased (Krause & Weis, 1991). The ratio of maximum variable florescence F_v/F_m (where, $F_v = F_m - F_o$) is linearly correlated with the quantum yield of net photosynthesis (Krause & Somersalo, 1989) and thus may be a good measure of seedling vigour (Husen, 2009). Chlorophyll fluorescence has also been used as a diagnostic tool to study the various environmental stresses (Ashraf et al., 2004; Yin et al., 2005; Husen, 2010), genotypic variation (Janssen et al., 1995; Husen, 2009), altitudinal variation (Husen et al., 2004a) and species specific diurnal changes (Husen et al., 2004b). Therefore, chlorophyll fluorescence is a sound method to diagnose seedling stock quality (Mohammed et al., 1995; Barbagallo et al., 2003; Husen, 2009). There is no report available on screening of fast growing indigenous fodder and fuel wood tree species growing at nursery stage, considering the growth, biomass and physiological characteristic features. Therefore, this research paper focuses extensive attention on these parameters to identify the best and fast growing fodder and fuel wood tree species of Garhwal Himalayas with wider relevance and great potential for afforestation programmes.

Material and Methods

Experimental Site

The experiments were conducted at the Plant Physiology nursery of New Forest campus, Forest Research Institute (FRI), Dehra Dun, Uttarakhand (UK), India. The FRI campus is located in Doon Valley and is surrounded by Western Lesser Himalayan ranges in the North and Shiwalik ranges in the South. This campus covers an area of 4.45 km^2, and lies at an elevation of 640 m above mean sea level. It is situated on North Latitude 30°20'40" and East Longitude 77°52'12" on the northern limit of the Oriental region.

Plant Materials

For quality planting material, the seeds of *Bauhinia purpurea* L., *Bauhinia retusa* Roxb., *Bauhinia variegate* L., *Celtis australis* L., *Ficus nemoralis* Wall., *Ficus roxburghii* Wall., *Grewia optiva* Drummond, *Leucaena leucocephala* (Lam.) de Wit, *Melia azedarach* L., *Ougeinia oojeinensis* (Roxb.) Hochr., *Quercus leucotrichophora* A. Camus, *Terminalia alata* Heyne ex Roth. and *Toona ciliate* M. Roem. were collected from the phenotypically superior trees in different localities under re-

serve forest/plantation/agroforestry areas of Garhwal regions, Uttaranchal, India. The superiority percentage was calculated following Wright (1976) as: S = (C/A × 100) − 100; where, S is superiority percentages, C is value for Candidate Plus Tree's and is stand for average value of five comparison trees. The details of the selected superior tree species from which seeds were collected are mentioned in **Table 1**.

Nursery Development and Experimental Design

Nursery was developed to identify and screen fast growing fodder and fuel wood tree species at New Forest campus, FRI, Dehra Dun, UK, India. Seeds obtained from the superior tree were carefully examined for disease free status and shown in plastic tray for germination. When seeds were germinated; these were transplanted to the polybags in mid June, 2002. The polybags were filled with the mixture of soil, sand and farmyard manure in the 2:1:1 ratio and the total weight of the medium was 1.50 kg. The planting medium had 0.120, 0.024 and 0.31 percent nitrogen, phosphorous and potassium respectively. These polybags with plantlets were acclimatized in green house. The upper portion of the green house was covered with green plastic shade, while the other parts remained open. Following 2 months of acclimatization all polybags were shifted to an open environment prevailing at New Forest campus. These plantlets were maintained carefully by regular watering and weeding. Complete protection was provided against diseases and insects by foliar spray with insecticides and fungicides, as and when required. The completely randomized factorial design (CRD) was used for this experimentation. Five replications with ten seedlings (5 × 10 = 50 seedling per species) were used.

Growth and Biomass Studies

After one year and one month (July, 2003) observation on growth namely height (cm) was measured from the ground line of polybags up to the tip of each species. Similarly, the ground line basal diameter (mm), number of leaves (in number) and

Table 1.
Detail characteristics of phenotypically superior fodder and fuel wood trees species.

Species name	Height (in meter)	Girth at breast height (in meter)	Clear bole (in meter)	Crown area (in meter^2)	Place of collection
Bauhinia purpurea L.	13.10	0.95	2.78	8.58	New Forest campus, FRI, Dehra Dun
Bauhinia variegate L.	12.62	0.93	2.63	9.03	Kanwali Garden, Dehra Dun
Bauhinia retusa Roxb.	10.10	0.80	2.70	7.26	Chakrata Forest Division, Dehra Dun
Celtis australis L.	16.52	1.70	6.00	9.50	Chakrata Forest Division, Dehra Dun
Ficus nemoralis Wall.	8.30	0.88	0.66	8.29	Gopeshwar, Chamoli
Ficus roxburghii Wall.	11.60	1.14	1.19	8.05	Gauchar, Chamoli
Grewia optiva Drummond	9.60	0.78	1.95	7.56	Dhaulas, Dehra Dun
Leucaena leucocephala (Lam.) de Wit	14.28	1.19	3.78	6.57	New Forest campus, FRI, Dehra Dun
Melia azedarach L.	16.00	1.39	4.51	11.70	Kanwali Garden, Dehra Dun
Quercus leucotrichophora A. Camus	19.30	1.05	4.31	9.63	Chakrata Forest Division, Dehra Dun
Ougeinia oojeinensis (Roxb.) Hochr.	10.44	1.34	3.02	7.21	Ban Chetna Kendra, Mussoorie, Dehra Dun
Terminalia alata Heyne ex Roth	27.50	1.25	8.37	12.80	New Forest campus, FRI, Dehra Dun
Toona ciliate M. Roem	21.50	1.82	7.50	9.66	New Forest campus, FRI, Dehra Dun

number of branch (in number) were recorded. Thereafter, at the same time the destructive sampling were performed for biomass (the total dry weight of each seedling in gm). Total dry mass/or dry weight of each seedling was obtained after placing the root, stem and leaves in an oven at 70°C for 48 h.

Chlorophyll Fluorescence Measurements

The chlorophyll fluorescence of each species were recorded during 10:00 to 11:00 hrs on cloud free conditions over a period of 2 days with the help of portable Hansatech Plant Efficiency Analyser (*Hansatech*, King's Lynn, England) in first week of July, 2003 (**Table 2**). For chlorophyll fluorescence, data was analyzed taking three replication of each species. Approximately, the 5th leaf from top was darkened with leaf clips for twenty minutes before the measurement. Then values of F_v/F_m (chlorophyll fluorescence) at ambient temperature were noted (Husen, 2009).

Statistical Analysis

Statistical analysis was carried out with the Statistical Package for Social Sciences (SPSS) software package version 6.1.3. A fixed-effects model was used for statistical analyses. The data recorded on the growth, i.e. height, the ground line basal diameter, number of leaves, number of branches and biomass were subjected to one-way analysis of variance (ANOVA) to test the significant variation among the different fodder and fuel wood species. In the ANOVA, the mean values of each replication were estimated. Mean values were compared by using Tukey's test at the $p \leq 0.05$ level. Chlorophyll fluorescence measurement of each fodder and fuel wood species contained three replications, and each replicate had a mean value of five plants (3×5), therefore, 15 seedlings were measured from each species. As mentioned previously that the chlorophyll fluorescence was measured over a period of 2 days and therefore, the mean data for 2 days were pooled for one way ANOVA analyses. For the comparison of chlorophyll fluorescence means Tukey's test was used at the $p \leq 0.05$ level of significance.

Results and Discussion

All studied parameters have shown significant variation (**Table 3**). Seedling of *G. optiva* at nursery stage exhibited maximum height (119.34 cm) and basal diameter (16.84 mm) increment in comparison to other fodder and fuel wood species. However, minimum height and basal diameter increment were recorded in *O. oojeinensis* (41.54 cm) and *C. australis* (4.13 mm), respectively (**Figures 1** and **2**). In addition, *G. optiva* was also found best in terms of number of branches while *M. azedarach* and *T. ciliate* showed least number of branches; and *F. roxburghii*, have not shown any branch (**Figure 3**). Productions of maximum number of leaves were found in *O. oojeinensis* (40.46) followed by *B. purpurea* (32.47), *G. optiva* (27.47) and *F. nemoralis* (26.63), while minimum number of leaves was observed with *T. ciliate* (6.22) (**Figure 4**). Studies on total dry mass production recorded highest in *L. leucocephala* (24.97 g) and lowest in *Q. leucotrichophora* (8.18 g) seedlings (**Figure 5**). Performance on the growth characteristics of fodder and fuel wood tree species at nursery stage varied significantly and may be attributed to adaptation because the seedlings from all the sources were raised under identical conditions (Singh & Pokhriyal, 2000). Furthermore, Sniezko & Stewart (1989) reported that provenance variation in growth characteristics is essentially genetic in nature. Biomass production of studied seedlings varied significantly. This variation can be associated with the adaptation mechanism of different species and seed source origin. Similar observations have also been examined in many other species by several investigators (Salazar, 1989; Bindroo et al., 1990; Singh et al., 2006).

Chlorophyll fluorescence study exhibited significant variation in photochemical efficiency in PS II as was recorded as F_v/F_m ratio on fully developed leaves in sunny days. *Q. leucotrichophora* was found to be photosynthetically most efficient species while *B. variegate* was exhibited least F_v/F_m ratio (**Figure 6**). The extent of chlorophyll fluorescence variation in selected fodder and fuel wood tree species was probably due to the specific liking of individual species or/environmental conditions. Several explanation have been given for variation in chlorophyll fluorescence, such as it might be due to low temperature (Hardacre & Greer, 1989), low irradiance (Janssen et al., 1995) and/or other environmental conditions (Parker & Mohammed, 2000; Husen et al., 2004a, 2004b) were observed by several researchers. It was interesting to note that both total biomass and chlorophyll fluorescence was higher in *Q. leuco-*

Table 2.
Meteorological data during chlorophyll fluorescence (F_v/F_m) measurements.

Date	Temperature (°C)			Relative humidity (%)		Rainfall (mm)	Evaporation (mm)	Bright sunshine (hrs)	Mean wind velocity (Km/hr)
				hr.					
	Max.	Min.	Mean	719	1419				
July 28, 2003	34.1	24.9	27.9	92	78	1.6	4.1	4.0	1.7
July 29, 2003	32.0	23.0	26.6	95	70	22.8	2.1	5.3	1.8

Table 3.
Analysis of variance for height, basal diameter, number of leaves, number of branches, dry mass and chlorophyll fluorescence.

Source of variation	Df	Mean square					
		Height	Basal diameter	Number of leaves	Number of branches	Dry mass	Chlorophyll fluorescence
Species	12	3195.68[**]	60.84[**]	551.87[**]	49.96[**]	119.92[**]	0.0059[**]
Error	52	3.23	0.04	0.94	0.04	0.28	0.0002

Note: [**]reflect significant at the $p \leq 0.01$

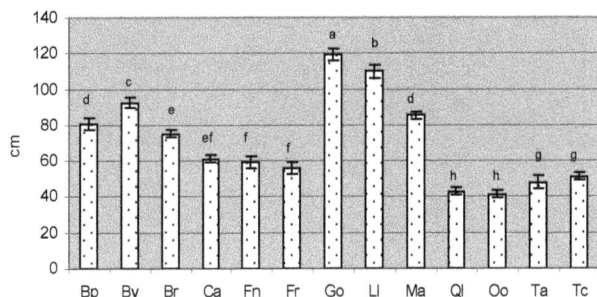

Figure 1.
Variation in height (in cm) of fodder and fuel wood tree species (Bp = *Bauhinia purpurea*, Bv = *Bauhinia variegate*, Br = *Bauhinia retusa*, Ca = *Celtis australis*, Fn = Ficus nemoralis, Fr = Ficus roxburghii, Go = *Grewia optiva*, Ll = *Leucaena leucocephala*, Ma = *Melia azedarach*, Ql = *Quercus leucotrichophora*, Oo = *Ougeinia oojeinensis*, Ta = *Terminalia alata* and Tc = *Toona ciliate*). Values followed by the same letter indicate no significant differences at $p < 0.05$ level according to Tukey's test. Each value represents the mean ± SE of five replicates.

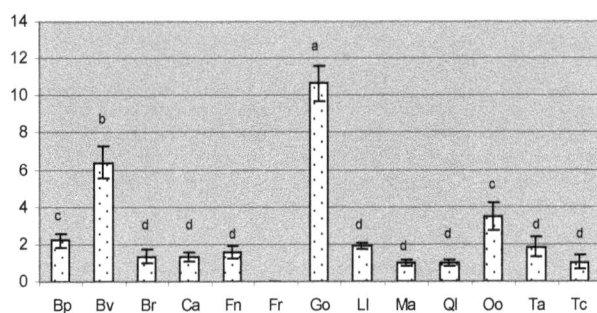

Figure 4.
Variation in number of leaves of fodder and fuel wood tree species (Bp = *Bauhinia purpurea*, Bv = *Bauhinia variegate*, Br = *Bauhinia retusa*, Ca = *Celtis australis*, Fn = Ficus nemoralis, Fr = *Ficus rox- burghii*, Go = *Grewia optiva*, Ll = *Leucaena leucocephala*, Ma = *Melia azedarach*, Ql = *Quercus leucotrichophora*, Oo = *Ougeinia oojeinensis*, Ta = *Terminalia alata* and Tc = *Toona ciliate*). Values followed by the same letter indicate no significant differences at $p < 0.05$ level according to Tukey's test. Each value represents the mean ± SE of five replicates.

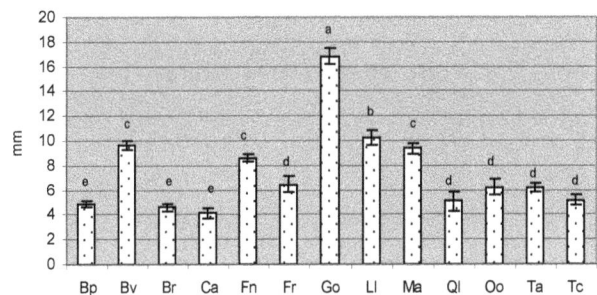

Figure 2.
Variation in basal diameter (in mm) of fodder and fuel wood species (Bp = *Bauhinia purpurea*, Bv = *Bauhinia variegate*, Br = *Bauhinia retusa*, Ca = *Celtis australis*, Fn = Ficus nemoralis, Fr = *Ficus roxburghii*, Go = *Grewia optiva*, Ll = *Leucaena leucocephala*, Ma = *Melia azedarach*, Ql = *Quercus leucotrichophora*, Oo = *Ougeinia oojeinensis*, Ta = *Terminalia alata* and Tc = *Toona ciliate*). Values followed by the same letter indicate no significant differences at $p < 0.05$ level according to Tukey's test. Each value represents the mean ± SE of five replicates.

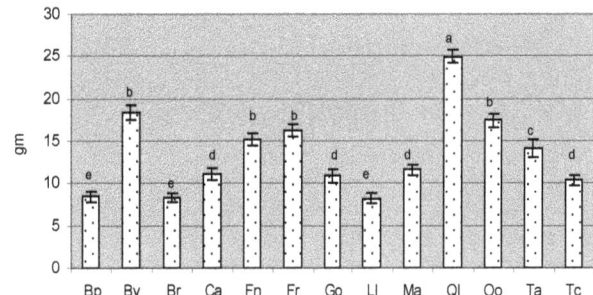

Figure 5.
Variation in total dry mass (in gm) of fodder and fuel wood tree species (Bp = *Bauhinia purpurea*, Bv = *Bauhinia variegate*, Br = *Bauhinia retusa*, Ca = *Celtis australis*, Fn = Ficus nemoralis, Fr = *Ficus roxburghii*, Go = *Grewia optiva*, Ll = *Leucaena leucocephala*, Ma = *Melia azedarach*, Ql = *Quercus leucotrichophora*, Oo = *Ougeinia oojeinensis*, Ta = *Terminalia alata* and Tc = *Toona ciliate*). Values followed by the same letter indicate no significant differences at $p < 0.05$ level according to Tukey's test. Each value represents the mean ± SE of five replicates.

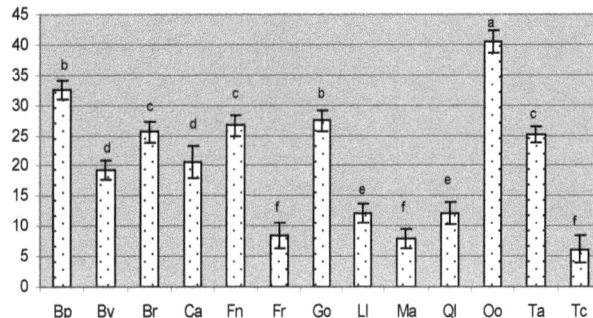

Figure 3.
Variation in number of branches of fodder and fuel wood tree species (Bp = *Bauhinia purpurea*, Bv = *Bauhinia variegate*, Br = *Bauhinia retusa*, Ca = *Celtis australis*, Fn = Ficus nemoralis, Fr = *Ficus roxburghii*, Go = *Grewia optiva*, Ll = *Leucaena leucocephala*, Ma = *Melia azedarach*, Ql = *Quercus leucotrichophora*, Oo = *Ougeinia oojeinensis*, Ta = *Terminalia alata* and Tc = *Toona ciliate*). Values followed by the same letter indicate no significant differences at $p < 0.05$ level according to Tukey's test. Each value represents the mean ± SE of five replicates.

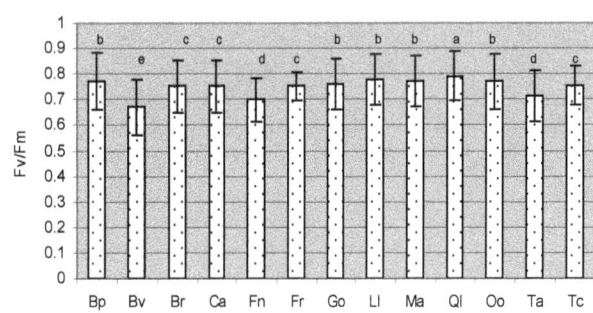

Figure 6.
Variation in chlorophyll fluorescence (F_v/F_m) of fodder and fuel wood tree species (Bp = *Bauhinia purpurea*, Bv = *Bauhinia variegate* Bv = *Bauhinia retusa*, Ca = *Celtis australis*, Fn = Ficus nemoralis, Fr = *Ficus roxburghii*, Go = *Grewia optiva*, Ll = *Leucaena leucocephala*, Ma = *Melia azedarach*, Ql = *Quercus leucotrichophora*, Oo = *Ougeinia oojeinensis*, Ta = *Terminalia alata* and Tc = *Toona ciliate*). Values followed by the same letter indicate no significant differences at $p < 0.05$ level according to Tukey's test. Each value represents the mean ± SE of five replicates.

trichophora; which is supported by more carbon assimilation. Physiological characterization (chlorophyll fluorescence) has been used more in stock quality assessment (Mohammed et al., 1995; Husen, 2009) because this test is noninvasive, nondestructive and rapid (Vidaver et al., 1989). However, in this finding *Q. leucotrichophora* was photosynthetically most active plants while *G. optiva* exhibited maximum growth in terms of height and basal diameter at the nursery stage. Hence, selection for plantation of *Q. leucotrichophora* and *G. optiva* around agricultural fields in different agroforestry systems of Garhwal Himalaya for rapid growth and higher biomass production can be advantageous. In addition, *Q. leucotrichophora* and *G. optiva* also recommended as a fast growing fodder and fuel wood tree species for Garhwal Himalayas which leads with wider relevance and great potential for extensive afforestation programmes.

Acknowledgements

Financial assistance by National Agricultural Technology Project (ICAR), New Delhi, India is gratefully acknowledged. I thank Dr. Laxmi Rawat, Forest Research Institute (FRI), Dehra Dun, India for meteorological data.

REFERENCES

Anonymous (1991). Census data of Garhwal region of Uttar Pradesh.

Barbagallo, R. P., Oxborough, K., Pallett, K. E., & Baker, N. R. (2003). Rapid, noninvasive screening for perturbation of metabolism and plant growth using chlorophyll fluorescence imaging. *Plant Physiology, 132*, 485-493.

Bhatt, B. P., & Sachan, M. S. (2004). Firewood consumption along an altitudinal gradient in mountain villages of India. *Biomass and Bioenergy, 27*, 69-75.

Bhatt, B. P., & Verma, N. D. (2002). *Some multipurpose tree species for agroforestry systems*. Umiam: ICAR Research Complex for NEH Region.

Bindroo, B. B., Tiku, A. K., & Pandit, R. K. (1990) Variation of some traits in mulberry varieties. *Indian Forester, 106*, 320-323.

Hardacre, A. K., & Greer, D. H. (1989). Differences in growth in response to temperature of maize hybrids varying in low temperature tolerance. *Australian Journal of Plant Physiology, 16*, 181-187.

Husen, A. (2009). Growth, chlorophyll fluorescence and biochemical markers in clonal ramets of shisham (*Dalbergia sissoo* Roxb.) at nursery stage. *New Forests, 38*, 117-129.

Husen, A. (2010). Growth characteristics, physiological and metabolic responses of teak (*Tectona grandis* Linn. f.) clones differencing in rejuvenation capacity subjected to drought stress. *Silvae Genetica, 59*, 124-136.

Husen, A., & Nautiyal, S. (2004). Growth performance of some fuelwood and fodder tree species at the three altitudes of Garhwal Himalayas. *International Conference on Multipurpose Tree in the Tropics: Assessment, Growth and Management*, Jodhpur, 22-25 November 2004.

Husen, A., Khali, R., & Nautiyal, S. (2004a). Altitudinal variation in chlorophyll fluorescence/photosynthetic efficiency in seedlings of some indigenous fodder species. *Indian Forester, 130*, 89-94.

Husen, A., Khali, R., & Nautiyal, S. (2004b). Chlorophyll fluorescence in relation to diurnal changes of three *Ficus* species. *Indian Forester, 130*, 811-818.

Janssen, L. H. J., Von Overen, J. C., Van Hassett, P. R., & Kuiper, P. J. C. (1995). Genotypic variation in chlorophyll fluorescence parameters, photosynthesis and growth of tomato grown at low temperature and low irradiance. *Photosynthetica, 31*, 301-314.

Khoshoo, T. N. (1987). Strategies for meeting the fire wood needs in the hills. In T. N. Dhar, & P. N. Sharma (Eds.), *Himalayan energy system* (pp. 11-19). Nainital: Nainital Gyanodaya Prakashan.

Krause, G. H., & Somersalo, S. (1989). Fluorescence as a tool in photosynthesis research: Applications in studies of photo inhibition, cold acclimation, and freezing stress. *Philosophical Transactions of the Royal Society, 323*, 281-293.

Mohammed, G. H., Binder, W. D., & Gillies, S. (1995). Chlorophyll fluorescence: A review of its practical forestry application and instrumentation. *Scandia Journal Forestry Research, 10*, 383-410.

Parker, W. C., & Mohammed, G. H. (2000). Photosynthetic acclimation of shade grow red pine (*Pinus resinosa* Ait) seedlings to a high light environment. *New Forests, 19*, 1-11.

Perschel, R. T. (1991). Pioneering a new human/nature relationship. *Journal of Forestry, 89*, 18-22.

Salazar R. (1989). Genetic variation of 16 provenances of *Acacia mangium* at nursery level in Turrialba Costa Rica. *Commonwealth Forestry Review, 68*, 263-272.

Shah, S. L. (1982). Ecological degradation and future of agriculture in the Himalaya. *Indian Journal of Agriculture Economics, 37*, 1-22.

Singh, B., Bhatt, B. P., & Prasad, P. (2006). Variation in seed and seedling traits of *Celtis australis*, a multipurpose tree, in Central Himalaya, India. *Agroforestry Systems, 67*, 115-122.

Singh, N., & Pokhriyal, T. C. (2000). Biomass distribution pattern in relation to seed source variation in *Dalbergia sissoo* seedlings. *Annals of Forestry, 8*, 238-249.

Singh, R. V. (1982). *Fodder trees in India*. New Delhi: Oxford & IBH Publication Co.

Singh, S. P. (1998). Chronic disturbance, a principal cause of environmental degradation in developing countries. *Environmental Conservation, 25*, 1-2.

Sniezko, R. A., & Stewart, H. T. L. (1989). Range wise provenance variation in growth and nutrition of *Acacia albida* seedlings propagated in Zimbabwe. *Forest Ecology and Management, 27*, 179-197.

Vidaver, W. E., Binder, W. D., Brooke, R. C., Lister, G. R., & Toivonem, P. M. A. (1991). Assessment of photosynthetic activity of nursery grown *Picea glauca* seedlings using an integrated fluorometer to monitor variable chlorophyll fluorescence. *Canadian Journal of Forestry Research, 19*, 1478-1482.

Wright, W. J. (1976). *Introduction to forest genetics*. New York: Academic Press.

8

Review and Progress of China's Forest Continuous Inventory System

Guozhong Lin, Xiaorong Wen, Chunguo Zhou, Guanghui She*

College of Forest Resources and Environment, Nanjing Forestry University, Nanjing, China

China was one of the earliest countries to set up a system to continuously inventory natural forest resources. From the beginning of the 1970s until today, seven forest resource inventories have been carried out. This research summarizes the progress of forest continuous inventories and analyzes the existing deficiencies of China's forest continuous inventory system and forest management plan inventory. As stated above, this research offers corresponding countermeasures and suggestions: establishing a sample plot system for comprehensive national forest inventory and monitoring with each province's continuous forest inventory based on the foundation of the national sample plot system, able to develop the province as a subset of the overall province-level forest resource inventory according to the actual conditions in each province. Through annual multi-resource/multi-benefit surveying of the forests, the monitoring of forest amounts, quality, functions and benefits will be assisted in its entirety. The further integration of the forest continuous inventory and the forest management plan inventory is also discussed. This research also proposes the varied probability sampling method with sub-compartments as the basic sampling unit (or combinations of sub-compartments). This will also satisfy the requirements of ecological inventory by region.

Keywords: Review; China's Forest Inventory System; Forest Continuous Inventory System; China Forest Resource

Introduction

Forest Inventory proceeds with relevant scientific methods and standards according to the needs of forest management, ecological environment construction, forestry production, scientific research and other factors. One of the important tasks of forest resource management is to collect, count, analyze and assess the data according to a defined scale of time and space, for forest resource distribution, amount, quality, and trends of development, as well as relevant natural and social economic factors (IUFRO, 1994).

The survey of forest resources carried out in order to understand the current situation of the forest resources and dynamic is known as Forest Continuous Inventory.

The Forest Continuous Inventory system includes the sampling method plot settings and survey index, time interval, and statistics and analysis etc. China's work on the forest inventory system actually took place after 1950. It experienced 4 stages: 1) the initial establishing forest inventory system in nationwide in 1950s; 2) Sampling techniques using mathematical statistics as a theoretical basis from 1960s to1980s; 3) Remote sensing (RS), geographic information systems (GIS), global positioning system (GPS) technology was gradually introduced forest continuous inventory technology from 1990s to 2000s; 4) The expansion of inventory contents towards a multi-resource format includes forest productivity, biomass, carbon storage, etc. from 2000s to now.

The national inventory system only operates once every 5 years. Through the review of China's forest resource continuous inventory of progress, to further improve the perfecting forest continuous inventory system, how to improve the efficiency of inventory expanded inventory results application, some useful advice put forward in this paper.

China's Forest Inventory System and Its Course of Development

The Main Types of Forest Inventory

The forest inventory system and its contents, methods, technology, form of results, etc. develop parallel to the rapid advance of information technology and the widespread use of mathematical statistics. These new technologies and techniques enable integration of the macro-requirements of national forestry construction and development with forestry production, and lead to the gradual establishment and perfection of management practice requirements (Xiao, 2005).

After nearly 60 years of development, China's large-scale forest continuous inventory took shape as a relatively perfect forest resource inventory system in view of the special characteristics and requirements of China's forestry development. According to the research goal, content, and technical methods, China's forest resource survey system can be divided into:

1) The nationwide forest inventory is referred to as a first class survey. Beginning in the 1970s, China's forest continuous inventory was gradually established at the province level (region, municipality), with the data from the provinces making up the main body of the overall forest Continuous inventory system. Every 5 years the inventory was repeated, using fixed sample plot surveys as the main method of carrying out regular monitoring. Within a unified time, and according to unified inventory requirements, a clear picture of the macroscopic

*Corresponding author.

situation of national forest resources and its succession of patterns was created. This provided scientific evidence for the forestry development strategy and adjustment of forestry policies in a timely and efficacious way. The national forest continuous inventory used a systematic sampling method, and the stratified sampling method was integrated into the sampling estimation. For the systematically extracted samples a stratified sampling estimation was completed according to forest volume or other survey factors, the principle being to make the factor variation as small as possible for all points within the layer and larger between the layers, in order to improve the accuracy of stratified sampling estimations. In addition, in order to improve the efficiency of sample plot extraction, double samples and double regression method estimates are frequently used. This takes advantage of easily measured auxiliary tree survey factors to realize estimates of forest accumulation. When the forest resource inventory also has other spatial data sources, such as aerial or satellite imaging, the multi-source data joint estimation method is usually still adopted.

2) The forest management plan inventory is referred to as a second class survey. The Forest Management Plan Inventory is one kind of forest resource inventory method. It uses state-owned forestry bureaus, nature preserves, forest parks and other areas of forest management units or administrative units above the county level as its departments, and is carried out with forest management sub-compartments as the basic unit. It is done once every 10 years, to provide evidence for the forest management plan, the overall plan and county-level or above forest planning, and the sub-compartment data files which take shape provide a data platform for forest management and tasks.

3) The forest task planning survey is known as a third class survey. This inventory is a carried out as a task-based survey with a particular scope or operational area as the basic unit. Usually it adopts the real measurement or sampling inventory method. For the forest resources in each operational area, site conditions and forest regeneration status are used to undertake a detailed survey, with the goal of satisfying the requirements of specific production tasks (e.g. afforestation, deforestation, forest tending etc.) arranged by basic level forestry production units, usually carried out in the production tasks of the previous year.

4) The annual special forest resource inventory is known as the verification survey. Its goal is to achieve a timely grasp of the situation regarding completion of the annual forestry production plan by forestry production units or project implementation units. It evaluates the effectiveness of forestry production assignments, so that the Forestry Department and other related departments can adjust their annual production plan, to provide a basis to boost the effectiveness of carrying out the Forest Management Unit's forestry production plans. The verification inventory method usually takes a locality (city) or county (forestry bureau) as its basic units, using the terrain verification survey method.

5) Specialized inventory. Specialized forestry inventory includes site-type surveys, forest soil surveys, forest regeneration surveys, forest disease and pest surveys, preparation of forestry tables, forest growth surveys, forest multi-benefit measurement surveys and assessments, wild animal resource surveys, and other specialized inquiries. Its findings provide basic data directly to main specialized surveys, district, planning, design and establishment of forestry production.

The Shape of the Forest Continuous Inventory System

China's work on the forest inventory system actually took place after the founding of the new China, beginning in 1950 (Yi, 1991). At that time, forest inventory technology from the former Soviet Union was comprehensively introduced. The Soviet technology was in reality originally from Germany, and in addition the previous knowledge in the country on forest inventory came from European and Japanese systems, so China's technology was close to the European system from the beginning (Goran & Hans, 1998; Kleinn et al., 1998; American Forest Council, 1992).

Table 1 shows the 4 stages of shape and progress of China's Forest Continuous Inventory.

In the early part of the 1950s, theodolites or compasses were principally used for measurement, and to regulate the survey area, using a grid method to divide the land into compartments and sub-compartments, set up banded standard plots, and measure each tree's diameter in the standard plot to calculate forest volume. Since the workload of the survey was too large, in the middle of the 1950s, aerial forest surveys were developed, along with aerial forest inventories and ground based comprehensive inventories. By the end of the 1950s, angle gauge measurement technology was also introduced (Scott, 1947; Ware & Cunia, 1962), since the method was simple, accurate and reliable, and it received a wide range of applications throughout the country. According to the needs of that time, year after year the whole country was surveyed piece by piece. Since there were many investigation methods in different areas, survey quality was not consistent, and the scope of the inventory was incomplete. The statistical calculation methods were also rather poor at that time and remained so until the early 1960s. In 1962 the Forestry Department, responding to the requirements of forestry production development, organized the provinces (regions, city) to further the development of statistical work in the forest resource inventory across the whole country. For the 12-year period from 1950 to 1962, all of the various forest resource survey data was systematized, statistically processed and finally brought into a nationwide summary.

During the 1960s, sampling techniques using mathematical statistics as a theoretical basis were introduced and large-scale experiments and measurement verification was organized. In terms of the survey's accuracy, quality, benefits and other aspects, remarkable results were achieved across the board. This laid the foundation for the sampling techniques we use in our forest inventories today.

In the 1970s, China began to explore a national forest continuous inventory system. This was due to the consideration

Table 1.

The 4 stages of shape and progress of China's Forest Continuous Inventory.

Stage	Period	Feature
Initial establishment	1950s	theodolites or compasses, angle gauge measurement
Exploration & Shape	1960s to 1980s	sampling techniques, fixed sample plots at 5-year intervals
Development	1990s to 2000s	RS, GIS, GPS technology application.
Optimization & Expansion	2000s to now	multi-resource monitoring includes forest productivity, biomass, carbon storage, etc.

that the former survey methods were all independent investigations and the results from before and after lack continuity and comparability. It was impossible to achieve precise information on trends of growth and decline of forest resources (Cunia & Chevrou, 1969; Newton et al., 1974). In 1973, the Ministry of Agriculture and Forestry (MAF) arranged a national forest continuous inventory project in each province (region, city), developing the administrative districts and counties (bureaus) as basic units in the inventory system. This was the first time since the founding of the new China that a comprehensive forest continuous inventory was carried out in a relatively unified time scale and on a national scope (Taiwan temporarily excluded). In this inventory, through the establishment of a pilot program, the country's forest continuous inventory utilized systematic sampling methods and hierarchical regression estimate methods. Beginning in 1978, this system was fully developed in every province (region, city). The forest continuous inventory system is in accordance with the systematic sampling principle. Fixed sample plots were set up at ground level, the diameter of each tree in the standard plot was measured, and at 5-year intervals there was a review. This kind of inventory system can accurately obtain the current conditions of forest resources and uncover growth and decline trends. Regular patterns of resource development can be grasped, the effectiveness of resource management can be analyzed, and predictions of resource development trends can be predicted.

Thus China became one of the earliest countries to set up a forest continuous inventory system. From the 1970s until today, seven forest inventories have been carried out. The surveying period was divided into the years 1973-1976, 1977-1981, 1984-1988, 1989-1993, 1994-1998, 1999-2003, and 2004-2008. The 8th inventory is already underway.

The first country-wide forest inventory was the first forest continuous inventory with the counties (bureaus) as the basic units which reached a national scope (excluding Taiwan) since the founding of the People's Republic of China. Its main focus was on ascertaining the current forest resource conditions. Starting with the second national forest continuous inventory, the inventory system used the province (district, municipality) as the basic unit rather than the county. In subsequent inventories a review of previous inventory data, a gradual perfection of methods, and unceasingly enriched survey indices can be seen. At the end of the 1970s systematic sampling technology was first deployed, and by the 1990s remote sensing (RS), geographic information systems (GIS), global positioning system (GPS) technology was gradually introduced. Forest continuous inventory technology has shown continuous improvement. Before the 5th inventory, priority was given to timber production. For the most part, traditional ground level sampling was undertaken on the sample plots at this time. The 6th inventory increased forest ownership, plant disease and insect pest contents, and comprehensively introduced remote sensing technology, with 3S technology as the foundation. It used remote sensory monitoring and ground surveying technology combined with the double stratified sampling remote monitoring system (Wang et al., 2007; Gregoire, 1993; Lyncii, 1995; Martin, 1982). At present, the 7th national forest continuous inventory has incorporated the forest's ecological functions, forest healthy class and biodiversity into the measurement categories as reflections of ecological status. The forest continuous inventory system is moving in the direction of development of a comprehensive monitoring system for both resources and ecological conditions (Xiao, 2005).

In the 1980s, China's forest continuous inventory system was further improved, unifying the technological system for the entire country. As information technology developed, remote sensing technology began to be used during surveying, and satellite imaging and aerial sample plot imaging also made their appearance. Within the statistical work of the inventory, unified survey factor coding and unified data storage formats came into usage. A complete statistical summary was entered into the computer system, guaranteeing the comparability and accuracy of survey results. This summary provides reliable basic data for the current national forest resource situation and its growth and decline status. In the area of sampling technology, an accurate two-stage and multi-stage estimation method for forest continuous inventory was proposed (The Forestry Department of the People's Republic of China, 1983).

Forest continuous inventories before the 1990s were mainly undertaken with the forest resources themselves as the focus. Beginning in the 1990s, the country began the development of the forest resource monitoring system. In 1991 the United Nations Food and Agricultural Organization provided China with aid projects in order to support the establishment of the monitoring system. Research was begun on the establishment of a national Forest Resource Monitoring Body. This period of forest resource inventory saw the beginning of widespread usage of 3S (remote sensing (RS), geographic information systems (GIS), the global positioning system (GPS)) technology, and improved investigative factors in the areas of forest ecology, health class, ownership and intensity of management (The Forestry Department of the People's Republic of China, 1994).

After the year 2000, China gradually established a forestry development strategy that gives priority to ecological construction. China's "Forest Continuous Inventory Technical Regulations" (The Forestry Department of the People's Republic of China, 2004; Arner et al., 2004; Cochran, 1997), issued in 2004, embody the survey indices of forestry development with a focus on ecological construction. Beginning with the 8th national forest resource inventory, the surveying system will gradually develop into a comprehensive monitoring system which will give equal priority to the forest resources and the ecological situation. There are 75 investigative factors for sample plots in the 8th inventory which can be divided into six aspects: land usage and coverage; site and soil; stand characteristics, forest function; ecological situation, and other forest land management related survey indices. At present the 8th national inventory is underway, and the comprehensive monitoring system projects are divided into six projects and 133 monitoring factors (Ye et al., 2000): such as land cover and land usage, soil and site, land degradation, forest function efficiency, wood and other forest products, forest health conditions and disasters, and biomass and biological diversity.

The Progress of Continuous Forest Resources Inventory

Since the 1990s, China has carried out a series of optimization and improvement tasks on the current forest resource monitoring system. From 1993-1997, China received aid from the United Nations Development Program (UNDP) and other institutions and launched research on the "National Forest Resources Monitoring System". In 1997 the original forestry department proposed a conception of a monitoring and assessment system for national forest resources and the environment, and the State Forestry Bureau Planning and Design Institute completed the overall design. In 1998, the compilation of the technological plan and regulations was completed. On December 12, 2004, research was initiated on establishing a framework for a comprehensive monitoring system for forest resources and the ecological situation (Bu, 2005). The general ideas and basic framework for such a system were put forth. The monitoring of water, air and soil at a microscopic level can accurately assess the status of forest resources and lead to the discovery of regular patterns of forest growth, and can also find the external factors influencing natural forest growth. This environmentally-focused monitoring will represent a new direction of development for the comprehensive monitoring of forest resources, and also mirrors international development trends.

The characteristics of information on forest resources and ecological conditions come from many sources, and in many forms, shapes, grades and structures (Lie, 2006). Under the guiding idea of systemic integration, computer network technology, database technology, 3S technology, model simulation technology, and resources will be comprehensively integrated. Establishing a public service platform for comprehensive monitoring of forest resources, and promoting the further development, application and integration of information resource for forestry development and ecological construction to provide a real-time, dynamic, open information service will improve evaluation and forecasting ability to help avert disasters and ensure the success of engineering projects.

Remote sensing, aerial remote sensing, global positioning system, global information systems, database technology and computer network technology are widely applied in the modernization of forest resource monitoring. In particular, the use of remote radar sensing data to carry out local surveys, such as coastal shelter forest surveys, has been gradually increasing.

The expansion of inventory contents towards a multi-resource format includes forest productivity, biomass, carbon storage, forest products, temperate area resources, desertification, rocky desertification of soil resources and wild animal and plant resources. The development towards a multi-benefit format mainly includes investigation of the ecological conditions and functions of the forest, relating to sample plot biodiversity, forest health, and information output on ecological functions and forestry engineering effectiveness. Multi-resource, multi-benefit monitoring will guarantee comprehensive monitoring of the forest volume, quality, functions and benefits.

In recent years, due to China's forest resource management requirements, each province hopes to have forest resource and ecological conditions information for every year. One key trend of variation in forest resource monitoring is how to use the 5-year cycle of forest continuous inventories to realize annual forest resource monitoring. The method mainly used is to add detail to the original inventory sampling grid and review every year. Due to the heavy workload and high costs this method is not adopted by many. Another method is to utilize the data from the inventory, and carry out an update using the growth and consumption model. A third method is to correct and control the original inventory data through the use of remote sensing to provide annual data to combine with the original data (Roesch & Reams, 1999; Reosch et al., 1993; Scott et al., 1999).

In 1999 Zhejiang province took the lead in performing an annual forest resource monitoring trial taking the requirements of the new situation into account. The method they adopted involves tightening the original forest inventory sampling grid and reviewing every year. Minor forest zones utilize modeling to update their data as an auxiliary method. Annual forest resource monitoring not only improves the usefulness of the results of the forest continuous inventory, but also provides annual forest resource data for resource management. Every year Zhejiang Province will make their annual monitoring of forest resources and ecological conditions into a public announcement, helping the society to understand the forest, and increasing public participation and enthusiasm for forestry.

Discussion

Modern forest inventory monitoring technology is not a single-source technology. Its development is also not an independent development of one branch of science. It is rather a comprehensive integration of many types of technology and even many different scientific fields (Zhu, 2001). At present, spatial remote sensing, aerial remote sensing, global positioning, global information systems, database technology and computer network technology, along with other high-technology products, have received different degrees of national usage in forestry (Wynne et al., 1999). Looking at each kind of technology in turn, it can be seen that ground surveying can't satisfy the needs of a macroscopic analysis; aerial digital imaging has too large an information content to be useful in the monitoring of forest resources over vast areas; and satellite monitoring can't satisfy the demands for level of detail needed in resource monitoring information. Therefore, as far as monitoring technology, it is necessary to organically combine ground surveying, aerial and satellite imaging as the information content, accuracy and time period requires. The technique of combining 3S technology and ground survey methods, promoting space/ground/sky-integrated monitoring, receiving information from many channels and on many levels, can be an effective forest resource monitoring method.

The rapid growth of the national economy and the need for sustainable forestry development has given rise to some contradictions. On the one hand there is a requirement for the connotation of forest resource inventory to broaden, resource data to be updated year by year, and to develop forest resource monitoring in the direction of multi-resource comprehensive monitoring. On the other hand, although the establishment of a forest continuous inventory system played a very important role in promoting digital forestry and has become an important foundation for forest resource management, people have questioned the spending of such enormous sums of money to develop forest continuous inventories. How to further expand the sphere of application of the forest resource inventory results, and research low-cost continuous forest inventory methods remains a crucial research topic today.

China's forest continuous inventory system is set up so the provinces are the basic units providing data, with all of them using the systematic sampling method. Due to the limitations of that method, the results of monitoring only apply to the provinces, and break down into local municipalities. At the county level there is no guarantee of sampling precision. Another factor is that the national inventory system only operates once every 5 years, and in reality it is very hard to synchronize the inventory activities of the various provinces within that 5-year period. National statistics come with certain difficulties for the continuous forest resource inventory, and there is no way to utilize the forest inventory system to directly provide data on annual forest resource status and growth and decline status. To this end, some specialists have proposed surveying 1/5 of the provincial sample plots every year, and leaving the remaining 4/5 to use data updating methods to take the place of annual ground surveys. But what is urgently needed to solve this problem is 1) the basis for this statistical theory; 2) a method to extract 1/5 of the sample plots and 3) a concrete method to update the data for the remaining 4/5.

Moreover, the further integration of the forest continuous inventory and the second class (sub-compartment) forest resource inventory is also an important purpose of research in this paper. Under normal conditions, on the basis of the fixed sample plots usually used for the province level continuous inventory, the county level inventory sampling system can be established to act as a control on the sub-compartment survey precision. For this reason, realization of the data of the forest inventory can make the sub-compartment management more practical. The actual situation shows that costs and workload will be greatly increased by establishing such a system. Many counties have abandoned the increased-precision sample plot system that depends on county level data because it brings unavoidable problems of precision control in sub-compartment survey. At the same time, new requirements have been put forward for the progress of forest ecology monitoring which use mountain massif or even river basin areas (combined sub-compartments) as basic units. Some specialists (She, 1998; She et al., 2007) have proposed the varied probability sampling method with the county-level data making up the main body of data, and sub-compartments as the basic sampling unit (or combinations of sub-compartments). This proposal is directed at the existing problems in second-class forest resource inventory, specifically in the south of China. With this type of sub-compartment sample plot a forest resource fixed angle gauge plot dynamic measurement system can be established in each province. At the same time, the statistical estimations of sub-compartment class sample plots can serve as a control on the precision of the entire second class forest (sub-compartment) inventory. This method not only uses the high efficiency of varied probability sampling established in the forest resource monitoring system, but also can combine the sampling estimation method with sub-compartment inventory. This will satisfy the requirements of ecological inventory by region. Sub-compartment class fixed sample plots can become a new forest resource monitoring system, according to the inventory being carried out.

First and second class forest resource inventories are both important components of the national and local forest resource monitoring system. The amount of sample plot data used in the nationwide province-based inventory estimates is rather large, and the corresponding workload and costs are prohibitive. On this basis it is suggested to establish a national forest continuous inventory system and sample plot monitoring system that takes the whole country as its basic unit. In this way, the amount of sample plots in each province can be greatly reduced, and also it will be beneficial to the development of an annual national forest inventory. At the same time the statistical data can control the precision of each province's own inventory data collection. Each province's continuous forest inventory can be implemented according to the method suggested above on a national sample plot basis according to the really existing conditions in the province, creating a more perfect forest resource

monitoring system. The new forest continuous inventory system Put forward by this paper not only can play a role in original system but also can realize two classes of survey combination.

REFERENCES

American Forest Council (1992). Report of the blue ribbon panel on forest inventory and analysis. Washington DC: American Forest Council.

Arner, S. L., Westfall, J. A., & Scott, C. T. (2004). Comparison of annual invertory designs using forest inventory and analysis data. *Forest Science, 50,* 188-203.

Bu, C. J. (2005). Research of the country's start of a framework for establishment of forestry resource and ecological condition comprehensive monitoring system. *Journal of China Forestry Industry, 1,* 4.

Cochran, W. G. (1997). Sampling techniques. Hoboken, NJ: John Wiley & Sons, Inc.

Cunia, T., & Chevrou, R. B. (1969). Sampling with partial replacement on three or more occasions. *Forest Science, 15,* 204-224.

Goran, K., & Hans, T. (1998). The Swedish national forest inventory 1983-1987. Ultuna: Department of Forest Survey, Swedish University of Agricultural Sciences.

Gregoire, T. G. (1993). Estimation of forest growth from successive surveys. *Forest Ecology and Management. 56,* 267-278.

IUFRO, S4.02-05 (1994). International guidelines for forest monitoring. Vienna: IUFRO.

Kleinn, C., Dees, M., & Polley (1998). Forest inventory and survey systems in Germany. Bonn: Federal Ministry of Food, Agriculre and Forestry.

Lie, X. Y. (2006). Forestry resources and ecological condition information resource integration structure analysis. *Journal of Forest Resources Management, 2,* 51-56.

Lyncii, T. B. (1995). Compatible estimation of components of forest growth from remeasured point samples with restricted generalized least squares. *Forest Science, 41,* 611-628.

Martin, G. L. (1982). A method for estimating in growth on permanent horizontal points. *Forest Science, 28,* 111-114.

Newton, C. M., Cunia, T., & Bickford, C. A. (1974). Multivariate estimators for sampling with partial replacement on two occasions. *Forest Science, 20,* 106-116.

Roesch, F. A., & Reams, G. A. (1999). Analytical alternatives for annual inventory system. *Journal of Forestry, 97,* 33-37.

Roesch Jr., F. A., & Deusen, P. C. (1993). Control variate estimators of survivor growth from point samples. *Forest Science, 39,* 66-77.

Scott, C. B. (1947). Permanent growth and mortality plots in half the time. *Journal of Forestry, 45,* 669-674.

Scott, C. T., Kohl, M., & Schnellbacher, H. J. (1999). A comparison of periodic and annual forest surveys. *Forest Science, 45,* 433-451.

She, G. H. (1998). Study on the applied theory and method of angle measuring in volume growing estimation. *Scientia Silvae Sinicae, 34,* 25-30.

She, G. H., Lin, G. Z., Wen, X. R. et al (2007). Improvement of forestry resources second class survey methods and the establishment of angle sampling monitoring system. *Journal of Nanjing Forestry University, 31,* 11-14.

The Forestry Department of the People's Republic of China (1983). Main technical regulations of forest resources investigation. Beijing: China's Forestry Press.

The Forestry Department of the People's Republic of China (1994). Main technical regulations of forest resources survey. Beijing: China's Forestry Press.

The Forestry Department of the People's Republic of China (2004). Main technical regulations of continuously forest resources inventory. Beijing: China's Forestry Press.

Ware, K. D., & Cunia, T. (1962). Continuous forest inventory with partial replacement of samples. *Forest Science Monographs, 3,* 415-422.

Wang, Y. H., Xiao, W. F., & Zhang, X. Y. (2007). Forest health monitoring and evaluation of the present situation and development trend domestically and abroad. *Journal of Forestry Science, 43,* 78-85.

Wynne, R. H., Oderwald, R. G., Reams, G. A., & Scrivani, J. A. (1999). Optical remote sensing for forest area estimation. *Journal of Forest, 98,* 31-36.

Xiao, X. W. (2005). China forestry resources inventory. Beijing: China's Forestry Press.

Ye, R. H., Zhou, W. D., Huang, G. S. et al. (2000). A technical scheme for national forestry resources and the ecological environment comprehensive monitoring and system evaluation. *Journal of Forest Resources Management, 3,* 17-21.

Yi, H. Q. (1991). China's forestry survey planning and design history. Changsha: Hunan Press.

Zhu, S. L. (2001). The present situation and the future development characteristics of forest resources monitoring survey abroad. *Journal of Forest Resources Management, 2,* 21-26.

9

Enhanced Structural Complexity Index: An Improved Index for Describing Forest Structural Complexity

Philip Beckschäfer[1], Philip Mundhenk[1], Christoph Kleinn[1], Yinqiu Ji[2],
Douglas W. Yu[2,3], Rhett D. Harrison[4]

[1]Chair of Forest Inventory and Remote Sensing, Georg-August-Universität Göttingen, Göttingen, Germany
[2]State Key Laboratory of Genetic Resources and Evolution, Kunming Institute of Zoology, Chinese Academy of Sciences, Kunming, China
[3]School of Biological Sciences, University of East Anglia, Norwich, UK
[4]Key Laboratory for Tropical Forest Ecology, Xishuangbanna Tropical Botanical Garden, Chinese Academy of Sciences, Menglun, China

The horizontal distribution of stems, stand density and the differentiation of tree dimensions are among the most important aspects of stand structure. An increasing complexity of stand structure is often linked to a higher number of species and to greater ecological stability. For quantification, the Structural Complexity Index (*SCI*) describes structural complexity by means of an area ratio of the surface that is generated by connecting the tree tops of neighbouring trees to form triangles to the surface that is covered by all triangles if projected on a flat plane. Here, we propose two ecologically relevant modifications of the *SCI*: The degree of mingling of tree attributes, quantified by a vector ruggedness measure, and a stem density term. We investigate how these two modifications influence index values. Data come from forest inventory field plots sampled along a disturbance gradient from heavily disturbed shrub land, through secondary regrowth to mature montane rainforest stands in Mengsong, Xishuangbanna, Yunnan, China. An application is described linking structural complexity, as described by the *SCI* and its modified versions, to changes in species composition of insect communities. The results of this study show that the Enhanced Structural Complexity Index (*ESCI*) can serve as a valuable tool for forest managers and ecologists for describing the structural complexity of forest stands and is particularly valuable for natural forests with a high degree of structural complexity.

Keywords: Forest Structure Index; Structural Complexity; Stem Map; Species Composition; NMDS

Introduction

The importance of ecosystem structure to species richness has been established through many studies. Already in the early 1960s MacArthur and MacArthur (1961) showed that the physical structure of a plant community was of greater importance than the composition of plant species in determining bird diversity. A meta-analysis by Tews et al. (2004) found that 85% of 85 reviewed studies on habitat heterogeneity and species richness conducted between 1960 and 2003 found a positive correlation between richness and structural variables. As plants play an important role in shaping the physical structure of many environments (Lawton, 1983; McCoy & Bell, 1991) the structural complexity of plant communities has frequently been used as an indicator of the diversity in other taxa (Whittaker, 1972; Franzreb, 1978; Temple et al., 1979; Aber, 1979; Recher et al., 1996; Moen & Gutierrez, 1997). Moreover, the habitat heterogeneity hypothesis (Simpson, 1949; MacArthur & Wilson, 1967) relates this positive association between species diversity and structural complexity by suggesting that more complex environments provide increased niche space and thus facilitate specialization and avoidance of competition through spatial segregation (Cramer & Willig, 2005). This further implies that the structural complexity of a forest is of great importance for the number and composition of species inhabiting it (Willson,

1974; Ambuel & Temple, 1983; Freemark & Merriam, 1986; Spanos & Feest, 2007).

The multidimensional character of forest stands makes it hard to describe structural complexity holistically. Canopy layering, the presence of particular understory species, trees with different bark types, and decaying logs and hollow trees (Doherty et al., 2000) have all been considered to be important components of structural complexity. However, three-dimensional stand structures are probably the most important of all characteristics (Pretzsch, 1997). According to Pretzsch (2009), the horizontal distribution pattern of trees, stand density, the differentiation of dimensions, and species intermingling constitute the most important aspects of stand structure that influence growth processes, habitats, species richness, and stability of forest ecosystems. Kimmins (2005) suggests the spatial arrangement of plants, both horizontally and vertically, the structure of tree canopies and the presence of canopy gaps, and snags, and coarse woody debris are the principal characteristics that influence the diversity of animals. While some of these attributes are hard to define and difficult to measure in the field, tree stem diameter and position are standard in measurement protocols of forest inventories in countries like, for example, the USA (United States Department of Agriculture Forest Service, 2011) and Germany (Polley, 2007). Hence, this study focuses on these variables and defines structural complexity as the spatial arrangement of plant dimensions, both horizontally and vertically (Zenner & Hibbs, 2000).

Several indices have been developed in the past decades to provide interpretable metrics of structural complexity and thus facilitate comparisons among stands (Pommerening, 2002; LeMay & Staudhammer, 2005; McElhinny, 2005). LeMay & Staudhammer (2005) identify three groups of indices: 1) indices based on tree attributes, 2) indices of spatial heterogeneity and 3) indices combining tree attributes and spatial heterogeneity. While indices in groups 1) and 2) focus on only one aspect of forest structural complexity, indices in group 3) intend to retain more information and thus may provide more comprehensive measures of structural complexity.

In this study we propose an index that integrates the horizontal distribution of trees, stand density, and the differentiation and intermingling of tree dimensions into one measure, while requiring only data on tree position and diameter-at-breast-height (*DBH*) for its calculation. Our index is a modification of the Structural Complexity Index (*SCI*) developed by Zenner & Hibbs (2000). In the first part of this study, we describe limitations of the *SCI* and propose two modifications to create an Enhanced Structural Complexity Index (*ESCI*). In the second part, the *SCI* and *ESCI* are calculated for forest inventory data from an upland landscape in Xishuangbanna, South China, and results are compared. In the last part, we compare how the *SCI* and *ESCI* correlate with turnover in insect species composition.

The Structural Complexity Index (*SCI*)

Zenner & Hibbs (2000) introduced the *SCI*, a formula that mathematically integrates both vertical (size differentiation) and horizontal (spatial position) components of forest structure. It is based on the position of trees whose *xy*-coordinates are complemented with a tree attribute, such as *DBH* or height, as a *z*-coordinate. By a spatial tessellation approach (Delaunay, 1934) each tree is connected to its neighbours such that traingles are defined that form a continuous faceted surface, i.e. a triangulated irregular network (TIN) (**Figure 1**). If tree height is selected as the *z*-coordinate, this TIN can be visualized as connecting the tops of neighbouring trees. Instead of tree height, any measured continuously or ordinally scaled tree attribute can be chosen as the *z*-coordinate. The *SCI* is defined as the surface area of the TIN in three dimensional space divided by the area covered by its projection on a plane surface (Equation (1)). If all trees have the same *z*-value (e.g. all trees have the same height or basal area as in an even aged plantation) the *SCI* equals 1, the lower limit of the *SCI*. For structurally more complex forest stands the *SCI* is >1.

$$SCI = \frac{\text{surface area of TIN}}{\text{projected area of TIN}} \qquad (1)$$

Limitations of the *SCI*

We illustrate basic characteristics of the *SCI* with a set of four simple forest stands which differ in their structural complexity but have the same *SCI* value. All stands cover the same area but vary in number of trees, range of tree height, and spatial mingling of trees with different heights (**Table 1**, **Figure 2**). Observe that a stand composed of 36 regularly planted trees with a range of tree heights between 14 and 18 meters (**Figure 2(a)**) has the same *SCI* value as a stand with 8 trees and a range

of tree heights between 1 and 12 meters (**Figure 2(c)**, **Table 2**). Intuitively, these two stands have very different structures, and it may be considered an undesirable property of the *SCI* that it is not able to differentiate these stands.

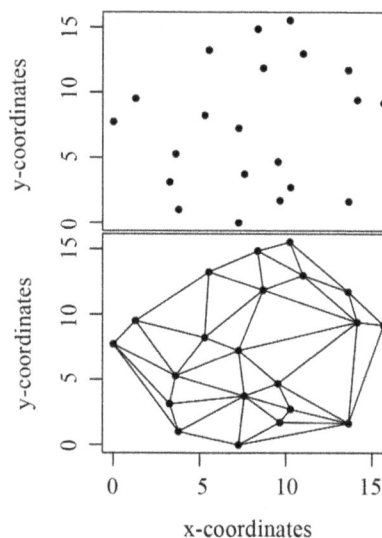

Figure 1.
Spatial distribution of stems (upper panel). Triangulated irregular network calculated for stems in the upper panel (lower panel).

Table 1.
Stand characteristics of example forest stands from **Figure 2**.

Stand	Area [m²]	No. of trees	Range of tree heights [m]	Mean tree height [m]	Standard deviation of tree heights	No. of individual heights
a)	1600	36	14 - 18	16.0	2.03	2
b)	1600	36	1 - 21	11.0	6.92	6
c)	1600	8	1 - 12	7.03	5.38	3
d)	1600	14	1 - 21	10.0	8.55	5

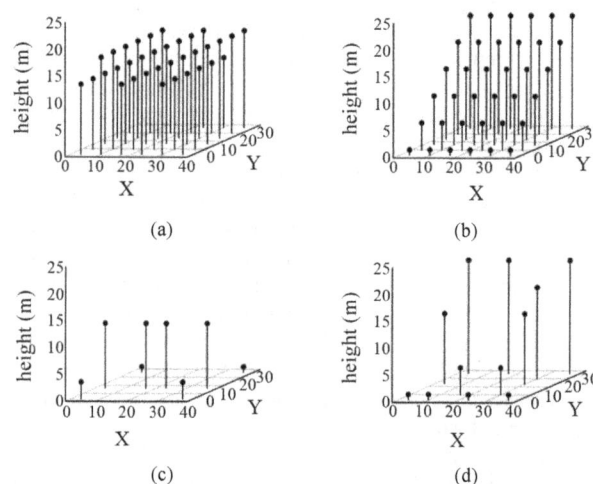

Figure 2.
Four different forest stands (a, b, c, d) having the same Structural Complexity Index (*SCI*) value. For stand characteristics see **Tables 1** and **2**.

Table 2.
Structural Complexity Index (*SCI*) and Enhanced Structural Complexity Index (*ESCI'* and *ESCI*) values of example stands from **Figure 2** and **Table 1**.

Stand	*SCI*	*ESCI'*	*ESCI*	TIN area [m²]	projected area [m²]	*VRM*	Trees/10 m²
a)	1.17	1.33	1.63	1295.77	1111.11	1.14	.23
b)	1.17	1.17	1.43	1295.77	1111.11	1.00	.23
c)	1.17	1.33	1.40	1295.77	1111.11	1.15	.05
d)	1.17	1.17	1.27	1295.77	1111.11	1.00	.09

The Enhanced Structural Complexity Index (*ESCI*)

We propose two modifications to the *SCI* to avoid the ambiguity described above and to better distinguish specific properties of structural complexity.

The two proposed modifications of the *SCI* towards the Enhanced *SCI* (*ESCI*) are:

1) Incorporation of triangle orientations;
2) Incorporation of triangle orientations + stem density.

The first modification enables the index to distinguish forest type a from type b, as in **Figure 2**. The two types do not differ in stem density, and trees in both stands are located on a regular grid with a spacing of 4 m. However, they differ in the range of tree heights and in the mingling of tree dimensions. In forest type b, there is a clear trend from small trees in the first row towards larger trees in the last row. This can be imagined as a forest edge in which tree height gradually increases towards the forest interior or as adjacent strips of clear cuts, each at a different age. In contrast, forest type a is composed of only two distinct tree heights; rows of 14 m high trees alternate with rows of 18 m high. The resulting TINs have the same surface area for both stands, resulting in the same *SCI* value.

To distinguish between these stand types, the orientation of triangles in the TINs is quantified by a vector ruggedness measure (*VRM*) (Equation (3)), adapted from a method proposed by Hobson (1972) and Sappington et al. (2007). Here, unit vectors are used to represent the orientation of a triangle (**Figure 3**). To centre a unit vector at triangle *i*, the cross product of triangle sides a_i and b_i is calculated. This results in a vector that is divided by its own length to standardize length to one. To ensure that all unit vectors are oriented upwards, unit vectors with a negative *z*-coordinate are mirrored. All unit vectors are summed, resulting in a new vector whose strength (*VSTR*, Equation (2)) is divided by the number of triangles in the TIN (*n*) and subtracted from 2 (Equation (3)). The resulting *VRM* is a dimensionless measure that ranges from 1 (non-rugged) to 2 (most rugged).

The *ESCI'* is calculated by multiplying the *SCI* with the *VRM* (Equation (4)). Based on the *ESCI'* forest types a (*ESCI'* = 1.6) and b (*ESCI'* = 1.3) can clearly be distinguished (see **Table 2**).

$$VSTR = \left| \sum_{i=1}^{n} \frac{a_i \times b_i}{|a_i \times b_i|} \right| \qquad (2)$$

$$VRM = 2 - \frac{VSTR}{n} \qquad (3)$$

$$ESCI' = SCI * VRM \qquad (1)$$

The second modification enables one to discriminate among forest types a and c and types b and d (**Figure 2**). Compared to forest types a and b, types c and d contain fewer trees (8 and 12 trees respectively) and cover different height ranges (**Table 1**).

Since stem density is considered an important structural characteristic (e.g. Pretzsch, 2009), the number of stems per unit area is included in the *ESCI* by multiplying the number of stems per 10 m² with the *ESCI'* (Equation (5)). Hence, the index calculated for stands with low stem densities is lower than for stands with high stem densities. To avoid too much weight being assigned to the stem density, a value of 1 is added to the number of stems per 10 m². The *ESCI* is a dimensionless measure >1 that increases with stem density, the intermingling of trees with different attributes, and differences between tree attributes.

$$ESCI = ESCI' * \left(1 + \text{No. of stems per 10 m}^2\right) \qquad (5)$$

Case Study from Mengsong, Xishuangbanna, Yunnan, China

Data Collection and Analysis

To investigate the *ESCI* and the *ESCI'* compared to the *SCI*, these indices were calculated for 28 plots of a forest inventory carried out in Mengsong township, Xishuangbanna, Yunnan, China (UTM/WGS84: 47 N 656355 E, 2377646 N, alt = 1600 m) (**Figure 4**). Plots cover the study site along a disturbance gradient from heavily disturbed shrub land, through secondary regrowth to mature montane rainforest stands. Each plot consists of nine circular 10 m-radius sub-plots (314.16 m²) arranged on

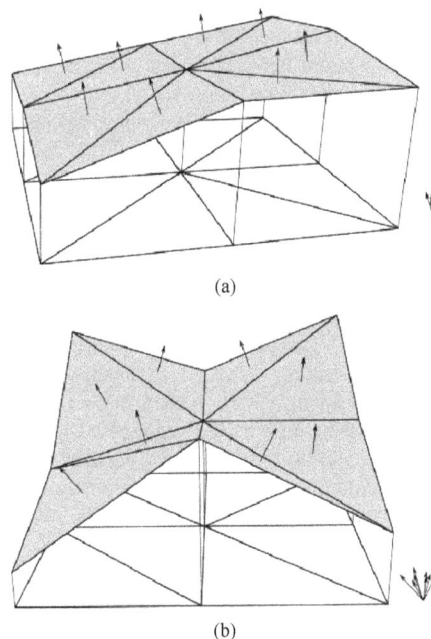

(a)

(b)

Figure 3.
Unit vectors centred at each triangle of a TIN. The vector ruggedness measure (*VRM*) quantifies the dispersion of these vectors: (a) Low *VRM* (low vector dispersion); (b) High *VRM* (high vector dispersion). *VRM* is used as a measure of the diversity of triangle orientations, hence a measure to describe the mingling of tree dimensions.

Figure 4.
Location of the study site Mengsong in Xishuangbanna, China. Black squares show plot locations
within the site.

a square grid with 50 m spacing (**Figure 5**). For sub-plots larger 250 m^2, mean and median *SCI* values have been shown to
be scale-invariant (Zenner, 2005).

In each sub-plot, *DBH* and position (azimuth and distance to
sub-plot centre) were recorded for all trees with a *DBH* ≥ 10 cm.
SCI, *ESCI'*, and *ESCI* were calculated for each sub-plot using
basal area of each tree as the *z*-coordinate. Sub-plots without at
least three trees were assigned *SCI*, *ESCI'* and *ESCI* values of
zero. The *R* statistical software (R Core Team, 2012) and the
geometry package (Grasman et al., 2011) were used for the
Delaunay triangulation. Per plot values were derived by averaging all 9 sub-plot values per plot.

In five sub-plots (No. 1, 3, 5, 7, 9) per plot, insects were collected using Malaise traps in April-May 2011. Malaise traps
were left in place for one week per plot and for each plot, species composition was determined using high-throughput DNA
metabarcoding of the cytochrome oxidase subunit I (COI) bar-
code gene (Yu et al., 2012). Insect species were approximated
with 97% threshold-similarity Operational Taxonomic Units
(OTUs), each representing a cluster of similar COI sequences.

Insect community composition was examined by non-metric
multidimensional scaling (NMDS) (Legendre & Legendre,
1998; Quinn & Keough, 2002). NMDS maps the position of
plots in species space, in our case represented by the Jaccard
dissimilarities of Hellinger transformed per plot OTU counts
(Yu et al., 2012), onto a predefined number of axes in an itera-
tive search for an optimal solution. NMDS is commonly regarded as the most robust unconstrained ordination method in
community ecology (Minchin, 1987). The *R* package *vegan*
(Oksanen et al., 2007) was used for the ordination analysis. To
investigate whether a relationship exists between structural
complexity and turnover in insect species composition we used
the function *envfit* in the package *vegan* to calculate the fit of
environmental variables to the ordination scores. *envfit* maxi-
mises the correlation of environmental variables to the matrix
of ordination scores and uses a permutation test to evaluate the
probability of obtaining the resulting or a higher r^2 value. *P*-
values stated here are based on 1000 permutations. To test whether differences in correlation coefficients between *SCI*, *ESCI'*

and *ESCI* are significant we took 1000 subsamples (without
replacement) of 26 observations from our field data. NMDS
ordinations were calculated for each subsample and the corresponding r^2 values with *SCI*, *ESCI'* and *ESCI* were calculated.
Subsequently, we applied Mann-Whitney *U* tests to investigate
whether differences of mean r^2 values of the replicated samples
were significant.

Results

From a total of 2890 trees that were recorded, each sub-plot
contained on average 11.47 trees, with numbers ranging from 0
to 34 trees per sub-plot. The basal area of individual trees varied from 78.61 cm^2 to 11,750 cm^2.

Across all sub-plots, *ESCI* values are consistently higher and
cover a larger range than do *ESCI'* values. The same is observed if *ESCI'* and *SCI* are compared (**Figure 6**).

In addition to the observed differences in index value range,
the indices treat special cases of tree arrangements differently.
For example, sub-plots 49_2 and 372_9 (**Figure 7**, upper panels)
have similar *SCI* values (1% difference), but their *ESCI'* values
differ substantially (22.75% difference) (**Table 3**). However, in
Figure 7, sub-plots 49_2 and 90_1 still have similar structural
complexities, as measured by the *ESCI'*.

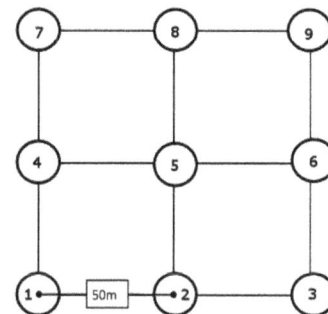

Figure 5.
Plot design: Cluster plot consisting of
9 sub-plots arranged on a square grid
with 50 meters spacing between sub-
plot centres.

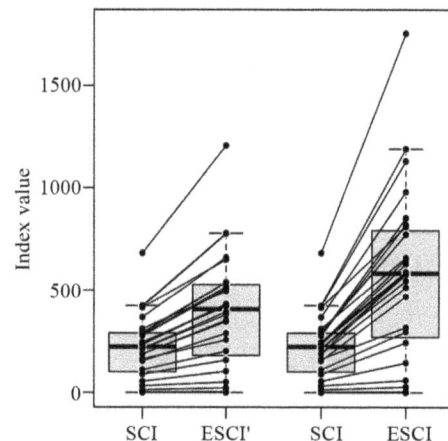

Figure 6.
Box and ladder plot comparing paired *SCI* and *ESCI'*
observations (left) and paired *SCI* and *ESCI* observations (right).

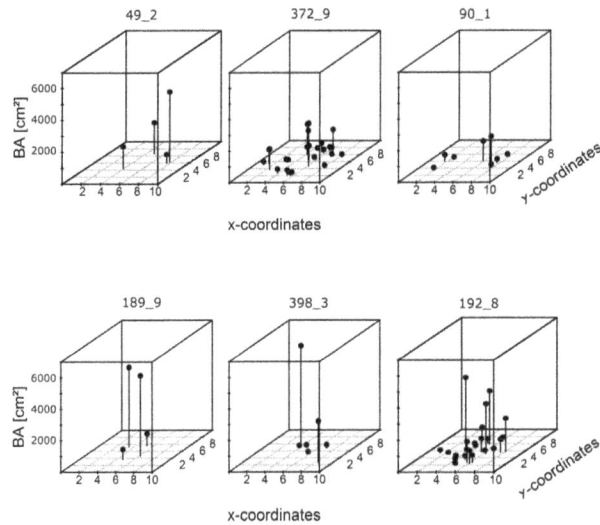

Figure 7.
Depiction of tree positions (*xy*-coordinates; in meters) complemented with basal area (*BA*) as the *z*-coordinates of sub-plots listed in **Table 3**. Upper panels: Sub-plot 49_2 has a *SCI* value similar to sub-plot 372_9. Based on the *ESCI* value 49_2 is similar to 90_1. Lower panels: Sub-plots with similar *ESCI'* values which would be distinguished by their *ESCI* values. Note: *ESCI* ranks the sub-plots in a different order than *SCI*.

Table 3.
Index values calculated for sub-plots shown in **Figure 7**.

sub-plot	projected area [m²]	TIN area [m²]	*SCI*	*ESCI'*	*ESCI*	*VRM*	No. of trees
49_2	47.53	24911.01	524.15	528.26	595.51	1.01	4
372_9	222.35	119133.61	535.79	1014.81	1790.07	1.89	24
90_1	113.11	33290.73	294.31	532.38	667.95	1.81	8
189_9	36.94	40896.36	1107.08	1445.78	1629.86	1.31	4
398_3	43.86	40417.48	921.44	1452.41	1729.81	1.58	6
192_8	166.83	133784.66	801.91	1473.12	2551.62	1.84	23

The effect of the incorporation of the stem density term becomes clear when the sub-plots in **Figure 7** (lower panels) are examined. According to their *SCI* values, sub-plot 189_9 has a higher structural complexity than does 398_3 or 192_8. *ESCI* ranks sub-plot 189_9 as the least structurally complex and 192_8 as the most structurally complex (**Table 3**).

The number of NMDS dimensions was fixed to three after visual examination of the scree plot of stress vs dimensions. Three dimensions provided a satisfactory stress value of 9.24%. There were highly significant ($p < .001$) correlations between all three structural complexity indices and the NMDS ordination. The *SCI* shows the lowest r^2 value of .51 followed by the *ESCI'* with $r^2 = .54$ and *ESCI* with $r^2 = .59$. Mean r^2 values based on 1000 sample replicates were significantly different among all combinations of indices ($p < .0001$).

Discussion

The number of stems per 10 m² is a critical factor in the *ESCI* formula. In the data set used in this study, we found that (1 + No. of stems per m²) results in a range of values from 1 to

2.02, for 0 and 32 trees per sub-plot respectively. This range does not assign a weight to the number of stems that outweighs the other terms of the equation. *SCI*, *VRM* and stem density are hence of approximately equivalent importance in determining the *ESCI* value. For other data sets with a higher number of stems per unit area, the relation to the base of 10 m² probably leads to density values that dominate the resulting *ESCI* value. A higher stem density might occur, for example, in different forest types, succession stages with high stem numbers, or if the *DBH* threshold is lower than 10 cm. In these cases it might be more informative to calculate *ESCI'* and stem density separately to describe structural complexity and to make inferences. The addition of 1 to the number of stems per 10 m² assigns a greater weight to low stem densities but has little influence if stem densities are high. This weighting takes into account that with increasing density, the potential for a spatial differentiation decreases.

Sub-plots in which stems were mapped for this analysis cover an area of 314.16 m², which is sufficiently large for the calculation of the *SCI* (Zenner, 2005). Nevertheless, due to the high scale dependency of forest structure (Franklin et al., 2002; Zenner, 2005), this plot size is probably at the lower bound of an adequate size. A further investigation of the variability of *ESCI'* and *ESCI* values at varying scales is recommended to achieve deeper insights regarding the minimum plot size for calculations of these indices.

In the structural assessment of forests, the inclusion of small diameter trees was found to enhance the detection of structure types (Zenner et al., 2011). In future studies it might be interesting to analyse whether *ESCI'* and *ESCI* show a similar response if trees with a *DBH* < 10 cm are considered.

Comparing *SCI*, *ESCI'* and *ESCI*, we find that the stronger emphasis on the intermingling of tree dimensions through the incorporation of the *VRM* as a measure of surface ruggedness improves the ability of the index to discriminate among stand conditions. This might result in a better utility of the index. For example, it could be used for a separation of forest edges with increasing tree height from other spatial arrangements of tree dimensions typical for forest interiors. Such discrimination makes sense from an ecological point of view since forest edges are characterized by a distinct microclimate and resource spectrum, with corresponding effects on species composition and abundance (McDonald & Urban, 2006). If in addition to the *VRM*, the stem density of a forest is taken into account, the ability to unambiguously characterize forest structural complexity is increased again. Ecologists and forest managers likewise consider stem density as an important aspect of structural complexity because it determines the mean growing space per plant and hence is an indicator of competition for resources within the stand (Pretzsch, 2009). Changing stem density is the principal way forest managers manipulate forests (Davis et al., 2001), and these changes alter forest habitats with consequences for forest organisms.

A possible ecological relevance of the modifications of the *SCI* is indicated by the correlation of the index values with the NMDS ordination, which describes turnover in insect species composition. Compared to the *SCI*, significantly stronger correlations with the NMDS ordination were observed for the *ESCI'* and the *ESCI*. The observed correlations support the habitat heterogeneity hypothesis by suggesting that the structural complexity of a habitat influences the insect species community composition. The higher r^2 values for *ESCI'* and *ESCI* suggest that these indices perform at least as well and possibly better in detecting this relationship. Nevertheless, the increase in r^2 values was only moderate, and hence it would be valuable to test the indices for structural complexity against other data sets from different geographic regions and taxa and to assess associations at multiple spatial scales.

Conclusion

The results of this study show that the suggested modifications to the *SCI* are valuable improvements that increase the ability to characterize the structural complexity of forests. *ESCI'* and *ESCI* allow for a more complete view of a forest structure than the *SCI*. This makes these indices relevant to ecologists, forest scientists, and forest managers who are interested in the relationship between ecosystem structure and biodiversity.

Acknowledgements

Above all, we are indebted to the Advisory Group on International Agricultural Research (BEAF) at the German Agency for International Cooperation (GIZ) within the German Ministry for Economic Cooperation (BMZ) for funding this research (project number 08.7860.3-001.00 "Making the Mekong Connected"—MMC). We are grateful to all members of the MMC-project for excellent support in coordinating and implementing research and field work, and in particular to the head of the project Prof. Dr. Xu Jianchu and the "various fathers" of the project including Dr. Horst Weyerhäuser and Dr. Timm Tennigkeit.

REFERENCES

Aber, J. (1979). Foliage-height profiles and succession in northern hardwood forests. *Ecology, 60,* 18-23.

Ambuel, B., & Temple, S. (1983). Area-dependent changes in the bird communities and vegetation of southern Wisconsin forests. *Ecology, 64,* 1057-1068.

Cramer, M., & Willig, M. (2005). Habitat heterogeneity, species diversity and null models. *Oikos, 108,* 209-218.

Davis, L., Johnson, K., Bettinger, P., & Howard, T. (2005). Forest management—To sustain ecological, economic, and social values. Prospect Heights: Waveland Press, Inc.

Doherty, M., Kearns, A., Barnett, G., Sarre, A., Hochuli, D., Gibb, H., & Dickman, C. (2000). The interaction between habitat conditions, ecosystem processes and terrestrial biodiversity: A review. State of the Environment, Second Technical Paper Series (Biodiversity), Environment Australia, Department of Environment and Heritage.

Franklin, J., Spies, T., Pelt, R., Carey, A., Thornburgh, D., Berg, D. et al. (2002). Disturbances and structural development of natural forest ecosystems with silvicultural implications, using Douglas-fir forests as an example. *Forest Ecology and Management, 155,* 399-423.

Franzreb, K. (1978). Tree species used by birds in logged and unlogged mixed-coniferous forests. *The Wilson Bulletin, 90,* 221-238.

Freemark, K., & Merriam, H. (1986). Importance of area and habitat heterogeneity to bird assemblages in temperate forest fragments. *Biological Conservation, 36,* 115-141.

Grasman, R., Gramacy, R.B., & Sterratt, D. C. (2011). Geometry: Mesh generation and surface tesselation. R Package Version 0.2-0. URL (last checked 5 June 2012).
http://CRAN.R-project.org/package=geometry

Kimmins, J. (2005). *Forest ecology: A foundation for sustainable forest management and environmental ethics in forestry* (3rd ed.). Upper Saddle River, NJ: Prentice Hall.

Lawton, J. (1983). Plant architecture and the diversity of phytophagous insects. *Annual Review of Entomology, 28,* 23-39.

Legendre, P., & Legendre, L. (1998). Numerical ecology. *Elsevier Science & Technology, 20,* 853.

LeMay, V., & Staudhammer, C. (2005). Indices of stand structural diversity: Mixing discrete, continuous, and spatial variables. In *Proceedings of the IUFRO Sustainable Forestry in Theory and Practice: Recent Advances in Inventory & Monitoring, Statistics, Information & Knowledge Management, and Policy Science Conference* (p. 4). Edinburgh, 5-8 April 2005.

MacArthur, R., & MacArthur, J. (1961). On bird species diversity. *Ecology, 42,* 594-598.

MacArthur, R., & Wilson, E. (1967). The theory of island biogeography. Princeton, NJ: Princeton University Press.

McCoy, E., & Bell, S. (1991). Habitat structure: The evolution and diversification of a complex topic (pp. 3-27). London: Chapman and Hall.

McDonald, R., & Urban, D. (2006). Edge effects on species composition and exotic species abundance in the North Carolina piedmont. *Biological Invasions, 8,* 1049-1060.

McElhinny, C., Gibbons, P., Brack, C., & Bauhus, J. (2005). Forest and woodland stand structural complexity: Its definition and measurement. *Forest Ecology and Management, 218,* 1-24.

Minchin, P. (1987). An evaluation of the relative robustness of techniques for ecological ordination. *Plant Ecology, 69,* 89-107.

Moen, C., & Gutierrez, R. (1997). California spotted owl habitat selection in the central Sierra Nevada. *The Journal of Wildlife Management, 61,* 1281-1287.

Oksanen, J., Kindt, R., Legendre, P., O'Hara, B., Stevens, M., Oksanen, M., & Suggests, M. (2007). The vegan package. Community ecology package. URL (last checked 5 June 2012).
http://www.R-project.org

Polley, H. (2007). Survey instructions for federal forest inventory II (2001-2002), 2nd corrected translation, February 2006, of the 2nd corrected and revised reprint, May 2001. Federal Ministry of Food, Agriculture and Consumer Protection, Bundesministerium für Verbraucherschutz, Ernährung und Landwirtschaft (BMVEL). URL (last checked 10 September 2012).
http://www.bundeswaldinventur.de/media/archive/513.pdf

Pommerening, A. (2002). Approaches to quantifying forest structures. *Forestry, 75,* 305-324.

Pretzsch, H. (1997). Analysis and modeling of spatial stand structures. Methodological considerations based on mixed beech-larch stands in Lower Saxony. *Forest Ecology and Management, 97,* 237-253.

Pretzsch, H. (2009). Forest dynamics, growth and yield: From measurement to model. Berlin: Springer Verlag.

Quinn, G., & Keough, M. (2002). Experimental design and data analysis for biologists. Cambridge: Cambridge University Press.

R Core Team (2012). R: A language and environment for statistical computing. Vienna: R Foundation for Statistical Computing.

Recher, H., Majer, J., & Ganesh, S. (1996). Eucalypts, arthropods and birds: On the relation between foliar nutrients and species richness. *Forest Ecology and Management, 85,* 177-195.

Sappington, J., Longshore, K., & Thompson, D. (2007). Quantifying landscape ruggedness for animal habitat analysis: A case study using bighorn sheep in the Mojave Desert. *The Journal of Wildlife Management, 71,* 1419-1426.

Simpson, E. (1949). Measurement of diversity. *Nature, 163,* 688.

Spanos, K., & Feest, A. (2007). A review of the assessment of biodiversity in forest ecosystems. *Management of Environmental Quality: An International Journal, 18,* 475-486.

Temple, S., Mossman, M., & Ambuel, B. (1979). The ecology and management of avian communities in mixed hardwood-coniferous forests. In *Management of north-central and northeastern forests for nongame birds* (pp. 132-153). USDA.

Tews, J., Brose, U., Grimm, V., Tielbörger, K., Wichmann, M., Schwager, M., & Jeltsch, F. (2004). Animal species diversity driven by habitat heterogeneity/diversity: The importance of keystone structures. *Journal of Biogeography, 31,* 79-92.

United States Department of Agriculture—Forest Service (2011). Forest inventory and analysis, national core field guide, volume I: Field data collection procedures for phase 2 plots, version 5.1. URL (last checked 10 October 2012). http://fia.fs.fed.us/library/field-guides-methods-proc/docs/Complete%20FG%20Document/core_ver_5-1_10_2011.pdf

Whittaker, R. (1972). Evolution and measurement of species diversity. *Taxon, 21,* 213-251.

Willson, M. (1974). Avian community organization and habitat structure. *Ecology, 55,* 1017-1029.

Yu, D., Ji, Y., Emerson, B., Wang, X., Ye, C., Yang, C., & Ding, Z. (2012). Biodiversity soup: Metabarcoding of arthropods for rapid biodiversity assessment and biomonitoring. *Methods in Ecology and Evolution, 3,* 613-623.

Zenner, E. (2005). Investigating scale-dependent stand heterogeneity with structure-area-curves. *Forest Ecology and Management, 209,* 87-100.

Zenner, E., & Hibbs, D. (2000). A new method for modeling the heterogeneity of forest structure. *Forest Ecology and Management, 129,* 75-87.

Zenner, E. K., Lähde, E., & Laiho, O. (2011). Contrasting the temporal dynamics of stand structure in even- and uneven-sized *Picea abies* dominated stands. *Canadian Journal of Forest Research, 41,* 289-299.

10

Fire and People in Three Rural Communities in Kabylia (Algeria): Results of a Survey

Ouahiba Meddour-Sahar[1], Raffaella Lovreglio[2], Rachid Meddour[1], Vittorio Leone[3], Arezki Derridj[1]

[1]Facultés Sciences Agronomiques et Biologiques, Department of Sciences Agronomiques B.P., Université Mouloud Mammeri de Tizi-Ouzou, Tizi-Ouzou, Algeria
[2]Department of Agriculture, University of Sassari, Sassari, Italy
[3]Department of Crop Systems, Forestry and Environmental Sciences, University of Basilicata, Potenza, Italy

This research was conducted to investigate about the causes of wildfires in three municipalities of the Tizi Ouzou and Bouira provinces, in Kabylia. Unknown forest fire causes account for 80% of total in Algeria, with a peak of about 99% in Kabylia, the most forested region in the country but also the most severely affected by wildfires. The three study areas exhibit a rather high forestry ratio (40% on average) but also a very high population density, up to 300 people per km^2, living in a multitude of small hamlets, near or inside the domanial forests and exerting an enormous pressure on them. Survey was carried out on people (N = 300) randomly selected in nine hamlets (*thaddarth*) through ad hoc questionnaires (134 possible replies) filled with face-to-face interviews. Factorial Correspondence Analysis was used to process data (300 × 134 replies in total). Fires result mainly voluntary (52.95%): pastoral fires to renew pastures (11.30%), political fires as security counter-terrorism measure (11.24%), uncontrolled garbage burning (6.83%). Negligent fires account for 41.79%: carelessly tossed cigarette butts (9.13%), agricultural works (burning of cut bush, stubble burning, 7.03%), restart of fire (6.83%) and forest activities in the forest (6.31%). Results depict a common core of fires due to the pressure on the domanial forests, on which traditional use of fire (pastoralism) and national security needs (counter-terrorism) dominate.

Keywords: Fire Motivations; Folk Crime; FCA; Kabylia; Pastoralism; Political Fire

Introduction

In the Mediterranean region more than 50,000 fires burn an estimated 600,000 - 800,000 hectares annually, about 1.5% of total Mediterranean woodlands (Rowell & Moore, 2000; WWF-IUCN, 2007; Cemagref, 2009). Fires are particularly abundant in the northern rim of the Mediterranean, where France, Greece, Italy, Portugal and Spain contribute with a yearly average of 49,838 fires and 471,644 burned hectares (1980-2010) (JRC, 2010).

In the southern rim fires are less abundant in terms of numbers and burned surfaces, certainly due to different socio-economic conditions (Velez, 1991), where the forest is a resource, for food, for fuel wood but mainly for grazing. Wildfires rarely naturally occur in the Mediterranean region where the only relevant natural cause of forest fires is lightning. Wildfires caused by lightning have a local character and are highly dependent on mesoscale atmospheric conditions (Petersen & Rutledge, 1998; Garcia-Ortega et al., 2011). Apart from those possible, but minimally important causes, fires reveal a strong impact of human actions.

Mediterranean region accounts the larger proportion of human caused fires in the world (95%) followed by South Asia (90%), South America (85%) and Northeast Asia (80%) (Leone et al., 2009).

Forest fire statistics are usually compiled by processing individual wild land fire reports collected after each event by the national Forest Services, Fire Departments or other similar governmental Agencies. Assessment of cause merely reflects the opinion of the reporting officer filling the fire report; secure determinations are therefore possible only when culprits are brought to justice (Leone et al., 2009). In the list of countries of southern Mediterranean rim affected by wildfires Algeria is the first best, with more than 1300 fires per year and 39,000 hectares burned (respectively 2.34% and 6.5% of the values for the whole Mediterranean; Meddour-Sahar, 2008). Algeria has one of the longest history of fire recording, together with Cyprus (Harris, 2007), since both coming from colonial experiences.

Changes in fire occurrence are evident throughout history of colonization, confiscation of communal land and the application of modern agricultural techniques that increased the amount of arable land (Davis, 2004; Bensaid et al., 2006; FAO, 2010).

The state-owned regime for forest and pasturelands and the settlement policies imposed by the colonial period, for instance, have brought extensive conflicts between pastoral groups and the public administration (Davis, op.cit). Rural incendiarism as protest against curtailing traditional use of common lands marked the second half of XIX century (Kuhlken, 1999), when collective fines, set against communities that continued to practice fire-driven agriculture despite a ban from French colonial authorities, were met with non-compliance and an increase in malicious fire raising (Holmes, 2007).

During the long and merciless independence war (1954-1962) fire was widely used in the scorch earth strategy by the French Army; more recently during the decade of severe political troubles of the Algerian Civil War, marked by terroristic activity, fire was among counter-terrorism measures and some forests have been destroyed to avoid giving refuge to armed groups (Bainem forest by Algiers, for example; Dridi, 2002). *"The situation in this area of the Mediterranean basin is particularly alarming and should be a priority for attention of the international community"* (Bariteau in: Baskaran et al., 2001).

Causes of fire in Algeria are the result of high density of rural population (ranging from 40 to over 600 inhabitants per km^2 in the North Central region of the country), of growing demography, of rural exodus and countryside abandonment, of urban sprawl and increasing demand of building areas along the coast, of increasing production of home waste and of traditional forms of land use dominated by pastoralism.

In many cases the degradation of forested areas reflects a population accustomed to using forest as a free-for-all with scarce concern for forest preservation (Thirgood, 1981), and for this in strong contrast with Forestry Administration (Berchiche, n.d.). This adds to an adverse climate, with recurring droughts and long, increasingly hot summers with prolonged, severe heat waves, such as in 2012. This scenario is rather similar to other M.E.N.A. countries' with similar climatic and productive features, where rural populations maintain excessively high pressure on wooded lands, overexploiting firewood and over-grazing (FAO, 2012).

In Mediterranean countries a large fraction of the total number of forest fires remains unexplained. Algeria holds the record of unknown causes with nearly 80%, followed by Tunisia, 65%, Morocco, 55% and Turkey, 48%. The high percentage of unknown causes as in Algeria, makes it difficult the implementation of a prevention policy targeted to specific social groups or activities or behaviours (Meddour-Sahar & Derridj, 2012). Fire causes in Algeria were object of researches in the past (Gravius, 1866; Thibault, 1866; Marc, 1916; IIA, 1933; Boudy, 1952), but more recent information about them is very scarce and dramatically reflects the extremely high incidence of unknown fires which, in some provinces in Kabylia, such as Tizi-Ouzou, is close to 99% (Meddour-Sahar & Derridj, 2010).

Methodology

In this paper we report the result of a survey on fire causes, carried out through interviews to people living in three fire prone areas in Kabylia (Algeria); they represent the powerless who usually have no voice and whose perspective or interpretation of the phenomenon could be different from foresters'. The 26 years time series analyzed (1985-2010), includes the "black decade" (1990-2000) of political instability which raged in the country.

Study Area

Survey was carried out in three different rural *baladiyath*[1] (Mizrana, Ain Zaouia, Haizer), in the *wilaya* (province) of Tizi Ouzou and Bouira in Kabylia (**Figure 1**), all of them marked by more or less severe fire history.

Kabylian communities live in the forest, which provides

Baladiah is the equivalent of municipality.

them with multiple resources: foodstuffs from poaching and gathering, honey, mushrooms, fodder, firewood, timber, cork, etc. (FAO, 2012), but rural populations maintain excessively high pressure on wooded lands. Official data from the Algerian Forest Administration of the two provinces (Tizi Ouzou and Bouira) permit to measure this high pressure: from 1999 to 2009, a total of 347 offenses were officially registered, even not always sanctioned to punish non compliant behavior.

Selected *baladiath* are representative of different situations (littoral, North part of Djurdjura massif, South face of Djurdjura massif). Each *baladiah*[*] is formed by different *douar* or *thaddarth* (village) (21 to 28), sometimes at the edge of forested areas, in many cases inside the forest. The *douar* under study are mountain and forest hamlets, all of them in territories having a forestry ratio >40%, thus among the most forested ones in the region.

Forests are highly degraded by fires and by the anthropic activities (illegal cutting and constructions, overgrazing, overexploitation, pollarding for fodder etc.).

Each *baladiah* covers a surface roughly ranging from 5000 to 9000 hectares and is synthesized, in terms of geographic indicators, in **Tables 1** and **2**.

Data and Methods

Survey was conducted in a traditional method (self-administered survey; Leones, 1998) by distributing the purposely prepared questionnaires. Surveyor filled every questionnaire with a face-to-face interview, directly contacting each one of the household components. The questionnaire used simple, familiar, and unambiguous words. It is a closed-ended or fixed-item type; questions are multiple-choice with unordered response choices. The questionnaire includes 37 questions allowing 134 responses and is structured in four sections:

- Demographic information (status, gender, residence and level of education, labor condition);
- Agro-forestry activities (type of activities, type and size of husbandry, livestock feeding);
- Knowledge about fires (size, damages, causes);
- Anthropic pressure on the forest (forests' condition, occupancy, garbage burning, needs for firewood, wood cutting, constructions, forest activities).

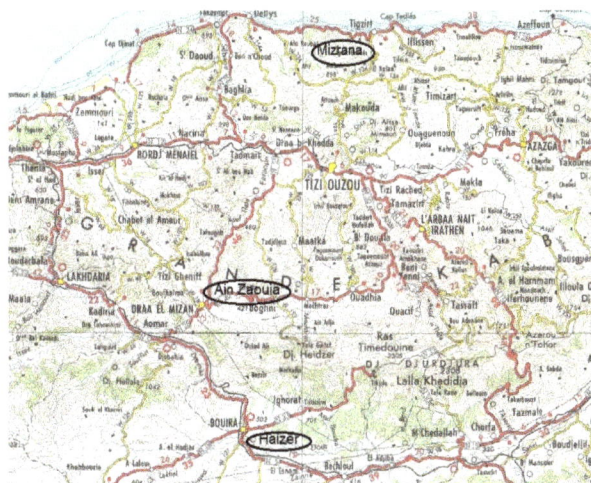

Figure 1.
Map of the study areas in Kabylia , North Central Algeria.

Table 1
The most relevant socio-economic indicators for the study areas.

Indicator/Baladiah	Mizrana	Ain Zaouïa	Haizer
Position	Littoral	North part of Djurdjura massif	South part of Djurdjura massif
Geographic coordinates	36°20' to 36°30'N 4°05' to 4°07'E		36°20' to 36°30'N 3°55' to 4°10'E
Surface (ha)	5784	5677	8900
Number of villages	28	29	21
Population	9488	17,372	18,371
Population density per km^2	168	305	206
Forest land (ha)	3595	2577	4827
Forestry ratio in %	72	48	56
Agricultural land (ha)	959	2211	3628
Pasture and rangelands (ha)	270	200	30
Unproductive land (ha)	200	400	80
Total agricultural land (ha)	5024	5388	8565
Sheep number	3128	2834	1632
Goats number	2395	1562	146
Cows number	326	345	993
Total livestock	8783	4741	2771
Livestock per km^2	151.80	83.33	31.13
Land-use	Agro-sylvo-pastoral	Agro-sylvo-pastoral	Agro-sylvo-pastoral

Table 2.
The most relevant forestry indicators for the study areas.

Forest	Domanial forest of Mizrana	Domanial forest of Boumahni	Domanial forest of Haizer
Forest description	Open forest	Mediterranean scrubland	Open degraded and fragmented forest
Vegetation	Cork oak, Zeen oak	Cork oak, Wild olive, Aleppo pine	Aleppo pine, Holm oak, Cedrus atlantica
Climate	Pluviometry: 800 to 1200 mm/yr Subhumid	Pluviometry: 800 to 1000 mm/yr Subhumid	Pluviometry: 700 to 1000 mm/yr Subhumid
Middle mountain terrain (400 - 800 m)	76.31%	53.21%	62.95%
Class of dominant slope (>25%)	60%	69%	47%
Forest fires number (1985-2010)	182	78	208
Burned surface (1985-2010) ha	286	143	104
Burned surface in % of forested area (1985-2010)	3.06	2.13	0.83

Survey was carried out in March-May 2010 in three villages for each *baladiah*, where we found the support of local authorities, including the transport on the field and presentation to the village chiefs (**Table 3**). Surveyor (O.M.S. and her team) was always accompanied by a local key-informant, usually a member of the village committee who helped the surveyor to get in touch with the relevant villagers and above all to establish a relationship of trust.

Table 3.
Number of sampled villages.

Wilaya (Province)	Baladiah (Town)	Thaddarth (Village)	Total household	Sample household
Tizi Ouzou	Mizrana	Azroubar	151	43
		Ouatouba	106	30
		Tamazirt Ourabah	96	27
		Total baladiah of Mizrana	353	100
	Ain Zaouia	Adbagh	29	18
		Ait Amar Moh	48	29
		Igharviyen	80	53
		Total baladiah of Ain Zaouia	157	100
Bouira	Haizer	Guentour	190	56
		El Mahsar	46	13
		Slim	105	31
		Total baladiah of Haizer	341	100
Total			851	300

At first some difficulties in contacting households arose and made necessary the intervention of village leaders to explain that surveyors were not sent by the government to investigate about offences in the forest, but emphasizing that their interest was directed to protect forest against fires.

The target population for the survey consists of 100 individuals aged 15 and older of both sexes, for each *baladiah*. The unit of observation is a household in the selected villages. Sampling plan and size for each village is reported in **Table 3**. At the household level, the sample of the interviewees was large enough in size in relation to the defined target population. The three *baladiah* encompass a total of 78 villages; the choice of three villages for *baladiah* (9 villages in total) therefore gave a sampling rate of 11.53%. Given the security constraints in the area (terrorism), the expected results and the specificities of the Kabylian village residents, a sample of nearly 300 individuals was decided, thus representing a sampling rate of 35.25 % (300 out of 851 households). Not having the availability of lists of inhabitants, a systematic sampling on the field was then applied in agreement with village leaders, by taking a household out of three for Mizrana and Haizer, and one out of two in Ain Zaouia. Sampling was very difficult in the absence of listing, and even on the field; contact also was difficult, since some families refused to participate in the investigation or were not at home. On the contrary, an open willingness came from University students, thus explaining the relatively high percentage of them among respondents (18.7%). A research whatever has never been carried out in those villages and their forests have no management plan.

Results

Data were processed using Sphinx plus V5 survey and statistics software (www.lesphinx-developpement.fr/).

Sample is mainly composed of male (81.7%), rather young individuals (65.7% under 39 years), mainly living in the forest (48.7%) or near it (19%). This is specific for Kabylia but not for Maghreb, since in Tunisia and Morocco the similar percentage is no more than 10% (Colin & Jappiot, 2001).

Labor status is strongly characterized by unemployment (51%), followed by employment in Government (16%) and work in agriculture (14%), be it self-employment or wage work; the number of retired (8%) and students (8%) is low. In general, all people, even in official status of unemployed or employed, work their land, as it is usual in rural societies. People under observation have a rather low education level (illiterate account for 25.3%; primary and college school level for 37%, secondary school for 19%, University level for 18.7%). The latter percentage, however, marks the interest for high levels of education also in small rural communities and, in addition, the relevant attractive power by University in the backcountry.

Main activities (more than 50%) are olive growing and husbandry, followed by growing fruit trees and vegetables. Beekeeping is rather important (14.7%) based on a reduced number of beehives (the modal value 46.7% is less than 4 beehives). Independently from species (cattle, sheep, goats) number of stock units is low: the modal class is 0 - 5 stock units. Products are mainly for self-consumption (78.7%). Livestock is mainly fed in the forest (27.2%) or fed with a mixed regime (fodder and grazing in forest (42.6%). Replies on occupancy of forest by local residents were bivalent, balanced ones: benefic for 41.7%, detrimental for 39.3%, no opinion for 19%.

Heating and cooking is mainly based on LPG (89.7%), followed by wood (76.7%) and electricity (27%). Wood is mainly harvested (illegally too) in the forest (71.7%) or purchased (19.3%). A large majority of respondents (88%) complains about scarce effort by Forest Administration in providing them with firewood. Domestic garbage is destroyed on site, arguably by fire (51.7%), abandoned in the forest (50%) and thrown into landfills 9.3%.

Illegal wood cutting and illegal constructions are considered important by 11.3%, but of low importance by 60.63% of respondents. Surprisingly, only 36.7% of respondents or their families, though living in the forest or rather close to it, are involved in forestry works: cleaning (21.7%), fire fighting activities (21%), cork harvesting (21.7)%. Forest is considered by far an advantage by 87.3% of respondents. Surveyed people express equivalent opinion about fires trend: increasing (45.7%). decreasing (45.3%) constant (only 9%). A great majority of surveyed people has seen a fire (90%), mainly of medium or large size (respectively 43.1% and 36.3%). Their opinion about fire damages mainly refers to crops (75.7%), fruit trees (52.3%), houses (22.3%) livestock (20.7%) and humans (only 2.3%). About forest fire causes, surveyed people were invited to declare presence or absence of the officially accepted causes, as proposed by Forest Administration. No question involved qualitative evaluation of the phenomenon.

Causes of Forest Fires

Percentage of causes results as follows:
- Natural: 0.74%
- Accidental: 4.47%
- Voluntary: 52.95%
- Involuntary: 41.79%

Percentages are well consistent with current literature, which often underlines an excess of emphasis given to voluntary fires (Velez, 2000; Franco Irastorza & Dolz Reus, 2007). The main causes are reported in **Table 4** here following (all *baladiath* merged):

As evident from **Tables 4** and **5**, a few motives have a fre-

Table 4.
Decreasing frequency of forest fires motives.

Main motives	%
Pasture renewal	11.30
Fires set for political reasons (security fires)	11.24
Cigarette remains	9.13
Pyromania	8.67
Agricultural works (burning of cut bush, stubble burning)	7.03
Illegal garbage dumping and burning	6.83
Restart of fire	6.83
Forest works (burning of cut bush)	6.31
All the others	<6

Table 5.
Fire motives in decreasing order of frequency in the three baladiah.

Ain Zaouia		Mizrana		Haizer	
Motives	(%)	Motives	(%)	Motives	(%)
Fires set for political reasons	23.29	Pyromania	19.63	Illegal garbage dumping and burning	8.56
Restart of fire	15.22	Pasture renewal	18.06	Cigarette remains	8.44
Pasture renewal	12.73	Cigarette remains	15.45	Pasture renewal	7.58
Illegal garbage dumping and burning	10.56	Fires set for political reasons	14.40	Forest works	7.33
Forest works	6.21	Agricultural works	11.26	Children's games	6.60
Agricultural works	4.66	Interest in land use changes	6.81	Agricultural works	5.99
Conflict related to land use	3.42	Conflict related to land use	5.76	Restart of fire	5.87
Cigarette remains	3.42	Forest works	4,19	Pyromania	5.75
Others	20.49	Others	1.57	Others	43.88

quency >6%, but they strongly differ from a *baladiah* to another, thus confirming that, at every scale, fire is site and culture specific (Leone et al., 2003).

Factorial Correspondence Analysis

Given the relevant amount of information gathered by the survey (300 × 134), forest fires causes were explored also with the help of FCA (Factorial Correspondence Analysis) a technique which belongs to the family of multidimensional descriptive statistics (Maniatis, 2010). We present only significant results of analysis, and some representative scatter plots.

Municipalities, Villages and Causes

Ain Zaouia is more distant from the average (axes origin), whereas Mizrana and Haizer are symmetrically opposite to it.

In Ain Zaouia, political fires are the more impressing motive, followed by fire restart; the latter motive, as already reminded, is closely related to security reasons which hamper the efficiency of fire fighting crews, thus confirming a sort of feedback with political fires.

Haizer is characterized by garbage burning, honey gathering, children's games, land use conflicts, followed by less important causes (brash burning, agricultural fires, tourists, machinery, power line arching, vehicle's muffling, hunting conflicts).

Mizrana has its distinctive trait in pastoralism, followed by cigarette remains, pyromania, land use change (it is a municipality rather close to coast, where housing boom is a reality).

Exploding the results in the nine villages, the mentioned matching is:

Ain Zaouia villages:

Ait Amar Moh and Igharbiyene are characterized by political fires and hunting interest, Adbagh by fire restart, conflicts with Administration.

Haizer villages:

Guentour stands out for garbage burning and conflict with Forest Administration, El Mahsar for land use conflicts and forest machinery, Slim for honey gathering and hunting conflicts.

Guentour and El Mahsar appear closer, therefore more similar and more involved with bush burning, whereas Slim is characterized by minor causes, such as machinery, hunting conflicts, land use conflict, agriculture and forest work by machinery.

Mizrana villages:

Mizrana, Tizi Ourabah and Outouba are involved with pastoralism and agricultural fires, in some opposition and rather distant from Azroubar which is characterized by pyromania, interest in land use and cigarette discarding.

For pyromania, present only in this context, probably the term was not well understood by respondents and merely mistaken for unknown as Franco Irastorza & Dolz Reus (2007) argue.

Location of Villages

People dwelling villages inside the forest mainly refer to pyromania, but also to cigarette remains, to interest and/or conflicts for land use changes; two main causes dominate: pastoralism, which pertains to their culture and political reasons. People living outside the forest have a less concerned image of problems: they recall rather obvious and banal reasons: children's games, power line arching, tourists, vehicle's mufflers. In a rather intermediate position the replies of people who dwell villages close to the forest or at the edge of it: they mainly refer to forestry works, to restart of fire, to honey gathering, but with strong emphasis on burning of garbage, which is probably part of their familiar scenario.

Educational Level

Municipalities and level of education form a rather compact cloud, rather distant from two causes which represent outliers: lightning, which is a rare event (within Algeria and all Maghreb) and interest in hunting, which is a forbidden activity now practiced only as poaching, therefore a rather risky one. In order of educational level, illiterate refer to restart of fire, tourists, forestry works, vehicles' mufflers, illegal dumping and burning of domestic waste. Primary and secondary school refer to honey gathering and change of land use. Middle school refers to political reasons and pastoralism and to less important leisure activities, cigarette remains, accident from agriculture and forestry works and machinery.

The University level clearly refers to pastoralism and pyromania, together with power lines arching, interest and conflicts in land use. In such results we can argue that some importance have mass media, since the motives are rather "high" and echo the frequent analysis circulated by them, sometimes containing not acceptable fire cause hypothesis such as glass refraction or magnification (Belgacem, 2012).

In the replies by the University level respondents we cannot also exclude a sort of reactivity in altering their performance, probably to conform to the expectations of the surveyors. At the opposite, the group of illiterate and low education summoned their experience of livelihood and traditional, rural, pastoral culture.

Pastoralism and political fires are quite coincident with axes origin and close to middle age classes, mainly 40 - 49, which in part arguably recalls personal experiences of the black period. Medium age class 30 - 39 seems attracted by land use changes and conflicts. More aged class 50 - 59 and 60 and over, clearly refer to livelihood activities: pastoralism, honey gathering, i.e. the traditional activities of aged rural people.

Age Classes

Analysis of age classes gives interesting result, though some classes are clustered: 30 - 39, 50 - 59, 60 and over are tightly grouped close to the origin of axes but under A1, whereas 20 - 29 and 40 - 49 are over the axe but symmetrically distant from axe A2. Rather distant from the cluster stays the younger class, >20, which appears as an outlier. Younger respondents refer to leisure activities, as expected; aged 20 - 29 are closer to pyromania, rather close to garbage dumping and cigarette remains, arguably under the influence of mass media.

Scatter Plots

Examples of scatter plots issued from FCA are here following (**Figures 2** and **3**).

Discussion

Survey results about causes are commented in their order of importance as in **Table 5** but only when exceeding a frequency of 4%. By far the most serious cause of forest fires attributed to the rural population is deliberate burning for grazing and land improvement.

The Most Relevant Motivations

Pastoralism: Pastoral fires are a traditional practice all over the Mediterranean basin in areas where fire is the cheapest way to regenerate pastures invaded by shrub (Cemagref, 2004; Pyne, 1997). Range burning could be related to the high number of sheep in the country as a whole: in Algeria (with more than 25 millions livestock, 77% of which represented by sheep; FAO, 2012) demand for red meat is high and growing. In the '90s the country imported around 20% of all that it uses (Homewood, 1993). Now sheep meat represents 30% of the total meat production of Algeria (Dutilly-Diane, 2006). In general fire is not

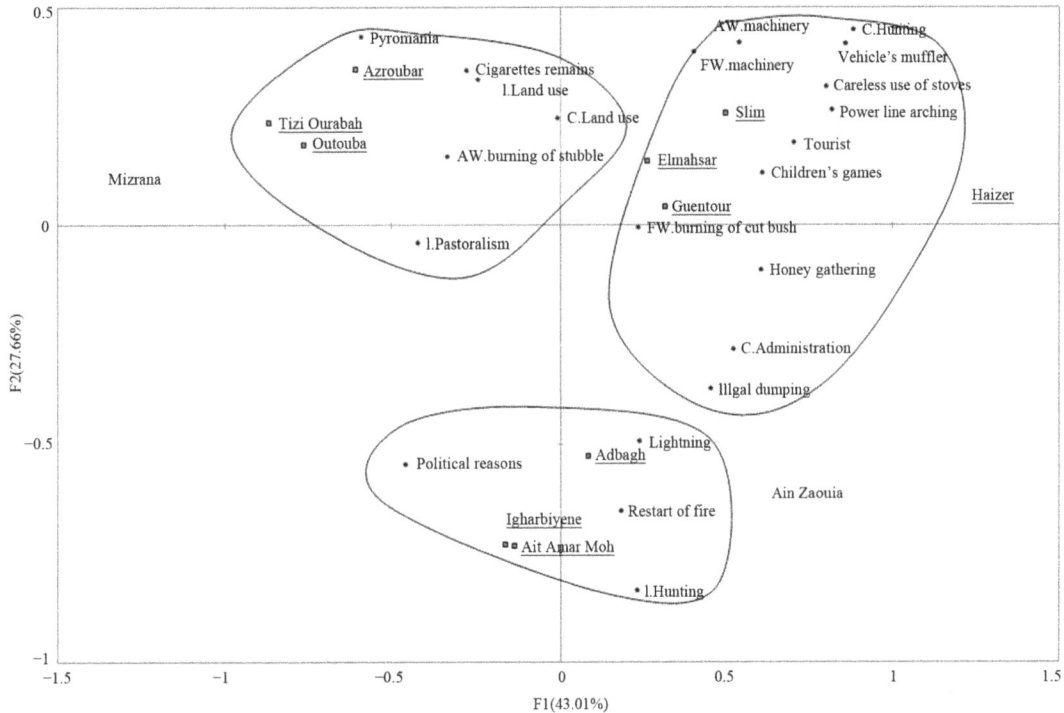

Figure 2.
FCA scatter plot related to the nine villages and causes. A1 axe explains 43% of variance, A2 explains 28%.Correspondence of motivation with municipalities is rather evident.

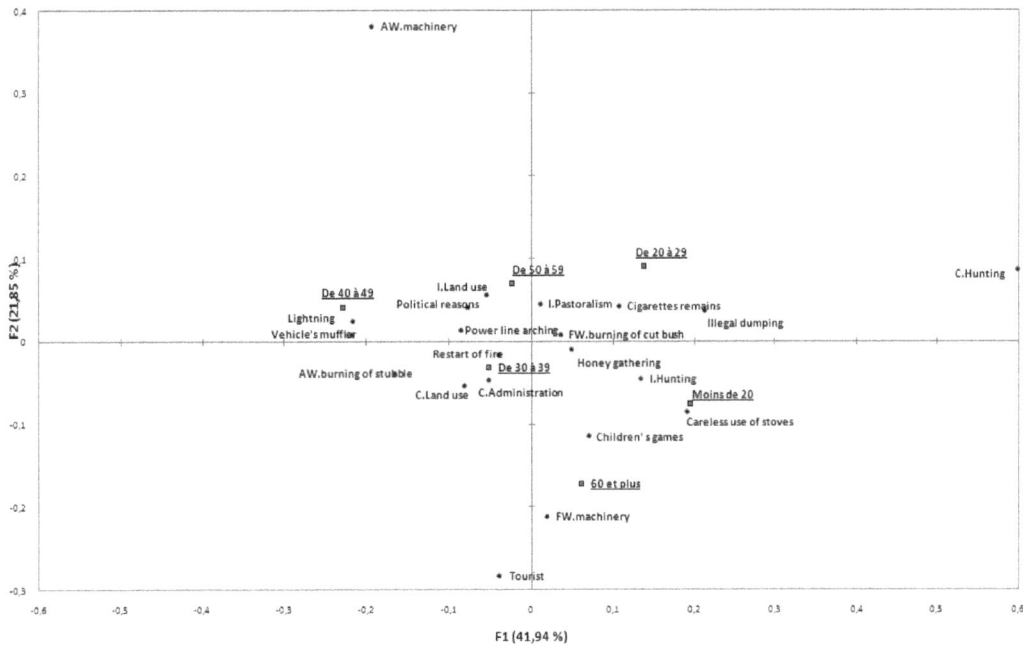

Figure 3.
FCA scatter plot related to age classes of surveyed and causes. A1 axe explains 37% of variance, A2 explains 27%.

directly applied to the forests, but rather to marginal rangelands and shrub lands close to them. When vegetation is stressed after a long summer drought, easily fires get out of hand affecting the forest boundary (Meddour & Derridj, 2012).The tendency to increase range areas at the detriment of degraded forest cover is a real problem, observed in other Mediterranean countries (e.g. in Sardinia, Italy). Shepherds do not refrain from use of fire, even if this means risk for forests (Meddour-Sahar et al., 2012). Fire setting normally follows traditional rules, such us procedures, timing, and time of return on same site; therefore this practice could be defined as a cultural trait of the local populations.

Political fires are the most site specific and disquieting cause of wildfires for Algeria, mainly in Kabylia. The term refers to forest fires as a tool used by the State in its fight against *jihadi* combatants (UNHCR, 2001; Dridi, 2002; Kervin & Gèze, 2004; Rahal, 2012). During the Algerian war (1954-1962) the French Army is reported having burned about 70% of forest in the Massif of Bou-Taleb (Madoui, 2000, 2002) also using napalm to destroy insurgents in Movis forest (Ouarsenis forests; Sari, 1976). Burned forests in Oued Lardjem, Theniet el Had, Ain Antar still exhibit scars of burning by napalm.

Cigarette remains: Carelessly tossed cigarette butts are responsible of roadside fires. Cigarettes, under normal conditions, do not start wild land fires unless the relative humidity is under 22%, it is windy, and a continuous, cured, finely-particulated fuel-bed exists (NWCG, 2005). Lack of roadside maintenance (through slashing or manual brush removal) increases fire hazard in summer, since roadsides are recognized as critical ignition start points (Romero-Calcerrada et al., 2008; Chapman & Balmain, 2004; Cardille et al., 2001) No specific roadside treatments (such as fire safety strips along roads i.e. "anti-cigarette strip"; FAO, 2001) to prevent the occurrence of fires and minimize the danger of fire spread are actually required for roads crossing forests or plantations in the area (Meddour-Sahar & Derridj, 2010).

Pyromania results with high frequency in general, and rather high, first best, in the *baladiah* of Mizrana. The term is often improperly used as a synonym of arsonist (Instituto Medio Ambiente, 2003; Dolz Reus & Franco Irastorza, 2005; Lovreglio et al., 2010), or just an improper way to express "unknown" (Franco Irastorza & Dolz Reus, 2007). In a recent survey in Sardinia, carried out by Delphi technique, pyromania resulted among the main causes of fires in the province of Cagliari (Lovreglio, unpublished data). It is probable that the number of authentic pyromaniacs, who suffer an addictive behavior, is over-estimated as many fire-raisers if caught "red handed" claim to be suffering from mental trauma in an attempt to escape a custodial sentence (Ashby, 2012).

Agricultural works refers to burning of cut bush or stubble burning, i.e. traditional systematic burning of stubble, for the preparation of the agricultural land for new sowing-purposes and the elimination of residues that restrict exploitation. Forest works are a rather common case of negligence, quite often related to old age of responsible, or imprudence; more rarely related to an overconfidence and boldness in manipulating fire.

Restart of fire can be caused by fire-fighters who do not ensure that the fire is completely out; the alarming conditions of insecurity in Kabylia, where terroristic groups are reported still active, could justify this apparently negligent behavior. Another reason could be the high number of fires which the fire crews must intervene on, therefore having no time enough for mopping-up any fire after it has been controlled (Meddour-Sahar et al., 2012).

Garbage burning is a direct consequence of extreme population density and type of settlements, which make garbage collecting rather difficult. The production of domestic waste in Algeria is estimated in 657,000 tons/year; (Benouar, n.d.), i.e. 0.50 Kg/hab/day (Dorbane, 2007). In the town of Tizi Ouzou the daily amount of domestic waste is estimated about 68.38 tons, corresponding to 0.81 kg/hab./day. In the wilaya of Tizi Ouzou a proliferation of illegal, uncontrolled dumps along roadside is reported, with the presence of only 30 controlled garbage dumping sites v. 1236 illegal ones (Meddour-Sahar &

Derridj, 2010; DPAT, 2011), i.e. an average of 18 sites per municipality. Dump sites are preferential points of fire start in forests (Djema & Messaoudene, 2009)

Interest in land use change mainly refers to fire used to change land use: from forested area to agricultural area or, more frequently, to building area. The latter is more evident in Mizrana, where demand of space for urbanization is very high. Such changes have no relevant obstacles given the lack of Land Cadastre and negligible fines for illegal behaviors. An increasing demand for housing can induce forest fires as an illegal means to increase the supply of available land (Gonzalez, 2007).

Children's games (with lighters, matches, small fireworks such as firecrackers or small amount of explosives) are currently reported in all countries, not therefore having any local character. Children cause fires out of curiosity or mischief; in the study area, fire is more probably a possible occasion to spend time together and relieve boredom, lacking other opportunities for free time.

Other fire causes exhibiting a frequency <4% are just listed, without a commentary: Tourists (3.09%), Honey gathering and related use of smoke (2.83%), Vehicle's muffler (2.76%), Barbecue fires (2.04%), Power line arching (1.71%), Conflicts with Administration (1.31%), Hunting conflicts (0.85%), Interest in hunting (1.25%), Lightning (0.79%).

Just some comments for conflicts with Foresters and Forest Administration seem useful. Conflicts related to forest policy could be interpreted as a kind of reaction, for example, when reforestation is carried out at the expense of traditional extensive grazing lands, against the lack of negotiation with the population (Meddour-Sahar et al., 2012). Similar reactions by local populations against national programs of reforestation (the so called *Green Belt* of the '70s) were already reported (Vallejo, 2005).

Conflicts with foresters. The policy adopted by the Forest. Administration often tends to further marginalize rural populations, which manifest themselves in the firing of forests against the application of a criminal jurisdiction and a colonial-style behaviour (Berchiche, n.d.). In those cases, though limited in frequency, fire can have a subtle but strong symbolism, because a patch of fire is a very visible and powerful claim.

Rural incendiarism is an expressive form of resistance, allowing a powerful statement to be expressed, with some livelyhood benefits through freshly cleared land or regenerated pasture, while having fewer constraints than other forms of protest because of its anonymity and difficulty to be identified (Holmes, 2007).

Causes Classification

For a classification of voluntary fires we adopted the classification by Douglas et al., (1992), which proposes six possible motives for arson: Vandalism, Excitement, Revenge or Protest, Crime Concealment, Profit, and Extremist. Douglas' classification was mainly conceived for urban fires but it well fits also with wild land fires (Leone & Lovreglio, 2003).

In the study case, ten possible motives after Douglas can be retrieved: Extremist: for political fires; Revenge or protest: for conflict related to land use, conflicts with foresters and Administration, hunting conflicts; Profit: for interest in land use and change in land use, for pastoral fires, for illegal garbage burning, for hunting.

For all the others, mainly involuntary fires, excluding pyromania classified as a mental addictive disorder (APA, 2000) a possible unifying category is folk crime (Bankston & Jenkins, 1982). Many law violations in rural communities such as the violation of gambling, hunting and game laws, and woods burning fall in such category. Woods burning, like poaching, is deeply rooted in some rural cultures (Bankston & Jenkins, 1982; Forsyth & Marckese, 1993; Forsyth et al., 1998).

As a result these violations became accepted as normal behavior in some cultural settings. Setting fire on woods is a "...*custom* (emphasis added by Authors) *that has been predicated upon the assumption that timber is either an undesirable barrier to land use or, at best, an expendable commodity secondary in importance to other land uses. In the course of time...it has become accepted as normal behavior... Although fire setters may be few in number, they can practice fire setting with the assurance or at least the toleration of their activities because community members subscribe in some degree the beliefs and/or attitudes that motivate the fire setter*" (Bertrand & Baird, 1975: pp. 6 and 11).

All this explained, behind the majority of motives for setting fire on woods in Kabylia some subsistence reasons exist (Emery & Pierce, 2005). Of course, the Penal Code of Algeria (Grim, 1989), is rather severe with fire setters, who are sentenced to hard labour jobs.

Conclusion and Final Remarks

Survey carried out in the three representative municipalities had the main scope to uncover motivations and drivers that lie behind wildfire in Kabylia. Of course we describe causes as perceived by local inhabitants and not by professionals of Forest Administration, since their statistics admit that the totality of fires is of unknown origin (99% in Tizi Ouzou wilaya).

We decided to give voice to the inhabitants, therefore admitting and accepting that their responses are frank and truthful. Our results therefore concern truth-as-observed, not exact science (Jollands et al., 2011).

It is the first time that a similar survey is carried out in Algeria; at our knowledge, this procedure is not so common also in more advanced countries, where statistic are always compiled on the basis of the subjective knowledge of professionals charged of this task. The fire scenario of our survey depicts relationship among people, land and fire rather different from the rhetoric image reported in some classical books on forest fires in Algeria, for instance Gravius (1866) and Thibault (1866).

In our results forested areas are degraded by human overuse and impact, swarming with a myriad of people using fire in a more or less legal way. The forest is a place where to use fire for getting fodder to animals; to illegally extract stones, accelerating the operations by burning used tires; to use fire to relieve boredom; to dispose garbage and wastes and burning them: in short, gradually consuming the forest or eroding it.

Results highlight a complex reality, issued from a new style of life where people live in the backcountry but work (if they do so..) in towns and are therefore in some way obliged to find new ways of problems solving, where time saving is an imperative, and fire is the privileged tool.

We can synthesize as follows: the causes, as perceived by dwellers (we insist that they are perceived but not real) can be distinguished in a "common core", represented by a cluster of negligent behaviors and scarcely important folk crimes; two more evident causes, represented by political fires and pastoralism, clearly dominate on them, whichever the criterion of analysis.

The negligent behaviors recall the anarchistic activities practiced in the Algerian forests (Djema & Messaoudene, 2009), results of the release of the rules of a former rural culture based on traditional behaviors, now no more consistent with changed style of life and age and/or education contrasts.

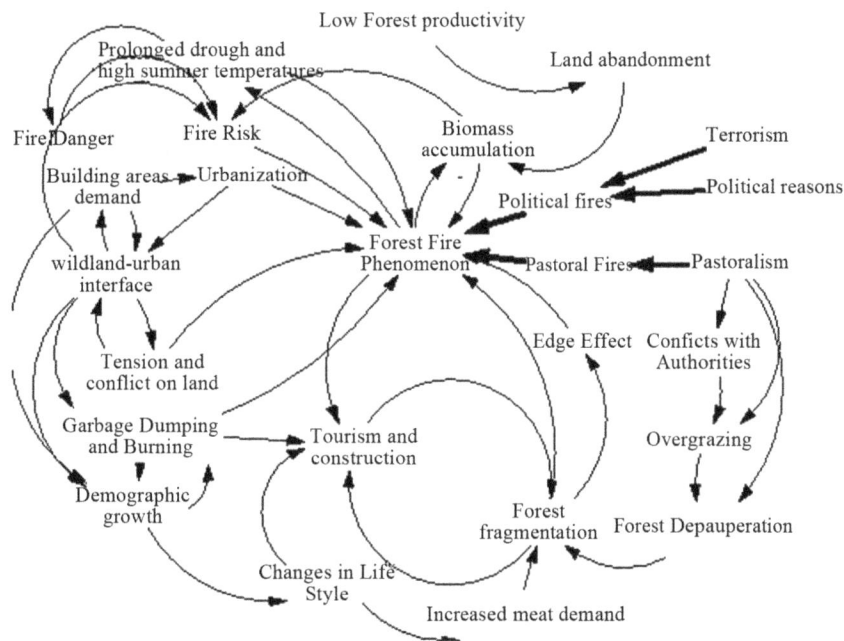

Figure 4.
CLD of forest fires and causative agents (factors) for Kabylia and Algeria: the number of arrows visualizes the multiplicity and complexity of interrelated factors.

We propose here as a conclusion (**Figure 4**) a CLD (Causal Loop Diagram; Richardson & Pugh, 1981; Kim, 1992), to visualize relationships of the most important variables which are behind causes in Algeria and in Kabylia.[2]

Though related to only three municipalities (i.e. 2.67% of the 67 municipalities of Tizi Ouzou and 45 municipalities of Bouira; 0.0019% of the 1541 municipalities of Algeria), our result is well representative of Kabylia.

We underline the importance of our survey, which for the first time measured the unknown causes, accounting for 99% in official statistics. Their importance must be considered in the scope of a new approach to fire fighting: from fire suppression, mainly oriented to emergency measures, therefore a "reactive" process, to a longer term policy of removing the structural causes (Birot, 2009; Velez, 2008), then a more "pro-active" approach.

But if wildfires are to be tackled conclusively, then there is a need (Jollands et al., 2011) to address the root causes which lead to fire setting, properly targeting the social groups and concentrating efforts consistently with fire risks.

Contribution

All authors contributed extensively to the work presented in this paper. O.M.S designed the study, collected and analyzed data; O.M.S. and V.L. wrote the manuscript; R.M., R.L., A.D. gave technical support and conceptual advice. All authors discussed the results and implications and commented on the manuscript at all stages.

REFERENCES

American Psychiatric Association (2000). *Diagnostic and statistical manual of mental disorders* (4th ed.). Washington DC: Author.

Ashby, P. (2012). Spain's burning forests-fire-raising and pyromania. URL (last checked 6 June 2012). http://suite101.com/article/spains-burning-forests-fire-raising-and-py romania-a405886

Bankston, W. B., & Jenkins, Q. A. L. (1982). Rural crime in the south: An overview of research traditions and theoretical issues. *The Rural Sociologist, 2,* 233-241.

Baskaran, K. D., Bariteau, M., El-Kassaby, Y. A., Huoran, W., Kagayama, P., Kigomo, B. N., Mesén, F., Midgley, S., Nikiema, A., Patiño, V. F., Prado, J. A., Sharma, M. K., & Ståhl, P. H. (2002). *Regional updates, prepared for the twelfth session of the FAO panel of experts on forest gene resources, Rome, Italy, 21-23 November 2001.* Forest Genetic Resources Working Papers, Working Paper FGR/34E. Rome: Forest Resources Development Service, Forest Resources Division. URL (last checked 4 June 2012). http://www.fao.org/DOCREP/005/AC646E/ac646e09.htm

Belgacem, D. (2012). Le DG de la Protection Civile à "Liberté" "nous avons engagé 15,000 hommes pour lutter contre les feux de forêt". URL (last checked 15 September 2012). http://www.liberte-algerie.com/actualite/nous-avons-engage-15-000h ommes-pour-lutter-contre-les-feux-de-foret-le-dg-de-la-protection-ci vile-a-liberte-184288

Benouar, D. (n.d.). The need for an integrated disaster risk reduction management strategy in North African cities: A case study of urban vulnerability in Algiers (Algeria). URL (last checked 1 June 2012). http://acds.co.za/uploads/jamba/vol1no1/benouer.pdf

Bensaid, S., Gasmi, A., & Benhafied, I. (2006). Les forêts d'algérie de Césarée la romaine a ce jour. *Forêt méditerranéenne, 3,* 267-274. URL (last checked 1 February 2012).

www.foret-mediterraneenne.org/fr/catalogue

Berchiche, T. (n .d.). *La politique forestière et le développement rural* (ronéotypé).

Bertrand, A. L., & Baird, A. W. (1975). *Incendiarism in southern forests: A decade of sociological research.* Washington DC: United States Department of Agriculture.

Birot, Y. (2009). *Living with wildfires: What science can tell us.* EFI Discussion 15, European Forest Institute.

Boudy, P. (1952). *Guide du forestier en Afrique du nord.* Paris: La Maison Rustique.

Cardille, J. A., Stephen, J., Ventura, S. J., & Turner, M. G. (2001). Environmental and social factors influencing wildfires in the upper midwest, United States. *Ecological Applications, 11,* 111-127.

CEMAGREF (2004). Haute-corse, pastoralisme et incendie. *Info DFCI, Bulletin du Centre de Documentation Forêt Méditerranéenne et Incendie, 52,* 1-8. URL (last checked 2 June 2012). http://www.irstea.fr/

CEMAGREF (2009). Repeated fire and drought: A menace for Mediterranean forests. *Science Daily, 26.*

Chapman, S., & Balmain, A. (2004). *Reduced-ignition propensity cigarettes. A review of policy relevant information.* Commonwealth Department of Health and Ageing. URL (last checked 1 June 2012). http://www.health.gov.au/internet/main/publishing.nsf/Content/1526 AD854C903649CA256F8E0011ACBE/$File/smoking_rip.pdf

Clément, V. (2005). Les feux de forêt en Méditerranée: Un faux procès contre Nature. *L'Espace Géographique, 4,* 289-304.

Colin, P. Y., & Jappiot, M. (2001). Politiques de prévention et de réhabilitation en réponse aux origines des incendies de forêt. Exemple de quatre pays du bassin méditerranéen (Chypre, Maroc, Syrie, Tunisie). *Info DFCI, Bulletin du Centre de Documentation Forêt Méditerranéenne et Incendie, 47,* 1-5. URL (last checked 2 June 2012). http://www.irstea.fr/

Davis, D. K. (2004). Eco-governance in French Algeria: Environmental history, policy and colonial administration. *Proceedings of the Western Society for French History, 32,* 328-345.

Direction Planification Aménagement Territoire (DPAT) (2011). Annuaire statistique de la wilaya de Tizi Ouzou, 2010. No. 27.

Djema, A., & Messaoudene, M. (2009). The Algerian forest: Current situation and prospect. In M. Palahi, Y. Birot, F. Bravo, & E. Gorrize (Eds.) *Managing Mediterranean forest ecosystems for non-timber goods and services* (pp. 17-28). EFI Proceedings No 57.

Dolz Reuss, M. L. (2005). State of the art of forest fire causes in Spain. *Proceedings of the II International Conference on Prevention Strategies of Fires in Southern Europe,* Barcelona, 9-11 May 2005.

Dorbane, N. (2007). Contraintes de la gestion des déchets dans les villes algériennes: Cas de Tizi-Ouzou. *Revue Campus, 6,* 4-16.

Douglas, J. E., Burgess, A. W., Burgess, A. G., & Ressler, R. K. (1992). *Crime classification manual.* New York: Lexington Books.

Dridi, D. (2002). The army: Friends of the scorched earth. URL (last checked 6 June 2012). http://www.algeria-watch.org/en/articles/2002/bainem.htm

Dutilly-Diane, C. (2006). *Review of the literature on pastoral economics and marketing: North Africa report prepared for the world initiative for sustainable pastoralism.* IUCNEARO. URL (last checked 6 June 2012). http://data.iucn.org/wisp/documents_english/north_africa_reports.pdf

Emery, M. R., & Pierce, A. R. (2005). Interrupting the "telos": Locating subsistence in contemporary US forests. *Environment and Planning A, 37,* 981-993.

FAO (2001). *International handbook on forest fire protection. Technical guide for the countries of the Mediterranean basin.*

FAO (2010). *Forests and climate change in the near east region. Forests and climate change.* Working Paper 9.

FAO (2012). *State of Mediterranean Forests (SoMF).* URL (last checked 5 April 2012). http://www.fao.org/forestry/225560f888476830f6931aba4b886baea8 8af1.pdf

Forsyth, C. J., & Marckese, T. A. (1993). Folk outlaws: Vocabularies of motives. *International Review of Modern Sociology, 23,* 17-31.

Forsyth, C. J., Gramling, R., & Wooddell, G. (1998). The game of

---

[2]A causal loop diagram consists of variables connected by arrows, denoting the causal influences among the variables. Diagram was drawn using *Vensim plus* software (www.vensim.com/software.html).

poaching: Folk crimes in southwest Louisiana. *Society and Natural Resources, 11*, 25-38.

Franco Irastorza, I., & Dolz Reus, M. L. (2007). Análisis de la percepción de la sociedad ante el problema de los incendios forestales: Metodología y resultados. *Actas de la IV Conferencia Internacional sobre Incendios Forestales*, Sevilla, 13-18 May 2007.

García Ortega, E., Trobajo, M. T., López, L., & Sánchez, J. L. (2011). Synoptic patterns associated with wildfires caused by lightning in Castile and Leon, Spain. *Natural Hazards and Earth System Sciences, 11*, 851-863.

Gonzalez, A. (2007). Libertad, the housing boom and forest fires. URL (last checked 16 May 2012). http://ssrn.com/abstract=1107120

Gravius, G. (1866). *Les incendies de forêt en Algérie leurs causes vraies et leurs remèdes*. Constantine: Chez Louis Marie librairie Paris.

Grim, S. (1989). *Préaménagement et protection des forêts contre l'incendie. Belgique: Le préaménagement forestier*. Ministère de l'Hydraulique d'Algérie & Unité des Eaux et Forêts de l'Université catholique de Louvain-la-Neuve, 271-289.

Harris, S. A. (2007). *Colonial forestry and environmental history: British policies in Cyprus, 1878-1960*. Ph.D. Dissertation, Austin: Graduate School of the University of Texas. URL (last checked 16 April 2012). http:// repositories.lib.utexas.edu/bitstream/handle/2152/3244/harris

Holmes, G. (2007). Protection, politics and protest: Understanding resistance to conservation. *Conservation & Society, 2*, 184-201.

Homewood, K. M. (1993). *Livestock economy and ecology in El Kala, Algeria. Evaluating ecological and economic costs and benefits in pastoralist systems*. Pastoral Development Network Papers 35a. URL (last checked 1 April 2012). http://www.odi.org.uk/resources/details.asp?id=4467&title=social-economic-ecological-cost-benefit-analysis-pastoralism

IIA (1933). *Enquête internationale sur les incendies de forêts*. Rome: Institute International d'Agriculture.

Instituto de Estudios del Medio (2003). *Estudio sociòlogico sobre la perceptión de la poblacion Española hacia los incendios forestales*. Associatión para la Promocciòn de Actividades Socioculturales. URL (last checked 10 April 2012). http://www.idem21.com/descargas/pdfs/IncendiosForestales.pdf

Jollands, M., Morris, J., & Moffat, A. J. (2011). *Wildfires in Wales*. Farnham: Forestry Commission Wales. URL (last checked 16 June 2012). http://www.forestry.gov.uk/fr/wildfiresinwales#finalreport

JRC (2010). *Forest fires in Europe 2010*. Report No. 11. EUR—Scientific and Technical Research, 1831-9424. URL (last checked 16 March 2012). http://effis.jrc.ec.europa.eu/reports/fire-reports

Kervyn, J., & Gèze, F. (2004). *The organization of the forces of pressure*. Justice Commission for Algeria at the 32nd Session of the Permanent Peoples' Tribunal on Human Rights Violations in Algeria (1992-2004). URL (last checked 16 April 2012). http://www.algeria-watch.org/pdf/pdf_en/forces_repressure.pdf

Kim, D. H. (1992). Toolbox: Guidelines for drawing causal loop diagrams. *The Systems Thinker, 3*, 5-6

Leone, V., & Lovreglio, R. (2003). Human fire causes: A challenge for modeling. In E. Chuvieco, M. Pilar Martín, & C. Justice (Eds.), *Innovative concepts and methods in fire danger estimation 4th international workshop on remote sensing and GIS applications* (pp. 89-98). Ghent, 5-7 July 2003.

Leones, J. (1998). A guide to designing and conducting visitor surveys. URL (last checked 6 May 2012). http://ag.arizona.edu/pubs/marketing/az1056/az1056.html

Leone, V., Lovreglio, L., Pilar Martín, M., Martínez, J., & Vilar, L. (2009). Chapter 11 human factors of fire occurrence in the Mediterranean. In E. Chuvieco (Ed.), *Earth observation of wildland fires in Mediterranean ecosystems* (pp. 149-170). Berlin, Heidelberg: Springer-Verlag.

Lovreglio, R., Marciano, A., Patrone, A., & Leone, V. (2012). Le motivazioni degli incendi boschivi in Italia: Risultati preliminari di un'indagine pilota nelle Province a maggiore incidenza di incendi. *Forest@, 9*, 137-147. URL (last checked 26 June 2012).

http://www.sisef.it/forest@/contents/?id=efor0693-009

Madoui, A. (2000). Forest fires in Algeria and the case of the domanial forest of Bou-Taleb. *International Forest Fire News, 22*, 9-15.

Madoui, A. (2002). Les incendies de forêt en Algérie. Historique, bilan et analyse. *Forêt Méditerranéenne, 1*, 23-30. URL (last checked 1 February 2012). www.foret-mediterraneenne.org/fr/catalogue

Maniatis, P. (2010). Greek census 1981. A case of factorial correspondence analysis. *International Research Journal of Finance and Economics, 35*, 99-111.

Marc, P. (1916). *Les incendies de forêt en Algérie. Notes sur les forêts de l'Algérie*. Alger: Typographie Adolphe Jourdan. Imprimeur libraire-Editeur.

Meddour-Sahar, O. (2008). *Contribution à l'étude des feux de forêts en Algérie: Approche statistique exploratoire et socio-économique dans la wilaya de Tizi Ouzou*. Thèse de Magister, Alger: INA El Harrach.

Meddour-Sahar, O., & Derridj, A. (2010). Le risque d'incendie de forêt, évaluation et cartographie (Wilaya de Tizi Ouzou, période 1986-2005). *Sécheresse, 21*, 187-195.

Meddour-Sahar, O., & Derridj, A. (2012). Bilan des feux de forêts en Algérie: Analyse spatiotemporelle et cartographie du risque (période 1985-2010). *Sécheresse, 23*, 133-141.

Meddour-Sahar, O., Meddour, R., & Derridj, A. (2008). Les causes des incendies de forêt en Algérie. *Technologie et Environnement*. URL (last checked 1 February 2012). www.recy.net/actualites/20080615-incendies-algerie

Meddour-Sahar, O., Meddour, R., Leone, V., Lovreglio, R., & Derridj, A. (2012). Analysis of forest fires causes and their motivations in north Algeria: The Delphi technique. *iForest-Biogeosciences and Forestry* (accepted).

NWCG (2005). Wildfire origin & cause determination handbook 1. URL (last checked 23 April 2012). http://www.nwcg.gov/pms/pubs/pubs.htm

Petersen, W. A., & Rutledge, S. A., (1998). On the relationships between cloud-to-ground lightning and convective rainfall. *Journal of Geophysical Research, 103*, 14025-14040.

Pyne, S. (1997). *Vestal fire: An environmental history, told through fire, of Europe and Europe's encounter with the world*. Seattle and London: University of Washington Press.

Rahal, M. (2012). Fused together and torn apart stories and violence in contemporary Algeria. *History & Memory, 24*, 118-151.

Richardson, G. P., & Pugh III, A. L. (1981). Introduction to system dynamics modeling *with DYNAMO*. Cambridge MA: Productivity Press.

Romero Calcerrada, R., Novillo, C. J., Millington, J. D. A., & Gómez Jiménez, I. (2008). GIS analysis of spatial patterns of human-caused wildfire ignition risk in the SW of Madrid (central Spain). *Landscape Ecology, 23*, 341-354.

Sari, D. (1976). *L'homme et l'érosion dans l'Ouarsenis (Algérie)*. Alger: Editions SNED.

Secretariat of the Convention on Biological Diversity (2001). *Impacts of human-caused fires on biodiversity and ecosystem functioning, and their causes in tropical, temperate and boreal forest biomes*. CBD Technical Series No. 5, Montreal: SCBD.

Thibault, R. (1866). *Des incendies de forêt en Algérie: De leurs causes et des moyens préventifs et défensifs à leur opposer*. Constantine: V Guende Librairie, Place du Palais, Paris: E. Galette Librairie.

Thirgood, J. (1981). *Man and the Mediterranean forest: A history of resource depletion*. New York: Academic Press.

UNHCR Algeria (2001). Country report UNHCR/ACCORD. *7th European Country of Origin Information Seminar*, Berlin, 11-12 June 2001.

Vallejo, R. (2005). Chapter 45 Restoring Mediterranean forests. In S. Mansurian, D. Vallauri, & N. Dudley (Eds.), *Forest restoration in landscapes: Beyond planting trees* (pp. 313-319). New York: Springer. URL (last checked 8 April 2012). http://www.bf.unilj.si/fileadmin/groups/2716/downloads/%C4%8Clanki_vaje/2.VS%C5%A0/Mansurian_Forest_restoration_landscapes.pdf

Vélez, R. (1991). Los incendios forestales y la politica forestal. *Revista de Estudios Agro-Sociales, 158,* 83-104.

Vélez, R. (2000). *La defensa contra incendios forestales: Fundamentos y experiencias.* Madrid: McGraw Hill.

Vélez, R. (2008). Europe: Development and fire. *Proceedings of the Second International Symposium on Fire Economics, Planning, and Policy: A Global View.* Pacific Southwest Research Station General Technical Report, US Department of Agriculture, Forest Service.

WWF-IUCN (2007). Forest fires in the Mediterranean. URL (last checked 4 May 2012). http://www.uicnmed.org/web2007/documentos/Background_med_forest_fires.pdf

11

The Effective Ecological Factors and Vegetation at Koh Chang Island, Trat Province, Thailand

Nathsuda Pumijumnong, Paramate Payomrat

Faculty of Environment and Resource Studies, Mahidol University, Nakhon Pathom, Thailand

This study aims to characterize the tropical rain forest present in the Chang Island, Trat Province, Thailand, and to analyze the environmental factors to determine its composition and structure. Thirty one plots were sampled, plant cover was measured in 20×40 m^2 plots, and the importance value index was calculated. A total of 78 species belonging to 32 families were identified. Twenty soil samples were analyzed, and cluster analysis was employed to classify the vegetation communities. Floristic and environmental data were evaluated and ordered using canonical correspondence analysis. The results showed that the vegetation communities could be divided into 4 types and were significantly ($p < 0.05$) controlled by a secondary distribution according to elevation and the topographic wetness index (TWI). Mixed plant communities were more likely to distribute in regions with moderate to low levels of TWI, which were divided by levels of elevation into lowland multi-aged stands (Type 1) or a *Calophyllum thorelii* Pierre community (Type 2). The Dipterocarpus (*Hopea pierrei* Heim) community (Type 3) was more likely to occur in regions with moderate to high levels of TWI, but the result from cluster analysis showed that some of the plot samples from the Dipterocarpus community were separated by characteristic importance value index (IVI) values. There was also evidence that the area was impacted by an old disturbance created by a rubber plantation. This impact was referred to as a secondary succession community (Type 4).

Keywords: Chang Island; Vegetation Community; Canonical Correspondence Analysis; Ecological Factors

Introduction

Chang Island is one of only a few islands where a tropical rain forest is distributed over 70% of the inland area, which has been preserved in its present form (Environmental Research Institute, 2007). The island covers an area of 212.947 sq km and is a major island of the Mu Koh Chang National park in the Gulf of Thailand, which became Thailand's 45th National Park in 1982. During the Pleistocene epoch (Tougard, 2001; Esselstyn et al., 2009), large sea level fluctuations caused the sea level to fall by 50 - 150 m, creating land bridge connections. After this period, the sea level rose and the low-lying regions were covered with water forming many islands separated from the mainland. Various fossils have confirmed these events during this period (Bird et al., 2005). The research of Chamchumroon and Puff provides evidence to support this theory. It is remarkable that although a large proportion of rubiaceous taxa show disjunction between Koh Chang and peninsular Thailand, 20% of all rubiaceae taxa recorded from Koh Chang have been found to be distributed in peninsular Thailand (Chamchumroon et al., 2003). This indicates that biological species found on the mainland may also be found on the Chang islands with a greater diversity. The tropical rain forest of Chang island has important roles in storing bioelements and water for local communities and acts as a carbon bank that impacts a large global area. Since 1902, the records of Schmidt (1900) have shown that all of the hills of Koh Chang are entirely covered with the densest jungle. In addition, the Siamese and Chinese populations scattered along the coasts where the river debouches have had little influence on the forest cover of the hills. Over the past

century, tropical rain forest areas around the world and in Thailand now face loss due to human activities (Royal Forest Department, 1997, 2005). Chang Island is also facing the same problem as other forest resources and is in danger of losing forest cover by people focused on acquiring forest land for their settlements, farms, and plantations. Recently, land use change caused by tourism had led to construction on the island, including business buildings, luxury accommodations, roads, and harbors. This type of land use changes the ecological system of Chang Island. Although the Mu Ko Chang National Park was established at the end of 1982, negative human impact on the vegetation had already occurred in the more accessible coastal areas because the settled area of the old communities before the establishment of the National Park were not included in the preservation. Therefore, some species have become extinct, such as *Ixoradoli chophylla*, *Lasianthuso ligoneuron* and *L. schmidtii*, which according to the reports of Schmidt and Kerr (Chamchumroon et al., 2003; Schmidt, 1900), come from the Klong Nonsi area, a populated area near the Koh Chang District Office that has no traces of natural vegetation left. None of these species have ever been identified anywhere else in the Island. The effect of climate change is another problem because the area, size, and isolation of Chang Island make it less resilient to such change.

This study aimed to classify the vegetation communities on Chang Island and determine environmental factors that influence their composition and structure. Knowledge of these relationships may be critically important for planning appropriate adaptations to climate changes and disturbances due to human activities.

Materials and Methods

Study Area

The study area is located at Chang Island, Trat Province, in the southeastern region of Thailand close to the border with Cambodia. It is located at 132°51'57"N - 134°51'57"N and 20°10' 46"E - 22°08'89"E. The total area is approximately 212.947 square kilometers. Its length from north to south is 30 kilometers and the width is 14 kilometers. Chang Island is the largest island in Trat Province. It is close to the Cambodia border, approximately 300 kilometers southeast of Bangkok. Ninety percent of the total area is comprised of an extrusive and intrusive igneous rock mountain range interchanged with cliffs, steep hills, and cliffs reaching as high as 743 m above sea level (ASL), such as the Kao Yai Mountain. The plains along the east coast of the island are rich in sandy clay and the west coast is rich in a recent beach deposit (Tansuwan, 2007). Small rivers and streams on Chang Island are found in areas where the sea and the creeks meet, flowing through small mangrove forests around the island, with the exception of a fairly large forest located in a protected area on the south coast. The mountain slopes are covered by dense tropical rainforest forest. The principle tree species include *Dipterocarpus alatus*, *D. turbinatus*, *Anisoptera costata*, *Hopea odorata*, *Irvingia malayana*, *Podocarpus neriifolius*, *Diospyros* spp., *Castanopsis* spp., *Croton* spp., *Oncosperma horrida*, *Caryotamitis.*, *Daemonorops* spp., *Korthalsia grandis*, *Bauhinia bracteata*, *Freycinetia sumatrana*, *Platycerium coronarium*, *Amomum* spp., *Boesenbergia pandurata*, and *Kaempferia pulchra* (Royal Forest Department, 1997, 2005). The climate conditions are influenced by northeast and southwest Asian monsoons; the former bring dry air to Thailand during November through April separated by short periods of cool (November-February) and hot (March-April) seasons, and the latter bring moisture from May through October and account for 90% of the annual rainfall. The mean annual rainfall is 4902 mm and total rain fall in some years surpasses 6000 mm (2000 AD, 2006 AD). In addition, the average maximum, minimum, and mean temperatures are 31.8°C, 23.6°C, and 27.4°C, respectively (Thai Meteorological Department, 2010).

Field Sampling

A quantitative survey of the vegetation was conducted along the entire topographic gradient of the protected area from North to South (**Figure 1**). A total of 31 plot samples were divided within the 4 survey lines. All samples from the plots were collected in year 2011. A temporary sample plot method was used in each sample plot. A quadrat of 20×40 m^2 was set and used to investigate all tree plants higher than 1.3 m with a diameter at breast height (DBH) ≥ 4.5 cm. One 4×4 m^2 quadrat was used to sample plants with a DBH less than 4.5 cm and higher than 1.3 m, and one 1×1 m^2 quadrat was used to analyze seeding plants shorter than 1.3 m with a DBH less than 4.5 cm. Elevation, slope, aspect, topographic wetness index (TWI) (Sørensen et al., 2005) were measured in every plots. Twenty soil samples were randomly collected from 31 plots and correlated with the aspect and altitude change. For each site, the soil depth interval to collect samples was 30 cm from the top to a depth of 1 m or until the base rock was reached. These soil samples were air dried and passed through a 2 mm sieve to remove coarse gravels, roots, and debris. The soil texture, bulk density, soil moisture, pH value, soil organic carbon, total nitrogen, avail-

able phosphorus, and exchangeable potassium were subsequently analyzed in these samples (Office of Science for Land Development, 2005).

Data Analysis

A species list was created to further simply identification of species diversity. We also determined three values for each tree species in a given community, including the relative density, relative frequency, and relative dominance (dominance was defined as the mean basal area per tree times the number of trees of the species) (Curtis, 1959). The diversity within a community was calculated using the Shannon-Wiener index (H) (Hill, 2007). Multivariate analyses were performed on the floristic data matrices, species, where all species whose importance value was less than 5% were eliminated. Classification of plant communities was achieved using a two-way cluster analysis on the PC-ORD program (Finch, 2005; McCune, 2002). Thirty-one plots of plant communities were classified based on the important value of each species in each plot sampled. Ordination was performed within PC-ORD using canonical corresponded analysis (CCA) (McCune, 2002). CCA is a direct gradient analysis technique that relates community variation (composition and abundance) to environmental variation, thereby providing a determination of significant relationships between environmental variables and community distribution (terBraak, 1995). CCA axes were evaluated statistically using a Monte Carlo test. Four topographic variables and 11 soil properties were used in the CCA.

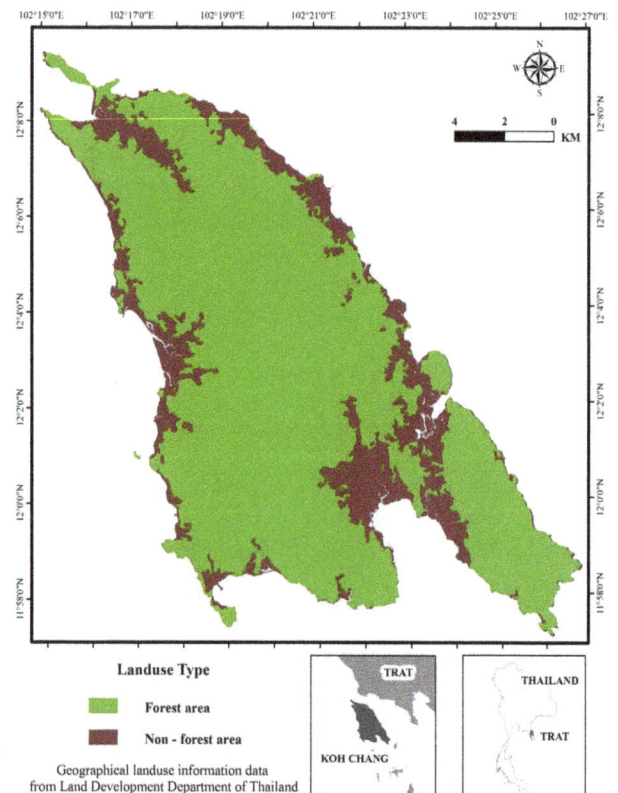

Figure 1.
Map of Chang Island obtained from a THEOS satellite image. The green indicates the study area.

Results and Discussion

Communities Classification

The classification of sub-communities characterized in this study was different from the classification by the Department of Forestry (Santhisuk, 2006). At the country scale, the forest area of Chang Island was defined only as a tropical rainforest of one of the fourteen sub-communities of Evergreen Forests, based on three major factors. First, the climatic factor present in East Thailand is such that the mean annual rainfall is between 1760 and 3140 mm. The number of days of rainfall is between 102 and 150 days, and the total rainfall in some years is more than 4000 mm. When this is compared to the climate of Chang Island (see in 2.1), we can see that the rainfall is clearly higher than in other parts of Thailand. This is because most areas of Chang Island are not far from the sea, and steep mountains are able to capture clouds and cause them to ascend. Second, the elevation of tropical rainforest is limited to 900 m asl and the highest peak of Chang Island is 743 m asl. The last factor is the native natural dominant species, which have developed over a long period of time into climax stage communities. Cluster analyses play a substantial role in analyzing the complicated characteristics of an area that has little change in environmental gradient. This type of analysis of Chang Island was quite successful in identifying distinct vegetation communities in our study area from the IVI of each species. Four community types were created with 50.76% - 76.66% of information remaining in at two-way cluster dendrogram (**Figure 2**), which indicated a good pattern in each community. Lowland multi-aged stands (Type 1) grouped from 6 samples and 60 species of stand trees in 27 families were recorded, and the IVI of each species was characterized. The species with IVI > 5% were *Tetrameles*

nudiflora R. Br. (Tetramelaceae), *Hopea pierrei* Heim (Dipterocarpaceae), *Madhuca pierrei* Lam (Sapotaceae), *Lithocarpus ceriferus* A. Camus (Fagaceae), and *Ficus callosa* Willd (Moraceae). The Shannon-Wiener index for diversity was 5.19 and the absolute density was 1287 tree ha^{-2}. The average tree height was 13.21 ± 7.28 m, and the average DBH was 47.03 ± 39.36 cm. This community is widely found at lowlands with moderate slopes and a relatively dry region in TWI. The *Calophyllum thorelii* Pierre community (Type 2) grouped from 7 samplings and 41 species of stand trees in 22 families was recorded, and the IVI of each spices was determined. The species with IVI > 5% were *Calophyllum thorelii* Pierre (Guttiferae), *Mesuaferrea* Linn(Guttiferae), *Hopea pierrei* Heim (Dipterocarpaceae), and *Cleistocalyx operculatus* Merr and Perry (Myrtaceae). Guttiferae is a family of tree species that grows from the coastal plains up to the mountain rain forest, and Myrtaceae is a major component of the understory layer. The Shannon-Wiener index for diversity was 4.43 and the absolute density was 2225 tree ha^{-2}. The average tree height was 12.31 ± 10.51 m, and the average DBH was 35.16 ± 26.06 cm. This community is widely found at medium to high elevations with moderate to steep slopes. The *Hopea pierrei* Heim community (Type 3) was classified first, with 78.66% of the information remaining while grouping due to the presence of the dominant *Hopea pierrei* Heim stand tree in each sampling (shown as the dark shade of matrix coding). The eight samples belonging to this community had 46 species of stand trees in 22 families characterized by *Hopea pierrei* Heim (family Dipterocapaceae), which is widely found in the forest area of Chang Island. However, the IVI ratio of each species of this community was occupied mainly by *Hopea pierrei* Heim IVI, rather than others species (IVI = 114.44 or 38.14%). Dipterocaps are inconspicu-

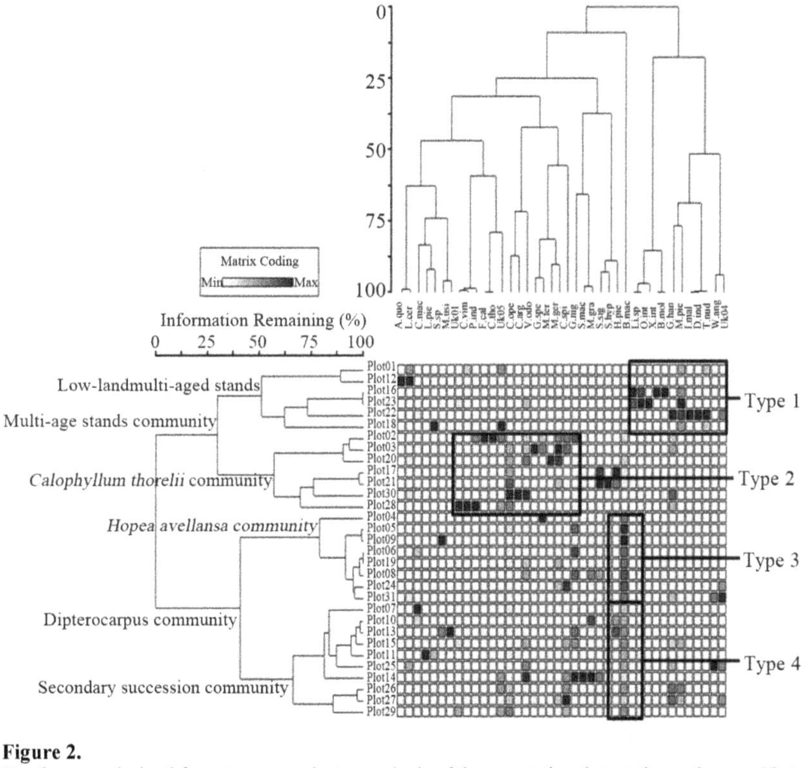

Figure 2.
Dendrogram derived from two-way cluster analysis of the vegetation data at the study area. Plots classification showing the indicator species for each division (see species code on **Table 2**).

ous in the Neotropics and Afrotropics, but are the dominant trees in the Indomalayan region in most tropical lowland evergreen forests (Göltenboth et al., 2007). The Shannon-Wiener index for diversity was 3.96 (high species evenness and richness) and the absolute density was 1807 tree·ha^{-2}. The average tree height was 13.72 ± 6.81 m, and the average DBH was 40.97 ± 49.13 cm. This community is widely found at mid elevations with moderate to steep slopes and in relatively wet regions with a high TWI (**Table 1**). The secondary succession community(Type 4) emerged as a group from 10 samplings and 41 species of stand trees in 22 families characterized by *Hopea pierrei* Heim. Almost all of the species were similar to the *Hopea pierrei* Heim community (Type 3), but differed in the species abundance as defined by the IVI ratio. The *Hopea pierrei* Heim (Dipterocapaceae) abundance decreased from 38.14% to 20.54% and the Shannon-Wiener index for diversity increased to 4.58. The absolute density was 1782 tree ha^{-2}. The average tree height was 13.37 ± 7.50 m, and the average DBH was 40.76 ± 34.65 cm. This community is widely found at low to medium elevations with flat to moderate slopes and a moderate TWI. Some evidence pointed to a defined stage of secondary succession, including the spreading of death rubber trees in the sampling plots, interviews from local people, and the location of the plots near agriculture areas.

Although the classification of forest communities using clus-

Table 1.
Mean ($\pm$) standard deviation (SD) of environmental variables for the plots associated with each of the six community type.

	Community types			
	Type 1	Type 2	Type 3	Type 4
Number of plots	6	7	8	10
Number of species	60	41	46	41
Shannon-wiener index	5.19	4.43	3.69	4.48
Elevations (m sal.)	210 ± 57	364 ± 179	320 ± 72	287 ± 133
Slope (°)	14 ± 9	19 ± 13	20 ± 14	14 ± 9
Topographic wetness index (TWI)	0.5 ± 0.6	1.29 ± 1.2	4.13 ± 2.5	1.70 ± 2
Soil depth (cm)	83 ± 15	68 ± 29	70 ± 15	60 ± 19
Organic matter (%)	4.1 ± 0.6	4.6 ± 1.4	3.9 ± 1	3.6 ± 1.2
pH	4.4 ± 0.5	4.5 ± 0.05	4.3 ± 0.2	4.2 ± 0.4
Total nitrogen (%)	0.2 ± 0.1	0.2 ± 0.1	0.2 ± 0.1	2.1 ± 0.1
CEC (cmol/kg)	8.9 ± 1.9	8.3 ± 1.5	9.4 ± 2.5	10.1 ± 3.2
Available phosphorus (ppm)	1.3 ± 1.2	1.8 ± 0.9	1.2 ± 0.7	1.9 ± 0.7
Exchangeable potassium (pmm)	51 ± 21.5	46 ± 16.7	32 ± 16.6	32 ± 9.3
Exchangeable calcium (pmm)	69 ± 64.6	34 ± 12	18.2 ± 9.4	16.7 ± 9.0
Exchangeable magnesium (pmm)	42 ± 42.6	27 ± 13	14.8 ± 9.6	16.6 ± 8.6
Bulk Density (g/cm^3)	1.1 ± 0.1	1.0 ± 0.1	1.2 ± 0.2	1.2 ± 0.2
Sand (%)	50.0 ± 11	58.7 ± 8	52.2 ± 13	48.4 ± 12
Silt (%)	18.9 ± 5	14.0 ± 3	16.5 ± 7	18.7 ± 6
Clay (%)	31 ± 8.3	27 ± 5.6	31 ± 7.4	32.9 ± 6

Table 2.
Recorded species and their Important Value Index (IVI) in each community type (con.).

Species	IVI				
	Sp. code	Type 1	Type 2	Type 3	Type 4
Aglaia cordata	A. cor	4.36	1.03	0.61	-
Antidesma bunius	A. bun	-	1.3	2.21	-
Antidesma laurifolium	A. lau	0.75	1.2	-	1.98
Aquilaria crassna	A. cra	-	2.54	-	-
Archidendron quocense	A. quo	7.11	1.16	-	7.09
Artocarpus lanceifolius	A. lan	0.76	-	-	-
Azadirachta var. *siamensis*	A. sia	0.85	-	1.41	-
Baccaurea ramiflora	B. ram	0.93	2.14	-	-
Barringtonia macrocarpa	B. mac	10.63	-	-	1.96
Barringtonia racemosa	B. rac	2.26	-	0.61	-
Bouea var. *microphylla*	B. mic	2.3	-	3.51	3.28
Bridelia affinis	B. aff	0.7	1.1	2.66	2.9
Brucea javanica	B. jav	0.7	-	-	-
Brucea mollis	B. mol	4.18	-	-	-
Calophyllum thorelii	C. tho	2.94	54.38	19.56	20.17
Carallia brachiata	C. bra	4.02	-	1.38	0.76
Carpinus londoniana	C. lon	8.35	8.23	5.72	10.41
Caryota bacsonesis	C. bac	-	-	0.62	-
Chaetocarpus castanocarpus	C. cas	3.45	1.09	0.87	2.97
Chisocheton siamensis	C.sia	8.2	2.23	2.04	4.87
Cinnamomum bejolghota	C.bej	-	0.97	-	-
Cleidion spiciflorum	C. spi	9.64	8.84	14.57	12.46
Cleistocalyx operculatus	C. ope	6.32	15.78	13.55	17.34
Cratoxylum maingayi	C. mai	0.8	1.86	-	-
Croton argyratus	C. arg	1.13	4	2.02	1.1
Dacrydium elatum	D. ela	-	3.13	-	-
Diospyros decandra	D. dec	0.77	-	0.66	2.58
Dipterocarpus sp.	D. sp	3.32	2.66	5.98	10.43
Dipterocarpus sp. 2	D. sp2	2.61	-	-	-
Dipterocarpus turbinatus	D. tur	0.74	-	1.69	-
Elaeocarpus robusus	E. rob	0.79	-	-	-
Eurycoma longifolia	E. lon	2.12	6.4	1.45	5.21
Fagraea fragrans	F. fra	-	1.49	-	-
Ficus callosa	F. cal	15.86	12.09	9.51	13.1
Firmiana colorata	F. col	0.7	-	-	-
Garcinia hanburyi	G. han	8.1	12.95	6.11	11.54
Garcinia nigrolineata	G. nig	3.29	2.01	7.88	9.12
Garcinia speciosa	G. spe	-	7.25	2.66	1.5
Heritiera javanica	H. jav	-	2.33	-	0.46

Continue

Species	Sp. code	Type 1	Type 2	Type 3	Type 4
			IVI		
Hopea pierrei	*H. pie*	23.59	30.55	114.4	61.63
Horsfieldia irya	*H. iry*	2.28	-	-	3.39
Iravingia malayana	*I. mal*	5.17	-	-	2.13
Ixora cibdela	*I. cib*	0.7	1.17	-	-
Knema linifolia	*K. lin*	2.48	-	-	-
Litchi chinensis	*L. chi*	1.47	-	-	-
Lithocarpus ceriferus	*L. cer*	16.11	6.35	1.02	3.56
Lithocarpus sp.	*L. sp*	2.85	-	-	-
Lithocarpus sp. 2	*L. sp2*	4.4	-	-	-
Litsea pierrei	*L. pie*	12.14	2.92	1.83	6.96
Lophopetalum duperreanum	*L. dup*	-	-	0.62	-
Madhuca grandiflora	*M. gra*	10.75	10.94	8.22	8.74
Madhuca pierrei	*M. pie*	21.47	7.04	6.93	11.21
Mangifera caloneura	*M. cal*	-	-	1.3	2.02
Maytenus marcanii	*M. mar*	0.7	1.12	0.64	-
Melanorrhoea usitata.	*M. usi*	-	-	6.33	9.54
Memecylon garcinioides	*M. gar*	5.26	12.46	12.61	15.76
Mesua ferrea	*M. fer*	2.01	34.7	6.09	1.58
Ochna integerrima	*O. int*	4.25	-	-	-
Oncosperma tigillaria	*O. tig*	-	-	-	2.69
Paeinari anamense	*P. ana*	1.5	-	-	-
Phyllanthus emblica	*P. emb*	-	-	0.96	-
Scaphium macropodum	*S. mac*	7.39	2.39	4.41	7.91
Semecarpus Sp.	*S. sp*	6.82	2.01	5.26	-
Shorea hypochra	*S. hyp*	8.62	9.31	6.42	8.23
Sloanea sigum	*S. sig*	-	4.67	2.78	2.67
Spondias bipinnata	*S. bip*	2.22	-	-	-
Sterculia foetida L.	*S. foe*	-	-	1.22	-
Syzygium diospyrifolium	*S. dio*	0.7	-	-	-
Tetrameles nudiflora	*T. nud*	24.96	-	0.7	1.94
Vatica odorata	*V. ode*	-	12.87	6.91	5.68
Walsura angulata	*W. ang*	3.57	-	1.95	2.69
Xerospermum noronhianum	*X. nor*	5.82	1.01	1.46	-
Unknow 1	Uk1	-	-	-	0.47
Unknow 2	Uk2	1.39	-	-	-
Unknow 3	Uk3	1.2	-	-	-
Unknow 4	Uk4	0.7	-	-	-
Unknow 5	Uk5	-	2.32	-	-
Unknow 6	Uk6	0.78	-	0.64	-

Note: *The gray shading for some IVI values indicate IVI values ≥ 15 (5%) for each community.

ter analysis on a small island tends to overlap (a result of the CCA ordination graph shown in **Figure 3**), it clearly showed a difference in the abundance of dominant species. *Hopea pierrei* Heim (Dipterocapus) was found in all the communities, including the *Hopea pierrei* Heim community, secondary succession community, *Calophyllum thorelii* Pierre community, and lowland multi-aged stands, but with different IVIs, which were 38.15%, 20.54%, 10.18%, and 7.86%, respectively. A similar pattern was obtained when a cluster analysis was performed on the literal evergreen forest of Samesan Island (Payomrat, 2011). Ordination diagram and species composition of five sub-community types showed an overlap of *Memecylon plebejum*, which is a dominant species between four communities, but different IVI ratios were observed. However, some studies of small islands have identified distinct vegetation communities using cluster analysis. Phra Thong Island has a large alluvial deposit sand plain with sinking seashore and productive mudbeaches on estuaries of many long rivers, which is different than the terrain of Chang Island. This island covers the Cajuput forest, Cajuput swamp, Peat swamp forest, Beach forest, grasslands, and mangroves (Pumijumnong, 2005). Cluster analysis of the sub-forest community (Thaisatuen, 2010) and Poaceae and Cyperaceae (Boutrat, 2009) on Phra Thong Island were successfully classified as communities with 60% - 95% remaining information, which is similar to the present study, with the exception of the grassland type.

Relationship between Communities and Topographic Factors

Elevation and TWI were similarly important factors controlling the community distribution of Chang Island. The eigenvalues in an ordination analysis represent the relative contribution of each axis to the explanation of the total variation in the data. The CCA eigenvalues of the 3 two ordination axes were 0.303, 0.253, and 0.172, respectively (**Table 3**). A Monte Carlo test confirmed the statistical significance of the axes ($P = 0.05$). The first three axes explained 12.5% of the variance in the species data. From the **Table 4**, the highest intraset correlations with axis 1 was TWI (0.819) and axis 2 was the elevation (−0.858).

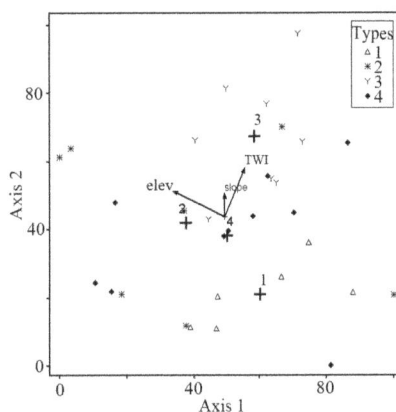

Figure 3.
CCA plot ordination with the community types derived from cluster analysis. Type 1 = lowland multi-aged stands; Type 2 = *Calophyllum thorelii* Pierre community; Type 3 = *Hopea pierrei* Heim community; and Type 4 = secondary succession community.

Table 3.
Summary statistics table for the CCA ordination.

	Axis 1	Axis 2	Axis 3
CCA			
Eigenvalues	0.303	0.253	0.172
% of variance explained	5.2	4.3	2.9
Cumulative % explained	5.2	9.6	12.5
Pearson correlation	0.981	0.88	0.894

Table 4.
Canonical coefficients and intraset correlations of environmental variables with the first three axes of the CCA.

	Canonical coefficients			Canonical coefficients (intraset)		
Variable	Axis 1	Axis 2	Axis 3	Axis 1	Axis 2	Axis 3
TWI	0.299	0.609	0.311	0.35	0.819	0.456
Elevation	−0.734	0.315	−0.199	−0.858	0.424	−0.291
Slope	0.002	0.29	−0.629	0.002	0.39	−0.921

Vectors representing environmental variables are shown in **Figure 3**. The length of the vector is proportional to its importance and the angle between two vectors reflects the degree of correlation between the variables. The angle between a vector and each axis is related to its correlation with the axis. Only variables with a correlation coefficient higher than 0.5 are represented in order to compare the classification between the cluster analysis result and CCA ordination. The derived community types are not clearly separated. Low-land multi-aged stands (Type 1) and *Hopea pierrei* Heim community (Type 3) grouped well on the upper right and lower right, respectively. The *Calophyllum thorelii* Pierre community (Type 2) and Secondary succession community (Type 4) have an overlapping distribution on the lower to mid-gradient.

TWI, which combines the local upslope contributing area and slope, is commonly used to quantify the control of hydrology, and therefore is used to define local topographic characteristics as a better representation than slope gradient alone. Many studies have showed a correlation between TWI, soil moisture, soil water level, ground water level, soil pH, and vascular plant species richness (Sørensen, 2005; Zinko et al., 2005; Giesler et al., 1998). Relative wetness in TWI tends to increase the richness in species and diversity. A comparison of TWI of the communities, species richness, and species diversity in this study showed that relative TWI tended to have a positive correlation with species richness (number of species)in three communities on the sloping area (*Calophyllum thorelii* Pierre community Type 2, *Hopeapierrei* Heim community Type 3, and secondary succession community Type 4). However, a negative correlation with species was observed. The community with the highest diversity and species richness, lowland multi-aged stands (Type 1), did not depend on the TWI value. This community is located on the flatland near the foot hill where the deepest soil affected the distribution of old trees with multi-layered strata. Soil moisture and humidity in the understory are high and much more stable compared to the canopy. TWI had a negative correlation with species diversity because of the success of *Hopea pierrei* Heim in the wetness area succession. As

shown in **Figure 4**, *Hopeapierrei* Heim was highly responsive to TWI, so much so that the others species had a reduced priority and the Shannon-Wiener index was decreased (**Tables 1** and **2**).

We selected key species by choosing species with an IVI value ≥15 or 5% from each community. Four species from the lowland multi-aged stands (Type 1) were *Tetrameles nudiflora* R. Br. (TETRAMELACEAE), *Hopea pierrei* Heim (DIPTEROCARPACEAE) *Madhuca pierrei* Lam (SAPOTACEAE), *Lithocarpus ceriferus* A. Camus (FAGACEAE), and *Ficus callosa* Willd (MORACEAE). Four species from the *Calophyllum thorelii* Pierre community (Type 2) were *Calophyllum thorelii* Pierre (GUTTIFERAE), *Mesua ferrea* Linn. (GUTTIFERAE), *Hopea pierrei* Heim (DIPTEROCARPACEAE), and *Cleistocalyx operculatus* Merr. and Perry (MYRTACEAE). Two species from the *Hopea pierrei* Heim community (Type 3) were *Hopea pierrei* Heim (DIPTEROCARPACEAE) and *Cleistocalyx operculatus* Merr. and Perry (MYRTACEAE). Four species from the secondary succession community (Type 4) were *Hopea pierrei* Heim (DIPTEROCARPACEAE), *Calophyllum thorelii* Pierre (GUTTIFERAE), *Cleistocalyx operculatus* Merr. and Perry (MYRTACEAE), and *Memecylon garcinioides* Bl. (MEMECYLACEAE). The IVI values of each species are shown in **Table 2**. A fit response curve created a turnover point of communities responding with TWI. Low land multi-aged stands (Type 1) and some species from the *Calophyllum thorelii* Pierre community (Type 2) occupied relative dry in TWI and turnover to Dipterocapus community (Type 3, 4) and *Calophyllum thorelii* Pierre community (Type 2) when TWI higher than 2, as defined by a moderate level of *Hopea pierrei* Heim abundance (key species of Type 3, 4) and *Cleistocalyxo perculatus* Merr. and Perry (key species of Type 2). However, the assumption of a high value of TWI occurred with the appearance of a creek based on field survey (TWI > 5), even if some of species, such as *Hopea pierrei* Heim, *Ficus callosa* Willd, and *Tetrameles nudiflora* R. Br. occupied this area.

Dipterocapace (*Hopea pierrei* Heim) were successfully distributed along all environmental gradients of Chang Island and were found in all strata. A characteristic feature of these plants is the amount of sunlight that can be tolerated during development. An emergent dipterocarp has a crown that receives full sunlight during the entire day, whereas as a seedling, it not only requires tolerant shade, but it will die when subjected to full sunlight during the entire day (Göltenboth, 2007). Thus, clear cutting of the forest will be a serious problem for dipterocarp, as seedlings will have less of a chance to survive. Buttress roots

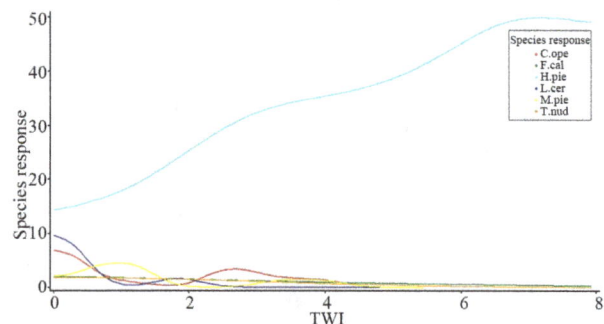

Figure 4.
Fit response curves between key species and TWI (the species codesare show in **Table 2**).

are another characteristic feature of dipterocarp (common name of *Hopea pierrei* Heim is TaKhianRak; "Rak" means root in Thai). It occurs when trees with a shallow root system prevail in areas where nutrient concentrations are largest near the soil surface or stand against a steep slope. Why these structure are formed is still not fully understood, as research has shown that buttresses do not increase the trees' resistance to mechanical stress, such as trunk snapping, or alleviate the pulling strain on the roots (Richard, 1996).

Elevation was positively correlated with tree density (r = 0.979). The *Calophyllum thorelii* Pierre community (Type 2), which was distributed along the high land until the peak, had the highest tree density (2225 tree·ha^{-1}), whereas the tree density of the *Hopea pierrei* Heim community (Type 3), secondary succession community (Type 4), and low land multi-aged stands (Type 1) were 1807, 1782, and 1287 tree ha^{-1}, respectively. This relation pattern is seen in the tropical rainforest of Phukhet Island as well (Kiratiprayoon, 1986). Normally, high density leads to a high percentage of basal area, but the basal area of Chang Island was not increased. TWI showed a positive correlation with the basal area in this study (r = 0.928). The average TWI of *Hopea pierrei* Heim community (Type 3), secondary succession community (Type 4), *Calophyllum thorelii* Pierre community (Type 2), and low land multi-aged stands (Type 1) was 4.13 ± 2.36, 1.70 ± 1.95, 1.29 ± 1.16, and 0.5 ± 0.55, respectively, while the percentage of basal area was 0.5886%, 0.4056%, 0.3396%, and 0.3845%, respectively. Many studies have shown the same results as our study, where moisture is a factor that can potentially control the above ground biomass in 6 forest types of Thailand: tropical evergreen rain forest, seasonal rain forest, lower mountain coniferous forest, upper mountain rain forest, mixed deciduous forest, and deciduous dipterocarp forest (Working group I: Scientific Basis of Climate Change, 2011; Rueangruea, 2009; Sunthisuk, 2006; Kiratiprayoon, 1986). Species response curve with regard to elevation showed six key species responding in three patterns (**Figure 5**). The *Hopea pierrei* Heim curve primarily showed a bell-shape curve (abundant at mid-elevation), which intersected the *Calophyllum thorelii* Pierre curve at an elevation at 550 m asl. It can be concluded that the Dipterocarpus (*Hopea pierrei* Heim) community tends to turnover to the *Calophyllum thorelii* Pierre community at elevations greater than 550 m asl, and turnover to the secondary succession community (Type 4) at low elevation with an increase of *Memecylon garcinioides* Bl., which is the other key species of this community. The response curve of *Madhuca pierrei* Lam and *Tetrameles nudiflora* R. Br., which are key species of lowland multi-aged stands (Type1), showed a negative correlation with elevation. A positive correlation of the *Mesua ferrea* Linn curve (key species of *Calophyllum thorelii* Pierre community, Type 2) intersected the *Madhuca pierrei* Lam curve at an elevation of 250 m asl. Moreover, the decrease of *Hopea pierrei* Heim at a low elevation suggests that the limit of distribution of lowland multi-aged stands (Type1) is 250 m asl. *Lithocarpus ceriferus* A. Camus, *Cleistocalyx operculatus* Merr. and Perry, and *Ficus callosa* Willd were excluded because these species did not respond to a change in elevation (the curves ran parallel with the x-axis).

Area of Community

Based on the fit respond curve for elevation and TWI, we conclude that the environmental conditions of each community

are as follows: Lowland multi-aged stands (Type 1): Elevation < 250 m asl and TWI < 2, *Calophyllum thorelii* Pierre community (Type 2): Elevation > 250 m asl and TWI < 5, *Hopea pierrei* Heim community (Type 3): 250 < Elevation < 550 m asl and 2 < TWI < 5, Secondary succession community (Type 4): Elevation < 550 asl and 2 < TWI < 5 and Creek representation: All elevation gradient and TWI > 5.

From the **Figure 6**, the total forest area was 17428.45 ha, which was comprised of low land multi-aged stands Type 1 (4807.99 ha or 27.59%), *Calophyllum thorelii* Pierre community Type 2 (5300.04 ha or 30.41%), *Hopea pierrei* Heim community Type 3 (2358.54 ha or 13.53%), secondary succession community Type 4 (3731.24 ha or 21.41%), and creek area (1230.65 ha or 7.06%). Some areas of the secondary succession community were complicated. These areas were composed of

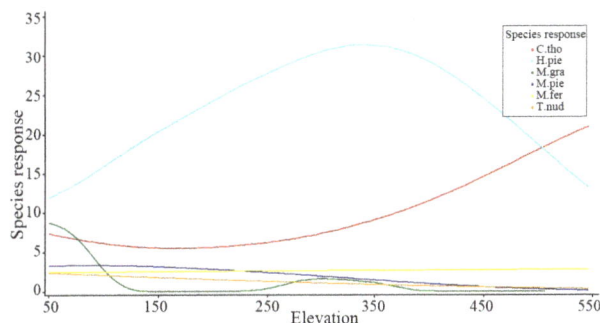

Figure 5.
Fit response curves between key species and elevation (the species codes are shown in **Table 2**).

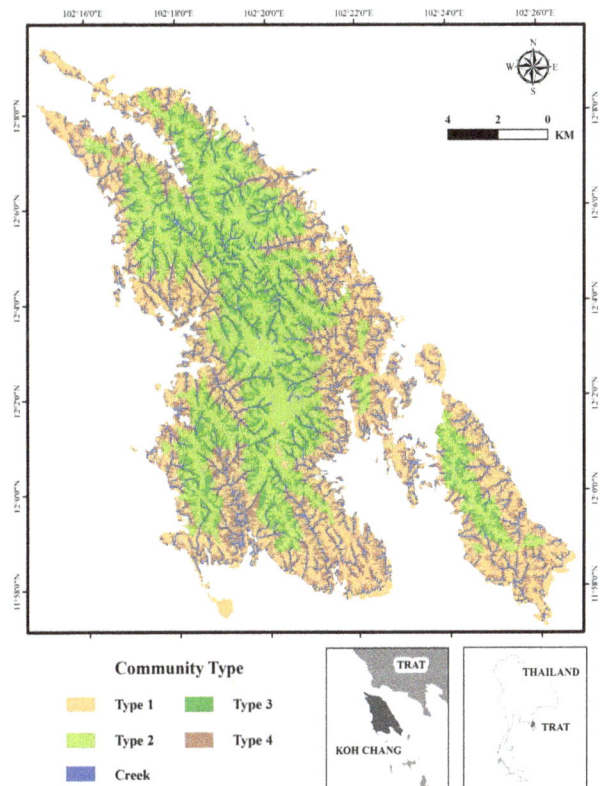

Figure 6.
Area of four community types (Type 1-4) and creek.

two probable communities: the natural Dipterocarpus community and secondary succession community, because one of the
factors controlling the secondary succession community was
human disturbance. Therefore, it was not included in this study.

Conclusion

Chang Island has a particularly high evenness and richness of
species. We investigated an area of 2.48 ha and recorde total of
78 species from a total of 32 families. Two-way cluster analysis
classified the tropical rain forest into four sub-communities:
lowland multi-aged stands (Type 1), *Calophyllum thorelii* Pierre community (Type 2), *Hopea pierrei* Heim community
(Type 3), and secondary succession community (Type 4). Elevation and the topographic wetness index (TWI) affected the
distribution of each community. Lowland multi-age stands,
which had the highest diversity and richness, correlated with
low elevation and relatively low TWI or dry conditions. The
Calophyllum thorelii Pierre community had the highest tree
density and correlated with high elevation and relative moderate to dry TWI conditions. The *Hopea pierrei* Heim community
had the highest percentage of basal area and the lowest diversity, and was extensively covered by the *Hopea pierrei* Heim
species. It correlated with a mid-elevation near creeks and had
relatively wet TWI conditions. Finally, the secondary succession community (Type 4) was related to mid to low elevation
and relatively moderate to wet TWI conditions.

REFERENCES

Bird, M. I., Taylor, D., & Hunt, C. (2005). Palaeoenvironments of
insular southeast Asia during the last glacial period: A savanna corridor in Sundaland? *Quaternary Science Reviews, 24,* 2228-2242.

Boutrat, W. (2009). *Relationship between soil properties with Poaceae
and Cyperaceae on the grassland at Phra Thong island, Kuraburi
district, Phangnga province.* Master Thesis, Nakron Pratom: Mahidol University

Chamchumroon, V., & Puff, C. (2003) The Rubiaceae of Ko Chang,
southeastern Thailand. *Thai Forest Bulletin (Botany), 31, 13-26.*

Curtis, J. T. (1959). The vegetation of Wisconsin. Madison: University
of Wisconsin Press.

Environmental Research Institute (2007). Study area and the index of
environmental quality to sustainable tourism and the island group
project (phase 2). Bangkok: Chulalongkorn University.

Esselstyn, J. A. & Brown, R. M. (2009). The role of repeated sea-level
fluctuations in the generation of shrew (Soricidae: Crocidura) diversity in the Philippine Archipelago. *Molecular Phylogenetics and
Evolution, 53,* 171-181

Finch, H. (2005). Comparison of distance measures in cluster analysis
with dichotomous data. *Journal of Data Science, 3,* 85-100.

Giesler, R., Högberg, M., & Högberg, P. (1998). Soil chemistry and
plant in Fennoscandian boreal forest as exemplified by a local gradient. *Ecology, 79,* 119-137.

Göltenboth, F., Langenberger, G., & Widmann, P. (2007). Tropical lo-
land evergreen rainforest. *Ecology of Insular Southeast Asia: The
Indonesian Archipelago,* 297-384.

Hill, D., Fasham, M., Tucker, G., Shewry, M., & Shaw, P. (2007).
Handbook of biodiversity methods. Cambridge: Cambridge University Press.

Kiratiprayoon, S. (1986) Comparative study on the structure of the rat-
tan bearing tropical rain forests. Master Thesis, Bangkok: Kasetsart
University

McCune, B., & Grace, J. B. (2002). Analysis of ecological communities.
Glenden Beach: MJM Software Design.

Office of Science for Land Development (2005). *Manual analysis of
soil, water, fertilizers, soil and plant analysis for certification* (2nd
ed.). Bangkok: Land Development Department.

Payomrat, P. (2011). *Effects of ecological factors on vegetation and
carbon stocks at Samesan island, Chonburi, Thailand.* Master Thesis,
Nakron Pratom: Mahidol University

Pumijumnong, N. (2005). *Effects of the tsunami on the Pra Thong island ecosystem.* Nakron Pratom: Mahidol University.

Richards, P. W. (1996). The tropical rainforest: An ecological study.
Cambridge: Cambridge University Press.

Royal Forest Department (1997). Master plan report Koh Chang marine
national park, Trad province. The land and forest resources. Conservation of natural resources.

Royal Forest Department. (2005). Master plan report Ton Nga Chang
wildlife sanctuary, Songkhla province.

Rueangrues, S. (2009). *Structure of Montane forests in Thailand.* Master Thesis, Bangkok: Kasetsart University

Santhisuk, T. (2006). *Forest of Thailand.* Bangkok: Department of National Parks, Wildlife and Plant Conservation.

Schmidt, J. (1900). *Flora of Koh Chang: Contributions to the knowledge of the vegetation in the Gulf of Siam.* Copenhagen: B. Luno.

Sørensen, R., Zinko, U., & Seibert, J. (2005). On the calculation of the
topographic wetness index: Evaluation of different methods based on
field observation. *hydrology and earth system sciences discussions,*
1807-1834.

Tansuwan, V., & Kitisarom, N. (2007). Study report of geology, Chang
island, Trat Province. Bureau of Geology, Department of Mineral
Resources.

ter Braak, C. J. F., & Verdonschot, P. F. M. (1995). Canonical correspondence analysis and related multivariate methods in aquatic ecology. *Aquatic Science, 57,* 255-289

Thai Meteorological Department (2010). Data of climate between the
years 1990-2009 at Khlong Yai station, Trat province.

Thaisatuen, T. (2010). *Multivariate analysis of relationship between
plant communities and soil properties on Phra Thong island, Kuraburi district, Phangnga province.* Master Thesis, Nakron Pratom:
Mahidol University.

Tougard, C. (2001). Biogeography and migration routes of large mam-
mal faunas in South-East Asia during the late middle pleistocene:
Focus on the fossil and extant faunas from Thailand. *Palaeo, 168,*
337-358.

Working Group I: Scientific Basis of Climate Change (2011). *Thailand's first assessment report on climate change 2011.* Bangkok: The
Thailand Research Fund.

Zinko, U., Seibert, J., Dynesius, M., & Nilsson, C. (2005) Plant species
numbers predicted by a topographhy based groundwater-flow index.
Ecosystem, 8, 430-441.

12

The Role of Physical and Political Factors on the Conservation of Native Vegetation in the Brazilian Forest-Savanna Ecotone

Henrique O. Sawakuchi[1], Maria Victoria R. Ballester[1], Manuel Eduardo Ferreira[2]

[1]Center of Nuclear Energy in Agriculture, University of São Paulo, Piracicaba, Brazil
[2]Social-Environmental Studies Institute/LAPIG, Federal University of Goiás, Goiânia, Brazil

The Araguaia River Basin covers a considerable extent of Brazilian Savanna (locally called Cerrado) and part of Amazon Tropical Rainforest, embracing high biodiversity and a vast flooding area. This region has been converted to agricultural lands since 1970s, for the past four decades, leading to a fragmented landscape that holds one of the few large remaining blocks of Cerrado primary vegetation. Therefore, to assess the degree of preservation of this area a 2007 primary vegetation map was derived through Boolean operations using land use and land cover maps from 1975, 1985, 1996 and 2007, from digital classification of Landsat MSS and TM images. To evaluate the role of driving factors on the presence of pristine vegetation, a logistic regression analyses was performed. Tested factors were: distance from roads and cities, terrain slope, land tenure, soil fertility and flooding. We found statistical significant values ($p < .05$) showing that distance from roads and cities, the increase in slope, the presence of protected areas, indigenous lands, wetlands and areas with low fertility have positive influence on the presence and maintenance of these pristine areas. The occurrence of original vegetation in many cases is associated with environmental constraints that difficult or do not allow agricultural use. Analysis of physical and political factors, which may have direct or indirect influence on the conservation and degradation of native vegetation are very important for the comprehension of the dynamics of regional land use, and provide supporting information for a more efficient and sustainable regional landscape planning.

Keywords: Amazon-Cerrado Transition; Pristine Vegetation; Driving Factors; Deforestation; Araguaia River Basin; Regional Planning

Introduction

The expansion of agricultural frontier in the tropics has been identified as one of the main drivers of deforestation (Alves et al., 2009; Geist & Lambin, 2001). In Brazil, since 1980s, although Savanna biome (locally known as Cerrado) has been converted to agricultural lands at higher rates than those found for the Amazon, most of the attention to the consequences of land use and land cover change has been focused in Tropical Rain Forest ecosystems (Alves et al., 1999; Batistella & Moran, 2005; Cardille & Foley, 2003; Fearnside, 2006; Laurance et al., 2004; Mahar, 1989).

Spatially, forest clearance in Brazilian Amazon has been mainly concentrated at the deforestation arc (Ferreira et al., 2005), encompassing the south, east and west boundaries of the Tropical Rain Forest and a transition zone between this biome and the Cerrado. This transition zone includes about 600 hundred km² of savannas located in the drainage basins of Tocantins and Araguaia Rivers. Since 1970s, intense farming activities, agricultural expansion and inappropriate land management have already resulted in active erosion at the head waters of the Araguaia River Basin, causing indirect impacts at the middle portion of the basin (Latrubesse & Stevaux, 2006; Coe et al., 2011). Due to its high biodiversity and vast flooding area, the alluvial zone of the Araguaia and the Bananal lowland regions were identified by the Brazilian Ministry of Environment as a priority area for conservation (MMA, 2010). Nevertheless, only few studies about the degree of conservation of the natural vegetation have been carried out in this region.

Conservation of the Araguaia-Bananal area, as well as other Cerrado large areas, is mainly related to regions where there is at least one factor preventing agriculture expansion, such as terrain slope, shallow soils and flooded areas (Machado et al., 2004). As a complex phenomenon, the origin of deforestation can be attributed to a series of factors (Alencar et al., 2004), related to local, regional and global scale processes. In the Amazon, many studies have shown that increasing in deforestation was related to tax incentives (Mahar, 1989; Nepstad et al., 2001), national economy (Alencar et al., 2004), wood industry, agriculture and cattle raising expansion (Alencar et al., 2004; Asner et al., 2009; Margulis, 2004; Walker et al., 2009), roads construction (Alves et al., 2009; Ballester et al., 2003), low slope, soil properties, among others. In this scenario, an important strategy to decrease deforestation is the public conservation policies aiming the creation of conservation units of integral preservation, sustainable use of natural resources and indigenous lands (Alencar et al., 2004).

Most studies about drive factors are focused on deforestation processes and there is a lack of information concerning what factors are responsible for preventing native vegetation losses. Therefore, the main objectives of this study were: 1) identify and quantify pristine vegetation in the middle Araguaia River basin and 2) analyze the influence of roads, cities, slope, land tenure, soil fertility, and flooding regime in the occurrence of primary vegetation remnants.

Methods

Study Area

The Araguaia River is part of the Araguaia-Tocantins Basin, draining areas of Tropical Rain Forest and Cerrado biomes, and considered one the South America most important riverine and wetland system (ANA, 2008). The study area covers 166,000 km^2, located in the central portion of the Araguaia River Basin, in central Brazil. This area encompasses part of the Tocantins, Mato Grosso, Pará and Goiás States, in the transition zone between Tropical Rain Forest and Cerrado, covering three quarters of the studied landscape (**Figure 1**). The natural heterogeneity of vegetation physiognomies, associated with increasing human disturbances, generated a complex dynamic change in native vegetation structure and configuration of the landscape.

The main Cerrado physiognomies found in the study area are: Grasslands, Shrublands and Forestlands, whereas dominant forest physiognomies are Open Rainforest (IBGE, 2004) and Transitional Forest, which is the most threatened physiognomy in Amazon. Its continuous loss has raised concerns, not only because of its ecological value (still poorly known), but also due to its open structure that makes it more susceptive to fires (Alencar et al., 2004).

The Araguaia is one of the main river basin draining the Cerrado biome, and includes the most important wetland of Central Brazil (Latrubesse et al., 2009), a quaternary alluvial lowland well developed and extending for more than 1100 km from Registro do Araguaia to Conceição do Araguaia (Carvalho et al., 2009; Latrubesse & Stevaux, 2006). In the northern part of this sedimentary basin is located the Bananal island, the largest fluvial island of the world. During the rainy season, a vast area of the Bananal plain is flooded. The vegetation is dominated by Grasslands and Riparian Forest, and in higher zones by Cerrado Woodland and Alluvial Forest (Diegues, 2002). Livestock is the dominant economic activity in this region, where productive areas can be planted or natural pastures. The native grasslands in the flooded areas of the Bananal lowland maintain a reasonable capacity for cattle rising during the dry season (Diegues, 2002).

The climate of the study area is warm and seasonally humid, with an annual mean precipitation of 1755 mm and mean monthly temperatures ranging from 25.1°C in January to 26.4°C in September. The dry season is from May to September (mean relative humidity of 40%), while the wet season is from October to April (mean relative humidity of 90%) (Borma et al., 2009). Plinth and Concretionary soils are the dominant soil types of the region, where can also be found Hydromorphic soils, Ox soils and Quartz sands (SEPLAN, 2008). The first two types present agricultural limitation due to low drainage and presence of ferruginous concretion, respectively (Coutinho, 2005).

Primary Vegetation Mapping

A 2007 primary vegetation map was derived through Boolean operations using land use and land cover maps from 1975, 1985, 1996 and 2007. These maps were obtained from Landsat Multispectral Scanner (MSS) for 1975 and Thematic Mapper (TM) for other years. Images were processed in ERDAS-IMAGE (version 9.2) using hybrid classification, composed by an unsupervised followed by a supervised methodology (Yu & Ng, 2006). The accuracy assessment was performed for 2007 by calculating overall accuracy and Kappa index using 287 ground truth points. The classification was considered good, with an overall accuracy of 85.02% and overall Kappa statistics of 0.75.

Native vegetation was separated in three different classes, defined according to the vegetation structure: 1) Forest, comprehending Dry Forest, Wooded Cerrado and Riparian Forest; 2) Cerrado Grassland, including all grassland dominated physiognomies and 3) Cerrado Woodland for shrubs dominated areas. To generate the final map, the three classes for each date were reclassified into one, called native vegetation. Finally, a simple Boolean operation was employed to select those areas of unchanged cover until 2007, eliminating re-growth regions. The 2007 native vegetation map was used to mask the different classes of native vegetation again, resulting in a map of pristine Forest, Cerrado Grassland and Cerrado Woodland remnants in 2007.

Driving Factors for Native Vegetation Conservation

The presence of native vegetation in Amazon is mainly motivated by isolation and existence of protected areas (Bruner et al., 2001), while deforested regions are related to roads and cities proximity (Alves et al., 1999; Ballester et al., 2003; Batistella & Moran, 2005). Therefore, to identify drivers of pristine vegetation remnants in our study area, we selected six factors: 1) roads distances; 2) cities distances; 3) terrain slope; 4) land tenure; 5) fertility; and 6) presence of flooding areas (**Figure 2**). Due to Landsat-MSS cell resolution we standardized the spatial resolution of all grid cell maps in 80 meters.

Roads, cities and fertility maps were obtained from a digital library available at IBAMA Remote Sensing Center (http://siscom.ibama.gov.br/sitecsr/). We used only state and federal roads, totalizing 3350 km, of which 1884 km are paved. Roads and cities distance maps were derived by calculating the Euclidean distance perpendicular to them (**Figures 2(a)** and **(b)**). Due to high distance values, their distribution were corrected by applying a logarithmic transformation (Serneels & Lambin, 2001).

The slope map, in percent, was derived from a Digital Elevation Model obtained by processing Shuttle Radar Topographic

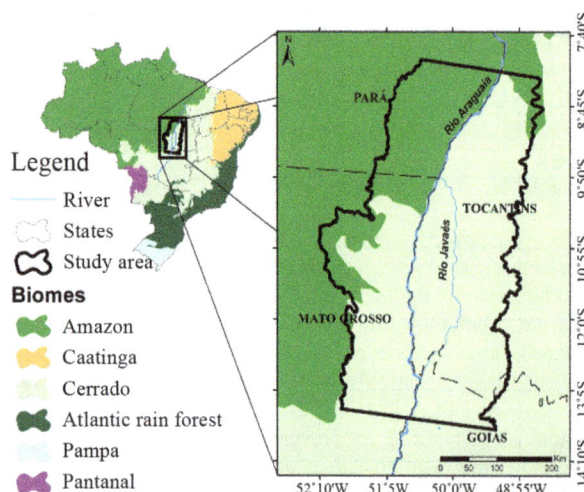

Figure 1.
Localization of the study area in the middle Araguaia River basin and the biomes that the area encompasses.

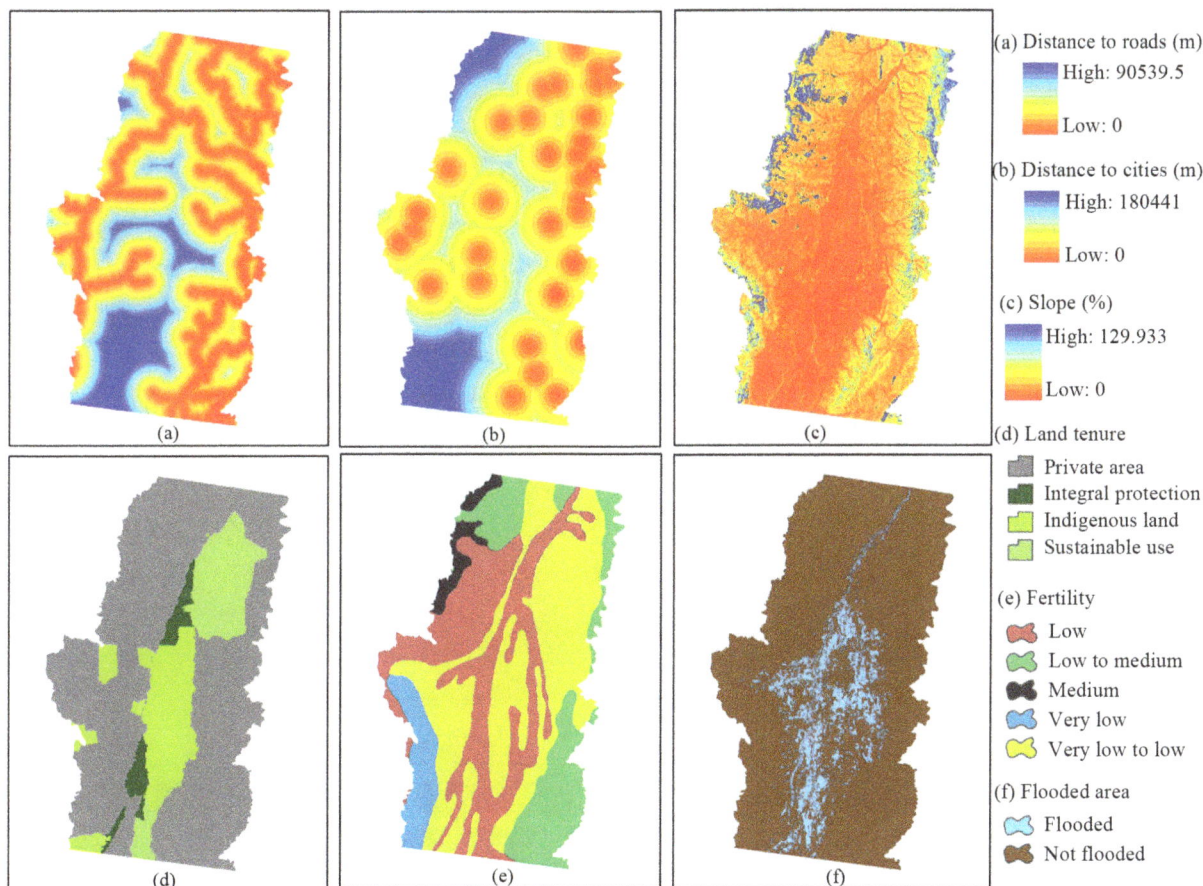

Figure 2.
Analyzed factors maps that can influence the occurrence of primary native vegetation in the middle Araguaia River basin: (a) Distance to roads; (b) Distance to cities; (c) Slope; (d) Land tenure; (e) Fertility; and (f) Presence of flooding areas.

Mission (SRTM) data from EMBRAPA satellite monitoring unit (www.relevobr.cnpm.embrapa.br). As shown in **Figure 2(c)**, there is an extensive flat area dominating the south and central part of the landscape, and areas with high slope concentrated in northwest and northeast regions.

Land tenure map encompasses private lands, indigenous land, state and federal conservation units—divided into integral protection (State and National Parks, and Wildlife Refuge) and sustainable for use (Environmental Protection Areas). Data were downloaded from a digital library available at the Brazilian Ministry of Environment (http://www.mma.gov.br/sitio/index.php).

The majority of conservation units and indigenous lands of the Araguaia basin are located in the study area and cover around 30% of the area (**Figure 2(d)**). To calculate coefficients of Land Tenure classes, private areas were used as reference.

Soil fertility data were obtained from the agricultural suitability map of Brazil (IBAMA dataset, http://siscom.ibama.gov.br/sitecsr/). In general, soils in the study area are dominated by low fertility classes. The dominant class was "Very low fertility" to "Low fertility", localized in the central zone from south to north. The second dominant class was "Low fertility", concentrated along the main rivers and in the west side of the study area (**Figure 2(e)**). The reference class used for soil fertility coefficients calculations was "Low", an intermediate class, allowing the visualization of higher and lower classes.

The flooded areas map was extracted from the South America Vegetation Map developed by Eva et al. (2002). These areas are concentrated in the central portion of the study region, encompassing over 17,480 km^2 (**Figure 2(f)**). In this case non-flooded area was used as reference for the calculation.

Statistical Analyzes

To determine the influence of each factor on the presence or absence of primary vegetation, a binary logistic regression analysis was performed using the maximum likelihood method in Minitab software, with p-value coefficients lower than 0.05. This analysis estimates the coefficient value, standard deviation and p value for each variable, indicating the effect (positive or negative) that each predictive variable has on a response variable. Furthermore, for categorical data are possible to verify the weight that each variable class has in relation to the presence of primary vegetation. As these analyses only test binary variables, the native cover classes were grouped in one class showing the presence or absence of native vegetation.

Data sampling was carried on using 50,000 random points distributed in the study area with a minimal distance of 100 m from each other, avoiding sampling in the same pixel (Chomitz & Gray, 1996; Mertens & Lambin, 2000; Serneels & Lambin, 2001; Ugon, 2004). These points were used to extract the information from all maps and build a matrix.

Results

Pristine vegetation in 2007 comprised an area of 86,800 km^2, which is equivalent to 52.3% of the study area. This vegetation was mainly concentrated at the central portion of the landscape, where an extensive area of Cerrado Grassland was found. Cerrado Woodland areas occurred mainly along the edges of Cerrado Grassland, and were concentrated at the east and southwest regions. Despite the fact that the original area of forest extended from the north to the southwest border, larger remnants patches were denser in the central part of the landscape (**Figure 3**).

Forest was the dominant physiognomy, encompassing 25% (41,430 km^2) of the landscape, followed by Cerrado Grassland and Cerrado Woodland, covering 16.6% (27,640 km^2) and 10.7% (17,732 km^2) of the study area, respectively (**Figure 4**).

Table 1 presents the results from the logistic regression model applied to evaluate how some regional factors influence

Table 1.
Coefficients values of the factors involved with presence and absence of primary native vegetation and its effect in the maintenance of this vegetation.

Predictor	Coefficient	p	Effect
Distance from Roads	0.314	0.00	Positive
Distance from Cities	0.130	0.00	Positive
Slope	0.019	0.00	Positive
Land Tenure			
Integral Protection	0.878	0.00	Positive
Indigenous Land	0.622	0.00	Positive
Sustainable Use	0.054	0.08	-
Fertility			
Low to Medium	−0.478	0.00	Negative
Medium	−0.638	0.00	Negative
Very Low	0.025	0.55	-
Very Low to Low	0.354	0.00	Positive
Flooded Area			
Flooded	0.291	0.00	Positive

For categorical data as Land tenure, Fertility and Flooded area the calculation of each class was done using private areas, Low fertility, and non-flooded areas, respectively, as reference.

the maintenance the amount of primary vegetation still remaining in the middle Araguaia River basin. Terrain physical characteristics, evaluated as slope and soil fertility, have an opposite effect on the presence of primary vegetation. A positive relationship was found between slope increase and presence of forest. While flatter areas are preferred for implementation of agricultural crops, remnants of forest tended to concentrate at steeper areas (Coefficient 0.019) (**Figure 5(a)**). In contrast, soil fertility had a negative impact on native vegetation cover. Areas with more fertile soil are preferable for agricultural practices and therefore, pristine vegetation remnants concentrate on very low to low fertility soils (Coefficient 0.354). These soils were cover by 22% of the remaining Forest, 25% of the Cerrado Grassland and 13% of the Cerrado Woodland areas (**Figure 5(b)**).

Natural flooding events affected about 10% of the study area and have a positive influence in the maintenance of pristine areas (Coefficient 0.291). Of the extent of flooded systems (17,400 km^2) 71% were cover by natural vegetation, with Forest encompassing 18% and Cerrado 53%. In private lands found in these ecosystems, agriculture and pasture areas covered 31%, since they are less suitable for agricultural practices (**Figure 5(c)**).

The evaluation of the role of land tenure showed that the presence of conservation units of integral protection and indigenous lands had a positive influence on primary vegetation and presented the higher coefficient values, 0.878 and 0.622, respectively. Of the total remaining natural vegetation in 2007, 27% of Forest, 41% of Cerrado Grassland and 13% of Cerrado Woodland were found in these areas (**Figure 5(d)**).

Distance from roads and cities showed an expected pattern of

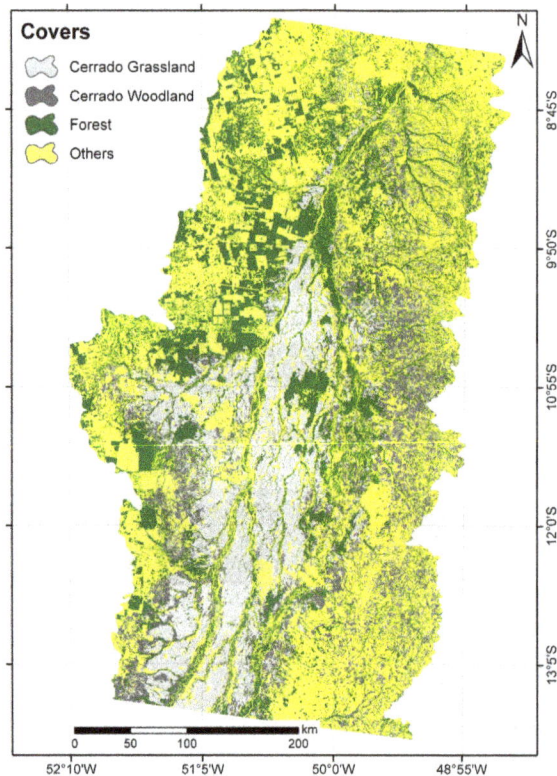

Figure 3.
Primary vegetation map in the middle Araguaia River basin in 2007.

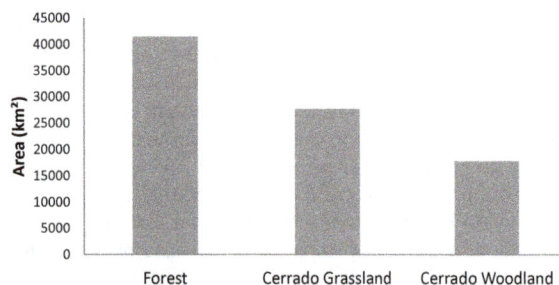

Figure 4.
Area covered for each physiognomy of native vegetation.

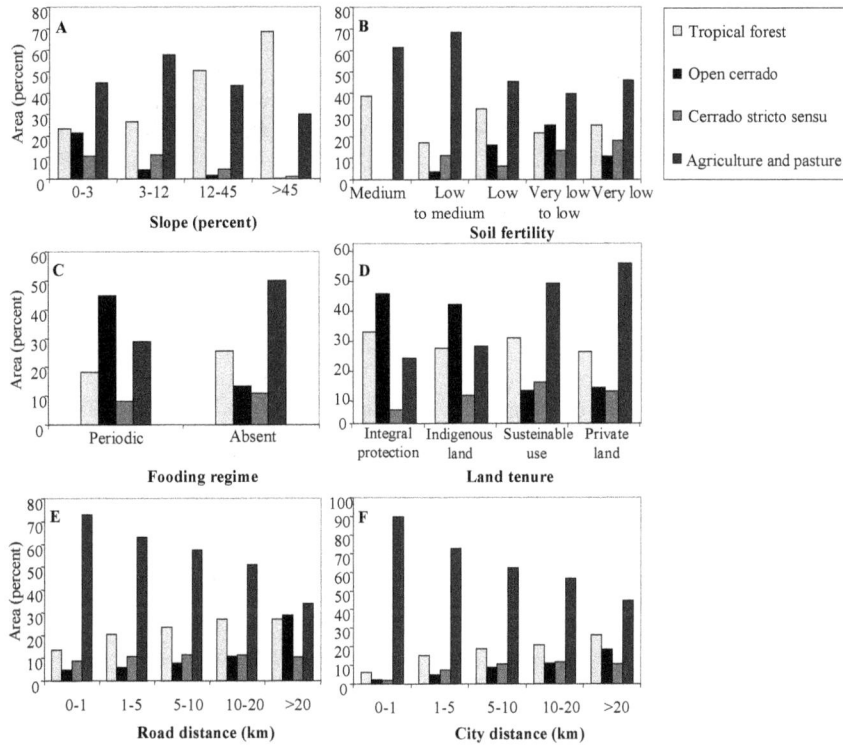

Figure 5.
Relative contribution of land cover and land use areas for each analyzed factor classes.

positive influence in the presence of primary vegetation. Larger areas covered with native forest are found as the distance from roads increases (Coefficient 0.314); of the total Forest remnants, 70% were at least at 10 km from a road, while for Cerrado Grassland and Cerrado Woodland this value was 86% and 66%, respectively. A similar pattern was found for distance from cities, but the lower coefficient value (0.130) indicates a less intense relationship (**Figures 5(e)** and **(f)**).

Discussion

In Brazilian Amazon, deforestation is primarily related to road development (Laurance et al., 2002) and governmental colonization programs (Alves et al., 1999; Batistella & Moran, 2005; Ferreira et al., 2007). Moreover, infrastructure implementation and government policy are the primary drivers controlling time and spatial scales of deforestation in most cases (Ballester et al., 2003). In general, this process begins with official or unofficial roads opening and paving in a preserved region, which in turn allow access to illegal wood exploitation, followed by conversion to pasture for cattle and crops (Ferreira et al., 2005; Yoshikawa & Sanga-Ngoie, 2011). Previous studies in the Amazon region have shown that deforestation is more intense and concentrated near roads, with up to 90% of forest clearing occurring within a 20 km buffer from a road (Ballester et al., 2003; Ferreira 2001; Gils & Ugon, 2006; Kirby et al., 2006; Ludeke et al., 1990; Nepstad et al., 2001), suggesting their influence in the increase of land cover conversion.

Road network construction and highway modernization are the main threats to preservation of large areas of pristine vegetation, facilitating access to logging, migration and farming (Primack, 2002). The simple announcement of roads opening or

improvements usually generates a speculative land rush that consequently can lead to a significant deforestation increase (Fearnside, 2006). Results from several modeling studies using this scenario (Ferreira et al., 2005; Laurance et al., 2001; Soares et al., 2004; Soares-Filho et al., 2005) have shown that public policy to control deforestation and land use planning play a key role in reducing the impact of roads construction and improvement on native vegetation.

Despite the fact that urban center proximity has a negative impact in forest preservation, its influence showed a different spatial effect on primary vegetation remnants distribution. While distance from roads presented a large effect in recent deforestation, our statistical analysis shows that distance from an urban center was more important at clearing initial stages, when nearer areas were preferred due to their proximity (Mertens & Lambin, 2000). This pattern emerges in areas located at cities closer surroundings. When these surroundings are occupied for agriculture and livestock activities, it is necessary to go increasingly far through roads, which in turn increases their influence in deforestation. In general, distance from roads and cities tend to have high correlation with deforestation and the presence of primary vegetation remnants, since the increase of migration, emergence and growth of urban areas are dependent on existing roads linking cities and facilitating agricultural production flow to other consumers markets.

The positive influence of slope can be associated with the major economic activity of the region, cattle rising, which dominates in planted and native pastures in low slope regions of the landscape. Nevertheless, during field surveys we were able to determinate that native pastures grow in moderate slope areas, as well as in flat flooded regions, resulting in the low observed coefficient value. Moreover, even in some cases in which land cover of these portions of the landscape has not changed, they have been impacted by human use and management practices of native pastures. A significant relationship between forest conversion in flat and not flooded areas was found for tropical forest conversion in Bolivia (Ugon, 2004). However, topography was irrelevant in forest conversion prediction models because road influence is stronger than the barrier effect of high slope (Gils & Ugon, 2006), reinforcing the weight that new roads have on deforestation.

Integral protection areas were the most effective for conservation. Similar pattern was observed for Peruvian Amazon (Oliveira et al., 2007). The low coefficient value of the "sustainable use" class was related to the agricultural and pasture use allowed in these conservation units. For the study area, the sustainable use class is equivalent to what is classified in Brazil as Environmental Protection Areas, which enables human occupation and is one of the less restrictive types of conservation unit in the National System of Conservation Units (Brasil, 2002). However, a high p-value (0.08) does not enable any consideration about the coefficient value of this class. The proportion of deforested areas within protected areas, such as conservation units of integral protection, sustainable use and indigenous land, was lower than outside them, indicating their importance as mechanisms to hold or slow down deforestation processes. The same pattern was found in other tropical areas. For instance, in Belize low rates of deforestation were found within the limits of national parks (Chomitz & Gray, 1996), while in Bolivia was verified that land with owners tend to be less invaded by settlers, preventing illegal occupation and consequently deforestation (Gils & Ugon, 2006).

As expected, logistic regression values showed a pattern in which the best fertility classes of the region have a negative influence in the presence of primary vegetation, while lower fertility classes have a positive influence in their maintenance. Although such pattern was expected, with the advance of agricultural techniques, large areas previously inappropriate for agriculture have now become available (Gils & Ugon, 2006). The class "Very low" was not considered due to its high p-value (0.55), which can be attributed to various mixtures of land cover classes in areas within this fertility class.

The Araguaia sedimentary plain is considered the most promising area for expansion of rice cropping in Brazil (EMBRAPA, 2008). The surroundings of Bananal Island, called Javaés Valley, were regarded as particularly suitable for this culture (Collicchio, 2008), and already have irrigation projects to produce different agricultural crops (Collicchio, 2006). Although all these projects have resulted in the conversion of large areas of native vegetation, which are adapted to the flooding regime, the occurrence of flooding have a positive influence in the maintenance of pristine vegetation. Nevertheless, this result can be related to land tenure, which could have a strong effect over this variable due to the overlap of indigenous land and integral protection conservation units with flooded areas, preventing the implantation of agricultural projects.

The effectiveness to preserve native vegetation within conservation units is not always easily achieved because deforestation will be greater if parks are closer to the capital cities, in sites closer to federal roads and on lower slopes (Pfaff et al., 2009). In some cases, the presence of pristine areas can be associated not only with conservation units but also with the local characteristics where they were created. Areas naturally protected from human exploration, such as remote places, steep areas, poor soils, flooded wetlands and any other factor that makes agriculture more difficult can be more important factors than the delimitation of conservation units. In these cases, it is not trivial to evaluate the real importance of the delimited protected areas (Joppa et al., 2008). In our particular case of study, the landscape was predominantly flat, with a large road network, and several protected areas, which are located mainly in wetlands and also on low fertility soils. The overlap of all these factors makes difficult to isolate the role of each factor, and hence its effectiveness in native vegetation conservation.

Other studies have shown a high and statistically significant influence of land tenure and distance from roads on the prediction of forest conversion. Distances from villages and topography have a smaller contribution while no significant predictive value was found for soil types (Gils & Ugon, 2006). Our results show that the presence of primary vegetation is linked in some extent to all tested drivers. Environmental constraints, inherent to the area, may be the factor responsible for making more demanding or precluding the conversion of primary vegetation into agricultural areas. Hilly slopes or wetlands areas can act as factors that hinder or even make impossible the transport of products for consumer markets due to high transportation costs (Chomitz & Gray, 1996). These local characteristics are also a limiting factor for mechanized agriculture, which has a preference for flat areas, where machinery access is facilitated. This feature is more important than soil quality, which can be improved with fertilizers (Serneels & Lambin, 2001).

Conclusion

Our analysis of all these variables showed that the presence of primary vegetation can be associated with different environmental constraints, making more demanding or preventing land conversion. The main factors responsible for the maintenance of pristine vegetation are the presence of protected areas and environmental constraints. Moreover, understanding the role of physical and political factors, which may have (direct and indirect) influence on the deforestation or maintenance of native vegetation are key elements to better accesses land use dynamic and provide supporting information for a more efficient and sustainable regional planning.

Acknowledgements

Funding and scholarship for this study were provided by FAPESP and Milenio/CNPq (proc. 2003/13172-2, 2007/01686-2 and 420199/2005-5, respectively). The third author received support from United States National Aeronautics and Space Administration (NASA)—Land-Cover and Land-Use Change Program (LCLUC) (NNX11AE56G).

REFERENCES

Alencar, A., Nepstad, D., McGrath, D., Moutinho, P., Pacheco, P., Diaz, M. D. C. V. et al. (2004). *Desmatamento na Amazônia: Indo além da emergência crônica*. Belém: Instituto de Pesquisa Ambiental da Amazônia (Ipam).

Alves, D. S., Morton, D. C., Batistella, M., Roberts, D. A., & Souza Jr., C. (2009). The changing rates and patterns of deforestation and land use in Brazilian Amazonia. In M. Keller, M. Bustamante, J. Gash, & P. S. Dias (Eds.), *Amazonia and global change* (pp. 11-24). Washington DC: American Geophysical Union.

Alves, D. S., Pereira, J. L. G., De Sousa, C. L., Soares, J. V., & Yamaguchi, F. (1999). Characterizing landscape changes in central Rondonia using Landsat TM imagery. *International Journal of Remote Sensing, 20,* 2877-2882.

Asner, G. P., Keller, M., Lentini, M., Merry, F., & Souza Jr., C. (2009). Selective logging and its relation to deforestation. In M. Keller, M. Bustamante, J. Gash & P. S. Dias (Eds.), *Amazonia and global change* (pp. 25-42). Washington DC: American Geophysical Union.

Ballester, M. V. R., Victoria, D. D., Krusche, A. V., Coburn, R., Victoria, R. L., Richey, J. E. et al. (2003). A remote sensing/GIS-based physical template to understand the biogeochemistry of the Ji-Parana river basin (Western Amazonia). *Remote Sensing of Environment, 87,* 429-445.

Batistella, M., & Moran, E. F. (2005). Dimensões humanas do uso e cobertura das terras na Amazônia: uma contribuição do LBA. *Acta Amazonica, 35,* 239-247.

Borma, L. S., da Rocha, H. R., Cabral, O. M., von Randow, C., Collicchio, E., Kurzatkowski, D. et al. (2009). Atmosphere and hydrological controls of the evapotranspiration over a floodplain forest in the Bananal Island region, Amazonia. *Journal of Geophysical Research, 114,* Article ID: G01003.

Bruner, A. G., Gullison, R. E., Rice, R. E., & da Fonseca, G. A. B. (2001). Effectiveness of parks in protecting tropical biodiversity. *Science, 291,* 125-128.

Cardille, J. A., & Foley, J. A. (2003). Agricultural land-use change in Brazilian Amazonia between 1980 and 1995: Evidence from integrated satellite and census data. *Remote Sensing of Environment, 87,* 551-562.

Carvalho, F. M. V., De Marco Júnior, P., & Ferreira, L. G. (2009). The Cerrado into-pieces: Habitat fragmentation as a function of landscape use in the savannas of central Brazil. *Biological Conservation, 142,* 1392-1403.

Chomitz, K. M., & Gray, D. A. (1996). Roads, land use, and deforestation: A spatial model applied to belize. *World Bank Economic Review, 10,* 487-512.

Collicchio, E. (2006). *Organização estadual de pesquisa agropecuária: Um instrumento de apoio ao desenvolvimento rural sustentável do Tocantins.* Palmas: Provisão.

Collicchio, E. (2008). *Zoneamento edafoclimático e ambiental para a cana-de-açúcar e as implicações das mudanças climáticas no estado do Tocantins.* Ph.D. Thesis, Piracicaba: University of São Paulo.

Coutinho, A. C. (2005). *Dinâmica das queimadas no estado do Mato Grosso e suas relações com as atividades antrópicas e a economia local.* Ph.D. Thesis, São Paulo: University Of São Paulo.

Diegues, A. C. S. (2002). *Povos e águas: Inventário de áreas úmidas* (2 ed.). São Paulo: Núcleo de apoio à pesquisa sobre populações humanas e áreas úmidas, USP.

EMBRAPA (2008). *Informações técnicas para a cultura do arroz irrigado no Estado do Tocantins: Safra 2008/2009.* Santo Antônio de Goiás: EMBRAPA Arroz e Feijão.

Fearnside, P. M. (2006). Desmatamento na Amazônia: Dinâmica, impactos e controle. *Acta Amazonica, 36,* 395-400.

Ferreira, L. V. (2001). Identificação de áreas prioritárias para a conservação da biodiversidade por meio da representatividade das unidades de conservação e tipos de vegetação nas ecorregiões da Amazônia brasileira. In J. P. R. Capobianco (Ed.), *Biodiversidade na Amazônia brasileira: Avaliação e ações prioritárias para a conservação, uso sustentável e repartição de benefícios* (pp. 268-286). São Paulo: Instituto Socioambiental.

Ferreira, L. V., Venticinque, E., & Almeida, S. (2005). O desmatamento na Amazônia e a importância das áreas protegidas. *Estudos Avançados, 19,* 157-166.

Ferreira, N. C., Ferreira, L. G., & Miziara, F. (2007). Deforestation hotspots in the Brazilian Amazon: Evidence and causes as assessed from remote sensing and census data. *Earth Interactions, 11,* 1-16.

Geist, H. J., & Lambin, E. F. (2001). *What drives tropical deforestation? A meta-analysis of proximate and underlying causes of deforestation based on subnational case study evidence.* Belgium: University of Louvain.

Gils, H. A. M. J. V., & Ugon, A. V. L. A. (2006). What drives conversion of tropical forest in Carrasco province, Bolivia? *Ambio, 35,* 81-85.

Joppa, L. N., Loarie, S. R., & Pimm, S. L. (2008). On the protection of "protected areas". *Proceedings of the National Academy of Sciences of the United States of America, 105,* 6673-6678.

Kirby, K. R., Laurance, W. F., Albernaz, A. K., Schroth, G., Fearnside, P. M., Bergen, S. et al. (2006). The future of deforestation in the Brazilian Amazon. *Futures, 38,* 432-453.

Latrubesse, E. M., Amsler, M. L., de Morais, R. P., & Aquino, S. (2009). The geomorphologic response of a large pristine alluvial river to tremendous deforestation in the South American tropics: The case of the Araguaia River. *Geomorphology, 113,* 239-252.

Latrubesse, E. M., & Stevaux, J. C. (2006). Características físico-bióticas e problemas ambientais associados à planície aluvial do rio Araguaia, Brasil central. *Revista UnG—Geociências, 5,* 65-73.

Laurance, W. F., Albernaz, A. K. M., Fearnside, P. M., Vasconcelos, H. L., & Ferreira, L. V. (2004). Deforestation in Amazonia. *Science, 304,* 1109b-1111b.

Laurance, W. F., Albernaz, A. K. M., Schroth, G., Fearnside, P. M., Bergen, S., Venticinque, E. M. et al. (2002). Predictors of deforestation in the Brazilian Amazon. *Journal of Biogeography, 29,* 737-748.

Laurance, W. F., Cochrane, M. A., Bergen, S., Fearnside, P. M., Delamonica, P., Barber, C. et al. (2001). The future of the Brazilian Amazon. *Science, 291,* 438-439.

Ludeke, A. K., Maggio, R. C., & Reid, L. M. (1990). An analysis of anthropogenic deforestation using logistic regression and GIS. *Journal of Environmental Management, 31,* 247-259.

Machado, R. B., Ramos Neto, M. B., Pereira, P. G. P., Caldas, E. F., Gonçalves, D. A., Santos, N. S. et al. (2004). Estimativas de perda da área do Cerrado brasileiro. Conservação Internacional.

Mahar, D. J. (1989). *Government policies and deforestation in Brazil's Amazon region.* Washington: World Bank Publications.

Margulis, S. (2004). *Causas do desmatamento da Amazonia Brasileira.* Brasília: Banco Mundial.

Mertens, B., & Lambin, E. F. (2000). Land-cover-change trajectories in southern Cameroon. *Annals of the Association of American Geographers, 90,* 467-494.

Nepstad, D., Carvalho, G., Cristina Barros, A., Alencar, A., Paulo Capobianco, J., Bishop, J. et al. (2001). Road paving, fire regime feedbacks, and the future of Amazon forests. *Forest Ecology and Management, 154,* 395-407.

Oliveira, P. J. C., Asner, G. P., Knapp, D. E., Almeyda, A., Galvan-Gildemeister, R., Keene, S. et al. (2007). Land-use allocation protects the Peruvian Amazon. *Science, 317,* 1233-1236.

Pfaff, A., Robalino, J., Sanchez-Azofeifa, G. A., Andam, K. S., & Ferraro, P. J. (2009). Park location affects forest protection: Land characteristics cause differences in park impacts across Costa Rica. *B.E. Journal of Economic Analysis & Policy, 9,* 26.

Primack, R. B. (2002). *Essentials of conservation biology* (3.ed.). Sunderland: Sinauer Associates.

SEPLAN (2008). *Atlas do Tocantins: Subsídios à gestão e planejamento territorial.* Palmas: SEPLAN.

Serneels, S., & Lambin, E. F. (2001). Proximate causes of land-use change in Narok District, Kenya: A spatial statistical model. *Agriculture, Ecosystems & Environment, 85,* 65-81.

Soares, B., Alencar, A., Nepstad, D., Cerqueira, G., Diaz, M. D. V., Rivero, S. et al. (2004). Simulating the response of land-cover changes to road paving and governance along a major Amazon highway: The Santarem-Cuiaba corridor. *Global Change Biology, 10,* 745-764.

Soares-Filho, B. S., Nepstad, D. C., Curran, L., Cerqueira, G. C.,

Garcia, R. A., Ramos, C. A. et al. (2005). Cenários de desmatamento para a Amazônia. *Estudos Avançados, 19,* 137-152.

Ugon, A. V. L. A. (2004). *A spatial logistic model for Tropical forest conversion: A case study os Carrasco province (1986-2002), Bolivia.* Master Thesis, Enschede: International Institute for Geo-Information Science and Earth Observation (ITC).

Walker, R., DeFries, R., Vera-Diaz, M. C., Shimabukuro, Y., & Venturieri, A. (2009). The expansion of intensive agriculture and ranching in Brazilian Amazonia. In M. Keller, M. Bustamante, J. Gash, & P. S. Dias (Eds.), *Amazonia and Global Change* (pp. 61-82). Washington DC: American Geophysical Union.

Yoshikawa, S., & Sanga-Ngoie, K. (2011). Deforestation dynamics in Mato Grosso in the southern Brazilian Amazon using GIS and NOAA/AVHRR data. *International Journal of Remote Sensing, 32,* 523-544.

Yu, X., & Ng, C. (2006). An integrated evaluation of landscape change using remote sensing and landscape metrics: A case study of Panyu, Guangzhou. *International Journal of Remote Sensing, 27,* 1075-1092.

13

A Wood Preservative Based on Commercial Silica Nanodispersions and Boric Acid against Fungal Decay through Laboratory and Field Tests

Sabrina Palanti[*], Elisabetta Feci

CNR IVALSA, Istituto per la Valorizzazione del Legno e delle Specie Arboree, Sesto Fiorentino, Italy

The paper is based on the development of a wood preservative without metal salts to be used in use classes 3 and 4 (EN 335), eco-friendly and harmless to humans and animals. Boric acid was used as a biocide, due to its effectiveness against fungi and insects. It is also known to be easily leached from wood exposed to weather action. Colloidal silica was therefore added in the formulations to guarantee the fixation of boric acid to wood. The different formulations were tested for the protective efficacy against decay fungi through laboratory tests (EN 113) and field trials (EN 252). The results were promising, especially those concerning boron fixation and efficacy against decay fungi through laboratory tests, where some formulations and retentions gave a durability class 1 (very durable) according to EN 350-1. The fourth evaluation, after 50 months of field trials showed only a slight difference between the treated samples and controls.

Keywords: Basidiomycetes; Boric Acid; Colloidal Silica; Wood Durability; Field Test

Introduction

Protecting wood from biotic decay is a necessity and also an obligation for producers of wooden artefacts. The wood preservatives have to ensure the resistance of wood to biotic agents for a long time in service conditions, so it is important that active ingredients are permanently fixed to the walls of wood cells.

Currently the most used products for use classes 3 and 4 (EN 335) are formulations based on copper salts and other co-formulates as azoles, organic compounds and boron (Forest Products Laboratory, 2010). The disadvantage of these compounds is due to utilization of metal compounds and to the facility of leaching them away when utilized in outdoor conditions.

Boric acid is a substance effective against fungi and insects xylophages and is also non-toxic to humans and the environment if used under the limit indicated in Biocides Directive 98/8/EC. Nevertheless, this substance alone cannot be applied to wood exposed to the atmospheric agents because it is easily washed away into the ground. The recent classification of boric acid and borates as toxic to the human reproduction, category 1B (EU Regulation 790/2009) means that they fall within the scope of the measures of risk reduction provided by the REACH regulation (licensing procedures and restrictions). The justification for the use of boric acid is linked to the fact that the repro-toxic effect occurs at a threshold concentration of 5.5%. Below this threshold this compound is considered safe.

There are some compounds that form complexes with boric

acid and thereby lock it when introduced into the wood. Their employment in the industry is hampered by some technical difficulties during the preparation.

In order to develop an environmentally friendly preservative, this research work has had as its aim the development of wood preservative formulations based on boric acid and colloidal silica nanodispersions with different weight percentagesin SiO_2 and size particles.

Silicic acid in combination with boric acid has been shown to be effective against wood decay fungi and termites, when applied to wood with vacuum-pressure impregnation cycle (Yamaguchi, 2001, 2003, 2005).

The theorized mechanism for the attachment of boron is due to the atomic structure of the colloidal silica performed by the compound SiO_4. This compound has tetrahedral structure with the silicon atom in the centre. Each molecule of silica is combined with others so that each silicon atom has available three oxygen atoms of the tetrahedron. It is assumed that the boron atoms, which have a structure similar to Si, may replace one atom into the tetrahedron (Ueda, 1993). Once heated, the mixture, boric acid and colloidal silica, takes a glassy appearance, blocking the vessels of wood, with flame-retardant property. In this case, the attachment of boron is due to the fact that boron atoms are replaced with those of silicon in the colloidal solution, once introduced into the wood by impregnation with vacuum-pressure cycle and hence is permanently blocked by the gelling of colloidal silica after heating wood.

In this research boric acid was added as the active biocidea-

---

[*]Corresponding author.

gainst fungi.

The evaluations of the experimental formulations were therefore concerned with:

- Antifungal efficacy through laboratory tests;
- Fixation of boric acid into the wood;
- Antifungal efficacy through in field tests.

Material and Methods

Treatments Based on Colloidal Silica and Boric Acid

The treatments were based on boric acid and nanodispersions of colloidal silica.

Some commercial silica dispersions (Eka Chemicals AB and Nissan Chemical Industrex) of different weight percentage and different silica nanoparticles size were utilized.

EN 113 wood blocks and EN 252 stakes were treated with a formulation based on 20.7% of a colloidal dispersion of silica particles in slightly alkaline water, Bindzil® CC 30 (Eka Chemicals AB) and 2.3% boric acid, technical grade.

Previous experiments to evaluate the fixation of boric acid into the wood were performed according to EN 84 on wood blocks treated with a formulation based on 20.5% colloidal silica (Snowtex C, Nissan Chemical) and 2.59% boric acid.

The treatment emulsions were impregnated into the wood according to the following cycle: vacuum 0.7 kPa for 30 minutes, pressure 200 kPa for 5 minutes, 400 kPa for 5 minutes, 600 kPa for 5 minutes and 700 kPa for 30 minutes.

Retentions were calculated as amount of dry residual expressed in kilograms per cubic meter of wood.

Wood Species

The wood samples were derived from Scots pine (*Pinus sylvestries* L.) sapwood.

Laboratory Decay Test (EN 113: 1996)

Laboratory decay test were carried out in accordance with European standard EN 113, 1996.

Treated wood samples, dimensions: $(50 \pm 0.5) \times (25 \pm 0.5) \times (15 \pm 0.5)$ mm^3, were exposed to fungal attack for 16 weeks in a conditioning room ($22 \pm 2°C$, $70 \pm 5\%$ RH). The test requires the inclusion of three certified fungi, one obligatory—the brown rot fungus *Coniophora puteana* (Schumacher ex Fries) Karsten (strain BAM Ebw.15) and the two additional brown rot fungi *Poria placenta* (Fries) Cooke sensu J. Eriksson (strain FPRL280) and *Gloeophyllum trabeum* (Persoon ex Fries) Murrill (strain BAM ebw.109). For each fungus, ten treated samples were tested. A treated sample was placed with an untreated reference one in a Kolle flask, containing one of the above described brown rot fungus grown on 20 ml of 4% malt and 2.5% agar medium. A modification of the EN 113 (1996), that consists in the division into two parts of the mycelium inside the Kolle flask before the exposure to wood samples (**Figure 1**), so that the fungal strain virulence (tested on control samples) was prevented from inhibition by the preservative washed away from the treated samples. In fact, the inhibition of the fungus by boric acid washed away from treated sample against control specimens, as resulted from preliminary tests, did not allow for demonstrating the effectiveness of the product, according to the criteria of the standard. This effect due to the leaching of boron from non-leached treated samples has been observed in preliminary

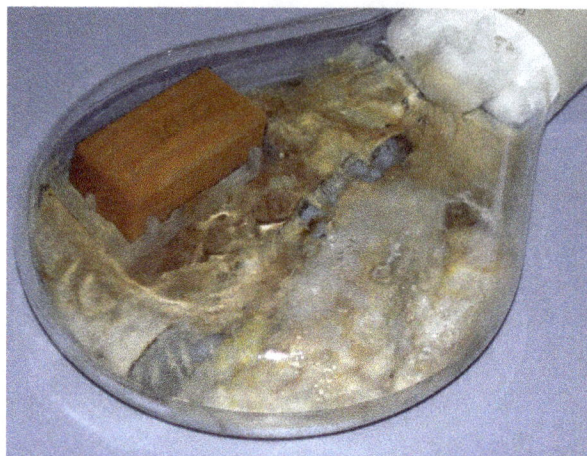

Figure 1.
Mycelium within the Kolle flask, divided into portions.

non-published experiments of the authors.

Furthermore, six treated wood blocks (called in EN 113 (1996) check blocks) were put in contact with culture medium without fungal strains for the determination of the correction coefficient, which was used to calculate the possible mass loss due to factors different from fungal decay. Another set of six untreated wood specimens for each tested fungus was used to check for virulence of fungus.

The resistance against fungi was evaluated through the measurement of the percent mass loss of wood, which was calculated for each individual block as the difference between the dry mass before the impregnation process and after the fungal exposure, corrected by correction coefficient. A minimum of 20% mass loss on control (untreated) samples was required for the test to be valid.

Leaching Tests (EN 84:1997)

Treated wood samples, dimensions $(5 \pm 0.1) \times (15 \pm 0.1) \times (30 \pm 0.5)$ mm^3, were subjected to leaching according to EN 84, 1997. This standard describes an accelerated ageing test of specimens treated with a preservative for simulating the service conditions, in particular exposure to rainfall. The cycle consisted of an initial 4 kPa vacuum-atmospheric pressure cycle with distilled water. Then, wood specimens were maintained in water (ratio of water to wood 5:1) for 14 days with 9 water changes, and then conditioned to constant mass.

The leaching waters were subjected to analysis by ICP (Inductive Coupled Plasma) for the determination of boron.

The percentage of residual boric acid in the wood was calculated by the difference with the amount obtained from the analysis on initial formulation, where its content is 100%.

Field Test (EN 252: 1989)

In ground test was performed in accordance with EN 252, 1989. This standard describes a method for the determination of efficacy of preservative applied by impregnation plant in ground condition. The method is also utilized for the determination of natural durability of wood in ground condition (EN 350, 1994).

The test sites were respectively situated in Cesa, Arezzo (lat 43.3°N; long 11.8°E), and Follonica, Grosseto (lat 42.9°N; long

10.8°E), Italy.

Duration of the test is a minimum 5-years period or until the stakes fail. The wood samples were placed in situ in October 2007 and in this paper the fourth evaluation was reported, after respectively 54 and 50 months for Cesa and Follonica.

Stakes, dimensions $(500 \pm 1) \times (50 \pm 3) \times (25 \pm 0.3)$ mm^3, were cut and conditioned before being impregnated and conditioned to constant mass.

The stakes were placed vertically in the soil leaving half of length exposed. Untreated stakes were also buried to half their length in the soil test site.

Ten replicates for treatment and reference controls were utilized.

The fungal decay was evaluated in the area in ground contact. Annual inspections were carried out by giving a light blow to the upper part of each stake followed by removal from the ground. Surface examination was performed with an awl. The fungal decay was evaluated on the basis of the depth of fungal softening and extension area according to a specific rating system: 0 sound, 1 slight attack, 2 moderate attack, 3 severe attack and 4 failure (stake breaks in the ground after blowing). The evaluation was performed in accordance with guideline for EN 252 of Nordic Wood Preservation Council (Borsholt & Henriksen, 1992). After evaluation, each stake was re-installed in the original position.

As preservative reference, ten stakes treated with Impralith KDS (Rütger Organics GmbH), with a retention use class 4 (EN 599-1, 2009), were used. As control reference ten untreated were utilized.

Conferred durability classes were calculated according to EN 350-1, 1994.

Results

Retention

The impregnation of EN 113 wood blocks gave the followings results: 201.37 ± 11.54 kg/m^3 (N = 36). Impregnation of wood mini-blocks for the leaching test gave as result 194.48 ± 4.03 kg/m^3 (N = 10).

The impregnation of EN 252 wood stakesgave the followings results: 55.6 ± 15.50 kg/m^3 (N = 20). The stakes impregnated with reference preservative Impralith KDS had an average retention of 8.65 ± 1.26 kg/m^3 (N = 25).

Decay Tests

Virulence of fungal strains confirmed the validity of the test (mass loss of untreated samples more than 20%).

Results of antifungal efficacy were expressed as average mass loss and standard deviation of treated samples with the formulation tested (**Table 1**). All reference wood blocks lost more than 20% mass, the minimum value required for the validity of the test according to standard EN 113.

The mean mass loss due to the check block was $16.11\% \pm 3.23\%$, this results was due to the leaching of boron from wood blocks into the medium.

Leaching Test

Results of leaching are reported in **Figure 2**, where the percentages of residual boric acid for the formulation impregnated into the wood are graphed. Number 0 corresponds to the

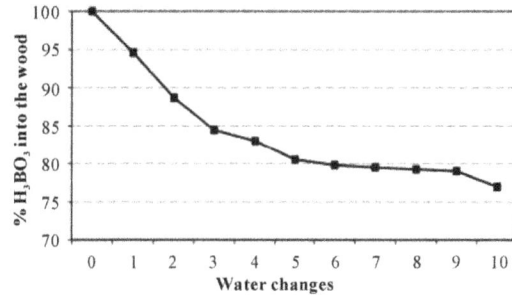

Figure 2.
Residual percentage of boric acid into the wood with respect to the initial concentration. The starting point (Water change number 0) represents the initial formulation and the corresponding boric acid content is 100%.

Table 1.
EN 113, decay test results; sd = standard deviation, N = number of samples.

Fungus	Wood blocks treated % mass loss average (sd) N = 10	Wood blocks references average % mass loss N = 10
C. puteana	6.87 (1.42)	59.16 (3.92)
P. placenta	6.65 (0.66)	48.25 (5.16)
G. trabeum	6.27 (1.02)	30.87 (5.45)

amount determined by ICP analysis in the impregnation solution, which was assumed to be entirely and homogenously penetrated into wood. This assumption can be considered realistic because at the concentration of 2.3% the boric acid is well solved into the water of silica dispersion.

The ICP boron analysis has shown a decreasing presence of the active ingredient in consecutively obtained leaching waters. The loss of boric acid was 23.0%. Compared to the amount present in the initial solution, boric acid decreased from 2.59% to 1.99%.

Field Test (EN 252)

Annual evaluations on stakes from the two experimental fields are shown in **Table 2**.

For each evaluation, the average decay values of the stakes treated with colloidal silica-boric acid formulation, the stakes treated with the reference wood preservative and the untreated stakes are reported.

Discussion

The different average retentions obtained for each test samples group were probably due to different size of wood specimens. It has been remarked that the different dimensions of specimens utilized in different standardized tests could have been influenced the final retention of formulations tested: the smaller the wood samples, the higher were the retentions, (e.g. EN 113, EN 84); the lower performance was obtained with the EN 252 stakes.

The expression of EN 113 decay test results as conferred durability of wood according to EN 350-1 (1994) led to durability class 1—very durable, with an expected service life exceeding 25 years (Eaton and Hale, 1993).

In **Table 1**, the masses losses indicate that the treatment gave

Table 2.
EN 252 Decay results.

Place: Cesa	15th month by installation	29th month by installation	40th month by installation	54th month by installation
	Average decay			
CC30 colloidal silica and boric acid	0.33	1.33	2.67	3.27
Impralith KDS	0.15	0.69	1.46	1.46
Untreated control	0.60	2.40	2.70	3.00
Place: Follonica	13th month by installation	24th month by installation	36th month by installation	50th month by installation
CC30 colloidal silica and boric acid	0.30	2.30	3.60	4.00
Impralith KDS	0	0	0.30	0.70
Untreated control	2.5	4	-	3.8 (the 2nd series)

protection against fungi with respect to untreated wood. Nevertheless, this preservative was not efficacy in accordance with the criteria of validity of EN 113 (1996), where is stated that a wood preservative is efficacy when its mass loss is below than 3%.

The amount of boric acid leached was probably an exceeding part that did not participate in the complex formation and consequently did not remain fixed in the wood. If it is assumed that after the impregnation the concentration of boric acid into the wood was the same as in the initial formulation, the fixation of boric acid was more than 75%. These results support the theory on the interaction between silica and boric acid and demonstrate the ability of silica to block the active ingredient in the wood.

Evaluation of EN 252 in two sites gave very similar results with the retention tested 55.6 kg/m^3.

In Cesa, an agricultural soil near Arezzo, during the fourth evaluation, the colloidal silica-boric acid treated stakes had an higher decay (3.27) than the untreated control stakes (3.00). During the first two evaluations there was a gap between the two sets of stakes, infact the decay grade were respectively 1.33 and 2.40 for silica-boric acid and untreated stakes, suggesting the treatment was better than untreated wood. During the third year this gap was plugged, the two series scoring the same grade of decay (2.67 and 2.70).

On the contrary, in Follonica, an agricultural soil close to the coast, the higher decay grade was reached by control stakes after 50 months from beginning and another series of control stakes has been installed. At the fourth evaluation (after 50 months) the decay grade of silica-boric acid was 4.00. In this case also there was not a substantial difference between treated and untreated specimens, suggesting the treatment is no better than untreated controls.

In both sites the copper salt wood preservative gave better results, reaching respectively 1.46 in Cesa and 0.70 in Follonica.

The fungi, principally white rots, found in Follonica, resulted more aggressive versus untreated and colloidal silica-boric acidtreated stakes with respect to those found in Cesa.

Conclusion

The formulations colloidal silica-boric acid tested in this work have been very successful at retention 201 kg/m^3 and 194 kg/m^3 respectively in terms of resistance to fungal decay through laboratory testand with regard to fixation into the timber.

The results obtained in EN 252 field test with the lower retention, 55 kg/m^3, support the idea that this retention value could be used in use class 3, not in contact with the ground, because this service condition is considered to severe, reaching a very high decay grade only after 50 - 54 months.

Acknowledgements

The authors thank the Regional Agency for Agriculture Developing and Innovation (ARSIA), TLF srl, Arezzo that supported this research and Mrs. Anna Maria Torniai from CNR Trees and Timber Institute who helped with fungi cultures.

REFERENCES

Borksholt, E., & Henriksen, H. K. (1992). Guideline for EN 252: Field test method for determining the relative protective effectiveness of wood preservatives in ground contact NWPC. Information No. 23/90 ISSN 0358-707X.

(1998). Directive 98/8/EC of the European Parliament and of the Council of 16 February 1998 concerning the placing of biocidal products on the market.

EN 252 (1989). Field test method for determining the relative protective effectiveness of a wood preservative in ground contact.

EN 84 (1997). Wood preservatives—Accelerated ageing of treated wood prior to biological testing—Leaching procedure.

EN 113 (1996). Wood preservatives—Test method for determining the protective effectiveness against wood destroying basidiomycetes—Determination of the toxic values.

EN 335 (2006). Durability of wood and wood-based products—Definition of use classes—Part 1: General.

EN 350:1 (1994). Durability of wood and-based products. Natural durability of solid wood. Guide to the principles of testing and classification of thenatural durability of wood.

EN 599-1 (2009). Durability of wood and wood-based products—Efficacy of preventive wood preservatives as determined by biological tests—Part 1: Specification according to use class.

European Commission (2009). Commission Regulation (EC) No 790/2009 of 10 August 2009 amending, for the purposes of its adaptation to technical and scientific progress, Regulation (EC) No 1272/2008 of the European Parliament and of the Council on classification, labelling and packaging of substances and mixtures.

Eaton, R. A., & Hale, M. D. (1993). Wood: Decay, pests, and protection. London: Chapman and Hall.

Forest Product Laboratory (2010). Wood handbook—Wood as an engineering material. General Technical Report FPL-GTR-190. Madison, WI: US Department of Agriculture, Forest Service, Forest Products Laboratory, 508 p.

Yamaguchi, H. (2001). Silicic acid-boric acid complexes as wood pre-

A Wood Preservative Based on Commercial Silica Nanodispersions and Boric Acid against Fungal Decay through Laboratory and Field Tests

87

servatives, IRG/WP01/30273.

Yamaguchi, H. (2003). Silicic acid: Boric acid complexes as wood preservatives. *Wood Science and Technology, 37,* 287-297.

Yamaguchi, H. (2005). Silicic acid-boric acid complexes as ecologi-cally friendly wood preservatives. *Forest Products Journal, 55,* 88-92.

Ueda, S. (1993). *Jkken kagaku kouza (in Japonese, text of experimen-thal chemistry) (16)* (4th ed.). Tokyo: Maruzen Co. Ltd., 588.

14

Evaluation of Broadleaf Tree Diversity at the Basin Scale—In Case of Artificial *Chamaecyparis obtusa* Forests

Sayumi Kosaka, Yozo Yamada

Graduate School of Bio-Agricultural Sciences, Nagoya University, Nagoya, Japan

In recent years, the various functions required of forests, especially the conservation of biodiversity, have been attracting increasing attention in Japan and worldwide. In Japan, 67% of national land is covered by forest, 41% of which is artificial forest (i.e., plantations). Therefore, forest biodiversity conservation efforts should also target artificial forests. In this paper, we seek to promote sustainable forest management that considers biodiversity conservation by examining indices that can be used by forest managers to evaluate the diversity of broadleaf trees. The result was that evaluation of broadleaf tree diversity in artificial forests at a basin scale was possible by combining several types of indicators.

Keywords: Artificial Forest; Forestry Management; Basin Scale; Species Diversity Index; Land Use Diversity Index

Introduction

In recent years, various functions have been required of forest ecosystems; these include not only the production of wood, but also the conservation of biodiversity, landslide prevention, cultivation of water sources, overall ecosystem health, and prevention of global warming. In particular, the conservation of biodiversity is considered a necessary function to promote sustainable land use and to conserve biodiversity of surrounding land-use types at a basin scale.

In Japan, approximately 67% of national land is covered by forest, approximately 41% of which is artificial forest (The Forest Agency, 2012). Therefore, in addition to natural ecosystems, conserving biodiversity within artificial forests is also crucial. To this end, management efforts within plantations must include thinning of the planted conifers and the maintenance of naturally regenerated broadleaf trees and understory vegetation. Forest managers must have a deep understanding of the unique species diversity of the broadleaf trees within their forests so as to promote forest management that considers tree conservation. However, clear guidelines or management approaches have not yet been established. Furthermore, appropriate studies are rare, both in Japan and worldwide, and tend to examine forests only at a small scale.

In this paper, we seek to promote sustainable forest management that considers biodiversity conservation by examining indices that can be used by forest managers to evaluate the diversity of broadleaf trees. We used several indices for evaluation, and we focused on the basin scale, which corresponds to the whole-forest level.

Survey Location and Methods

Survey Location

The survey was conducted at Hayami Forest located in Kihokucyou Kita, Mie Prefecture, Japan. Plots were established in Ootaga Forest (93 ha; 10 - 380 m above sea level; angle of inclination, 3 - 45 degrees) within Hayami Forest. Average annual temperature is 16.1˚C, and annual precipitation is 4200 mm (Meteorological Agency of Japan, 2012). Forest compartments vary in age and form a mosaic landscape.

Hayami Forest is currently managed using an intensive nurturing system aimed at the production of good wood over long-term cutting periods. *Chamaecyparis obtusa*, which is well suited to regional conditions, is the main production wood. Hayami Forest was the first forest to obtain forest certification from the Forest Stewardship Council A.C. (FSC) of Japan in February 2002.

Methods

Our research was conducted within five compartments that varied in age. We established 10 × 10 m plots within areas representative of the compartment, i.e., those lacking gaps, edge effects, and mountain streams. We examined all broadleaf and planted trees that were taller than 1 m. Broadleaf trees were chosen because they are affected by management efforts and they can be managed and investigated easily. For all broadleaf trees, we determined the number of species and the population size. We also measured tree height and diameter at a height of 50 cm from the ground, as we were unable to measure many individuals at heights of 1 - 1.3 m. For planted trees, we also measured height and diameter and counted the number of planted trees in 20 × 20 m areas to calculate stand density.

Analysis at the Compartment Scale

We analyzed the current conditions of compartments using stand density, which is an important measure of forest maintenance. For the analysis of broadleaf tree diversity within compartments, we used the number of species, population size, proportion of basal area (BA), and several species diversity indices. A 50-cm aboveground cross section was used to calculate BA as follows: BA = sum of 50 cm aboveground cross sections of a specific layer in the plot/sum of 50 cm aboveground cross sections of all trees in a plot. The proportion of BA of the natural forest was not calculated because no conifers exist within the natural areas of Hayami Forest, and thus the calculation result would be 100%. We used the Shannon-Wiener index (1), the inverse Simpson index (4), and the inverse logarithm Simpson index (5) as indices of species diversity (Yoshiaki & Kazunori, 2002):

Shannon-Wiener index (H')

$$H' = -\sum pi \log_2 pi \qquad (1)$$

where $pi = \dfrac{Ni}{N}$ is the relative frequency of i; Ni is the population size of I; N is population size.

Index of Simpson (D)

$$D = \sum pi^2 \qquad (2)$$

An unbiased estimator (D')

$$D' = \sum \frac{Ni(Ni-1)}{N(N-1)} \qquad (3)$$

An inverse Simpson index (L)

$$L = 1/D' \qquad (4)$$

An inverse logarithm Simpson index (L')

$$L' = \log \frac{1}{D'} \qquad (5)$$

Analysis at the Basin Scale

The land use diversity index (LUDI) was used for the analysis of forests at the basin scale. The LUDI is commonly used in landscape studies to improve the measurement of diversity at the landscape scale. This index enables the evaluation of all types of landscapes using mathematically weighted functions based on landscape structure and composition. The formula for the LUDI is as follows (6):

$$M' = \frac{2m\sum_{j=1}^{m}\left(j \cdot Wcj \cdot \left[kj - (kj-1)\dfrac{2\sqrt{\pi aj}}{pj} \right] \cdot aj \right)}{A(1+m)} \qquad (6)$$

where M' is the weighted land-use diversity index; m is the number of patch types present in a landscape unit

j is $1, \cdots, m$ is patch type;

Wcj is the compositional weight of patch type j;

kj is the upper limit of the structural weight of patch type j;

pj is the sum of the perimeter of patch type j;

aj is the sum of the area of patch type j;

A is the sum of patch areas in a landscape unit.

Here, we divided all compartments in Ootaga Forest into five

groups based on stand density. Thus, a value of 5 was used for m, and Pj, aj, and A were calculated using a geographic information system (GIS). In addition, evaluation results for compartments were used for Wcj, after correcting each value so that the maximum was 1. To determine the maximum value of the LUDI, we considered the maximum as M1max and compared these results with M'.

Results and Inquiry

Analysis of Compartments

Stand density exhibited a decreasing trend as the age of forest compartments within Hayami Forest increased (**Figure 1**).

Stand density within Hayami Forest was lower than forestry association of Kashimo and prefectural forest of Aichi, indicating that the management of Hayami Forest maintains low stand density in accordance with its forest management plan.

Number of Species

The total number of species peaked at a forest age of 49. After this age, species number declined with forest age (**Figure 2**). However, the total number of species did not largely differ among forest ages. On the other hand, when considering layer structure, the number of trees located lower in the canopy decreased with age, whereas the number of trees at intermediate levels increased with age. After age 67, an upper layer began to form, indicating that layer composition becomes more complex as the forest matures. In particular, layer composition within forests at age 99 approaches that of natural forests.

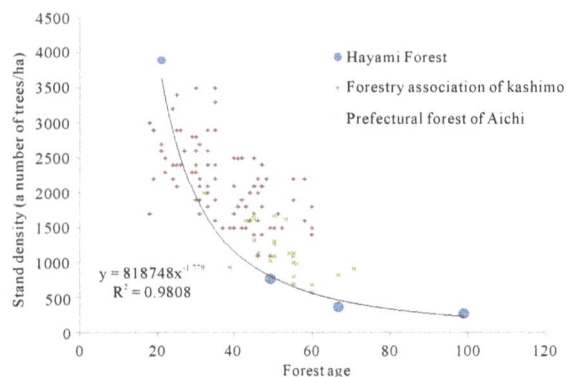

Figure 1.
Forest age and stand density.

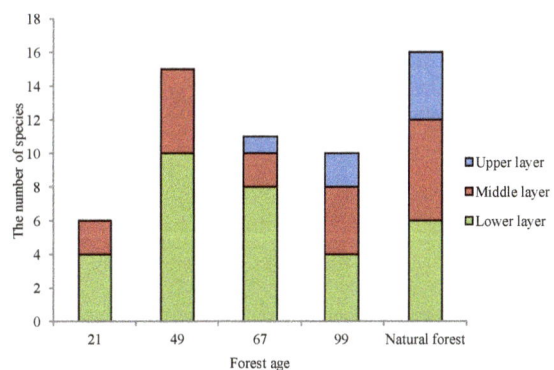

Figure 2.
Forest age and number of species.

The number of species is an effective index for forest management, as species richness is a good indicator of current forest conditions. However, the evenness of species must also be considered. Furthermore, the number of species tends to increase as plot size increases.

Population Size

Although population size exhibited a pattern similar to the number of species, the overall trend was different (**Figure 3**).

Compared to the number of species, the proportion of lower-layer trees increased within compartments that harbor large population sizes. Therefore, if the diversity of broadleaf trees is evaluated using population size, lower-layer trees are relatively dominant to middle and upper-layer trees.

The population size offers the advantage of assessing evenness between species. However, when several species are co-dominant, the index will overestimate the importance of those species.

Proportion of BA

Planted trees accounted for **88% - 99%** of the proportion of BA at all ages of forests, as the study was located within an artificial forest (**Figure 4**).

Broadleaf trees clearly continue to grow as the forest ages, as the dominance of broadleaf trees and the proportion of middle- and upper-layer trees increases with forest maturation. Because the proportion of BA is calculated from the sum of BA, lower-layer trees that exhibit low BA dominance will decrease even if the population size is large. In contrast, middle- and

upper-layer trees that exhibit high BA dominance will be abundant even if the population size is small. Considering this, the proportion of BA serves as an index that expresses the dominance of each layer and thus differs from the number of species. In addition, the equality of species can be accessed from this index in a manner different from population size. However, high dominance does not always indicate high diversity.

Species Diversity Index

All of the species diversity indices exhibited maximum values at a compartment age of 21, in which both the number of species and population size were lowest (**Figure 5**).

On the other hand, compartments of age 67, in which population size was largest, exhibited the lowest species diversity. These results suggest that evaluations using species diversity indices may be erroneous. In Hayami Forest, each compartment age varied little in species number, although population size varied greatly. Because species diversity indices are calculated using combinatorics, if species numbers vary little, plots with large populations will exhibit high values of species diversity.

Species diversity indices offer the advantage that results are easy to understand and to compare, as the result is just one number. However, the validity of the results must be assessed using the number of species or population size. Therefore, using species diversity indices alone is not a recommended approach.

Evaluation Using LUDI

The results of the four indices used to evaluate the diversity of broadleaf trees and the age of forests were used to evaluate the current distribution of compartments for Wcj. As a result, maximum M' was the number of species (**Table 1**) because the

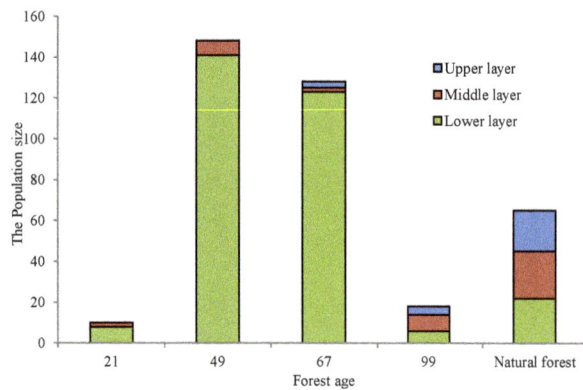

Figure 3.
Forest age and population size.

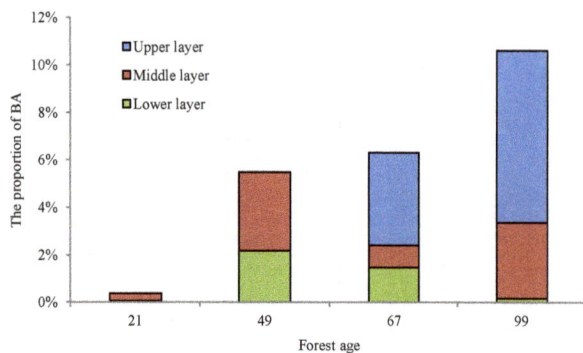

Figure 4.
Forest age and proportion of BA.

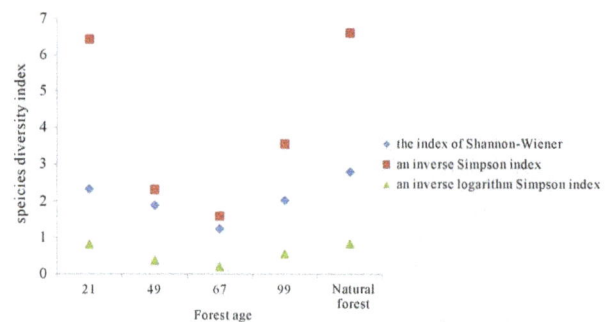

Figure 5.
Forest age and species diversity indices.

Table 1.
Results of LUDI.

Wcj	M'	$M'/M1max$
Forest age	3.262	0.393
Number of species	5.214	0.628
Population size	3.816	0.459
An inverse logarithm Simpson index	3.928	0.473
BA	3.577	0.431
1	6.710	0.808
$M1max$	0.000	-

number of species attained a high value regardless of forest age. On the other hand, other indices were almost equal in value for M' and were lower than that of the number of species. These indices exhibited low overall values because they were also low at the compartment level. However, the overall value of the LUDI cannot be determined by M' alone.

Therefore, to determine the maximum LUDI, we randomized the grouping of compartments, with the qualifications that the area, perimeter, and placement would not change and Wcj would be fixed at 1 (i.e., no weighting). We then calculated $M1$max for the maximum LUDI using 500 randomizations. The $M1$max results suggested that the ideal diversity of land use occurs at the basin scale. We also calculated $M/M1$max; that is, how much M' contributed to $M1$max. Furthermore, the ideal can easily be determined by depicting the results using GIS (**Figures 6** and **7**).

The $M'/M1$max of the number of species was high at 0.628,

suggesting that the current diversity of species is ensured at a certain level. The value of $M'/M1$max when Wcj was 1 was 0.808. Because the value of M' when Wcj is 1 represents the distribution of compartments, this value suggests a high evaluation of the distribution of compartments. Furthermore, diversity at the basin scale will be enhanced by declines in old forests (age 39 - 99) and increases in young forests (age 22 - 38), according to this analysis.

Originally, the LUDI was used as an index to evaluate land-use diversity and not for the evaluation of broadleaf tree diversity. However, by using the results of the four indices for Wcj, it is possible to evaluate diversity. Furthermore, this approach allows us to determine what proportion of $M1$max was achieved by M' by calculating $M'/M1$max. Using this index, the LUDI can be used as an index to evaluate the diversity of broadleaf trees. Furthermore, the LUDI can be employed by forest managers to determine the ideal distribution of compartments. One potential issue is that the method used to calculate maximum LUDI in this paper involved several qualifications; therefore, the results of LUDI alone are not adequate for designing a sustainable forest management approach.

Conclusion

Our results demonstrated that all tested indices were similar in their effectiveness, as they each evaluate different aspects of diversity. Moreover, we determined that the diversity of broadleaf trees at the basin scale can be evaluated using the LUDI. However, results obtained using individual indices must be evaluated with caution. Another issue is that we only used data for *Chamaecyparis obtusa* in Hayami Forest. It is unknown whether similar results would be obtained if data were used for forests of *Chamaecyparis japonica* and *Larix leptolepis* or for forests with an insufficient management regime. Future research should collect data within forests other than those of *Chamaecyparis obtusa*.

Although each index offers advantages and disadvantages, we found that the diversity of broadleaf trees at the basin scale was most effectively evaluated using these indices in combination. Alternatively, a method could be developed to use all indices in one model. We hope that this topic will be further explored and clarified through interdisciplinary research.

The present distribution of compartments

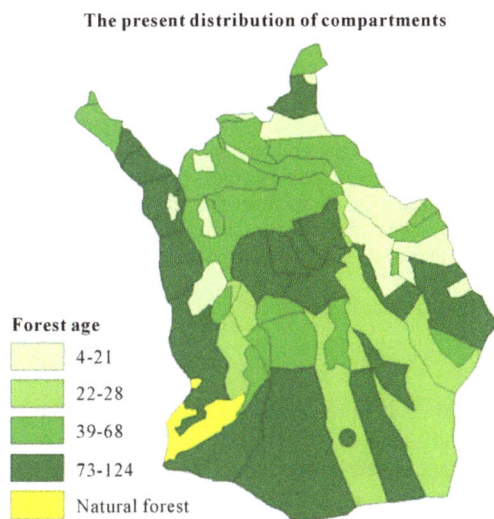

Figure 6.
The present distribution of compartments depicted using GIS.

The ideal distribution of compartments

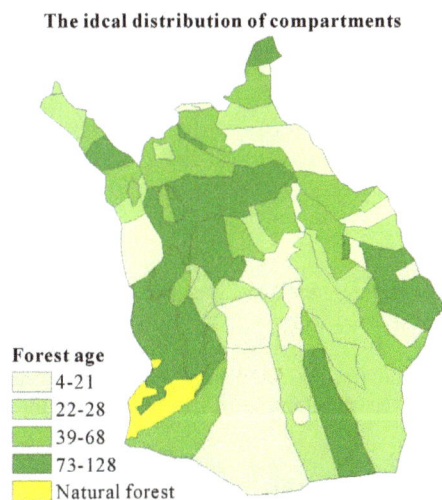

Figure 7.
Results of the LUDI depicted using GIS.

REFERENCES

Forest, H. (2011) Forest of Ootaga-the forest of FSC and a museum of forest and Owari Hinoki- 4 p.

The Forest Agency (2012). Proportion of forest and artificial forest within prefectural lands. URL (checked on 24 January 2012). http://www.rinya.maff.go.jp/j/keikaku/genkyou/sinrin_ritu.html

Yoshida, T., & Tanaka, K. (2005). Land-use diversity index: A new means of detecting diversity at landscape level. *Landscape and Ecological Engineering, 1,* 201-206.

The Meteorological Agency (2012). An online search of the data of past weather. URL (checked at 1 January 2012). http://www.data.jma.go.jp/obd/stats/etrn/index.php

Yoshiaki, I., & Kazunori, S. (2002). Problems around the indices of species diversity for comparison of different communities. *Landscape and Ecological Engineering, 53,* 204-220.

15

Evaluation of Durability Conferred by an Oleothermic Treatment on Chestnut and *Douglas fir* through Laboratory and in Field Tests

Sabrina Palanti

CNR IVALSA, Consiglio Nazionale delle Ricerche Istituto per la Valorizzazione del Legno e delle Specie Arboree, Sesto Fiorentino, Italy

The research evaluated two wood species, chestnut and *Douglas fir*, that are widespread in Tuscany, treated with an oleothermicprocess. Efficacy of this treatment against fungal decay was assessed through laboratory and field tests. The aim of this research was to investigate if an oleothermic process could add value to these natural resources when utilized in Use Class 4 (EN 335) for agricultural purposes such as vineyard poles. The treatment was effective on sapwood of both wood species in laboratory test, giving a protection similar to the untreated heartwood but in case of chestnut, it was easily washed away. In field test the leaching during the outdoor exposure reduced the resistance to fungal decay in both species.

Keywords: Conferred Durability; Chestnut; *Douglas fir*; Oleothermic Treatment

Introduction

The conventional wood protection for Use Classes 3 and 4 according to EN 335 (2006) is actually based on broad spectrum biocide formulations such as copper/organic biocides, copper-organometallics, metal free preservatives (Hughes 2004), and alkyl ammonium compounds (Pernak et al. 2004).

In the last decade, there has been a great interest within Europe in oils and water repellents for wood preservation princepally focusing on screening of natural and synthetic oils (Sailer and Rapp 2001; Van Eckeveld et al. 2001; Palanti et al. 2004), and on the processes for development of wood preservation technology (Thevenon 2001; Treu et al. 2001). Chestnut and *Douglas fir* are widely planted in Tuscany. Tuscany forest contains 862 hectares with 21% of chestnut and 4% conifers princcipally pine, *Douglas fir* and silver fir (Sanchez et al. 2005). Promoting the use of these species for agricultural purpose could add value to this natural resource.

Oleothermic treatment was explored for improving the natural durability of these species (Grenier et al. 2003). This process has been known as eco-friendly, biocide-free, cheaper, capable of being applied without drying the wood.

Oleothermic treatment also prevented the tannin leaching from chestnut which could maintain natural durability for a longer period (Grenier et al. 2003). While these results are promising, there is few data on the long-term performance of oleothermic treated chestnut and *Douglas fir* heartwood and sapwood. In order to evaluate these materials the following test were performed:

1) Laboratory tests of chestnut and *Douglas fir* sapwood, treated with oleothermic treatment, in according to EN 113 and EN 84.

2) Natural durability test in according to EN 252 (in field) of Chestnut and *Douglas fir*.

3) Resistance test in according to EN 252 of treated Chestnut and *Douglas fir*.

Materials and Methods

Wood Species

The tested wood species were chestnut (*Castanea sativa* Mill.) and *Douglas fir* (Pseudotsugamenziesii Mirb. Franco) of Tuscan origin, from Casentino, a valley near Arezzo. The reference controls were *Scots pine* (*Pinus sylvestris* L.) and European beech (*Fagus sylvatica* L.), sapwood, in accordance with European norms EN 350-1:1996 and EN 113:1996 and EN 252: 1989.

Treatment

The oleothermic treatment was performed at CIRAD in Montpellier, France using a two-step process (Grenier et al. 2003).

Chestnut and *Douglas fir* planks were conditioned at 20°C and 65% RH, cut in samples and sent to the plant for oleothermic treatment.

The treated samples were then conditioned to constant weight at 20°C and 65% RH.

Laboratory Test

Wood blocks $(50 \pm 0.5) \times (25 \pm 0.5) \times (15 \pm 0.5)$ mm^3, were

Evaluation of Durability Conferred by an Oleothermic Treatment on Chestnut and Douglas fir through Laboratory
and in Field Tests

93

cut from the boards.

Natural durability and the durability conferred by oil treatment of chestnut and *Douglas fir* were assessed in accordance with European standards EN 113 and EN 350. A comparison of resistance against fungi before and after leaching in accordance with EN 84 was carried out.

The fungi utilized for chestnut and *Douglas fir* were two brown rots, *Coniophora puteana* (Schumacher ex Fries) Karsten, strain BAM Ebw. 15 and *Gloeophyllum trabeum* (Persoon ex Fries) Murril, strain BAM Ebw. 19. The white rot fungus, *Coriolus versicolor* (Linnaeus) Quélet, strain CTB 863 A was only utilized for chestnut. The brown rot fungus *Poria placenta* (Fries Cooke) sensu J. Eriksson, strain FPRL280 was utilized for *Douglas fir*. These fungire present the certificated ones utilized in the European normative.

The wood samples were exposed to fungal attack for 16 weeks in a conditioning room (22˚C ± 2˚C, 70% ± 5% RH). For each set, a treated sample was placed with a reference untreated sample in a Kolle flask containing one of the test fungi growing on 20 ml of 4% malt and 2.5% agar medium. Another set of five untreated wood specimens was used to check for strain virulence for each tested fungus.

The resistance against fungi was evaluated using mass loss of wood, which was calculated for each individual block as the difference between the dry mass before and after the fungal exposure. The minimum of mass loss on reference wood blocks were 30% for *C. puteana* and 20% for *P. placenta, G. trabeum* and *T. versicolor*.

The natural durability class was calculated according to CEN/TS 15083, expressed as durability class, against wood destroying basidiomycetes fungi based on the higher median mass loss determined for all test specimens exposed to each of the three test fungi. Using the following scale: 1- very durable (mass loss ≤ 5%), 2- durable (mass loss >5% to ≤10%), 3- moderately durable (mass loss >10% to ≤15%) 4- slightly durable (mass loss >15% to ≤30%), 5- not durable (mass loss > 30%).

Leaching in Accordance with EN 84

Wood samples, see **Table 1**, were subjected to leaching according to EN 84: 1996. The cycle consisted of an initial 4 kPa vacuum - atmospheric pressure cycle with deionized water EN ISO 3696 (1996). The wood specimens were maintained in water (ratio of water to wood 5:1) for 14 days with 9 water changes, and then conditioned to constant mass before being exposed to one of the test fungus.

Table 1.
Median mass losses % of non-treated Chestnut heartwood or sapwood that was oleotheric treated. Samples were exposed to 3 fungi of EN 113 (H: heartwood, S: sapwood, T: Treated, L: leached, Y: yes, N:not, D class: durability class).

| H/ S | T Y/N | L Y/N | Median mass loss % (n = 5) | | | D class |
			C. puteana	*C. versicolor*	*G. trabeum*	
S	Y	N	1.24	1.49	-	1
S	Y	Y	12.91	12.50	12.92	3
H	N	N	1.33	1.17	0.92	1
H	N	Y	3.41	3.39	3.10	1

In Field Test

In ground performance tests were performed in accordance with EN 252: 1989 over a 70 months period (July 2005 to April 2011). The test site was on an agricultural soil situated in Tuscany, Cesa, Val di Chiana Arezzo.

Fifteen stakes for wood species, dimensions (500 ± 1) × (50 ± 3) × (25 ± 0.3) mm^3, derived from heartwood chestnut, heartwood and sapwood *Douglas fir* were conditioned and cut before being sent to the plant for oleothermic treatment.

The treated stakes were placed vertically in the soil leaving half of length exposed.

Untreated stakes of Scots pine and beech were also buried up to half their length in the soil test site as reference controls.

Fungal decay was annually evaluated in the area around the ground contact. Inspections were carried out by giving a light blow to the upper part of each stake to determine if the stake was close to failure, followed by removal from the ground. Surface examination was probed with an awl to detect weakness due to fungal rot. Fungal decay was evaluated on the basis of the depth of fungal softening and damaged area according to a specific rating system: 0 sound, 1 slight attack, 2 moderate attack, 3 severe attack and 4 failure (stake breaks in the ground after blowing). The evaluation was performed in accordance with guideline for EN 252 of the Nordic Wood Preservation Council (Borsholt and Henriksen, 1992).

It is possible to calculate the mean life as indicated in standard EN 350-1 when all stakes fail. The mean life, where not all stakes have reached the end of their life, can be predicted using the procedure of Purslow, 1976.

The number of replicates utilized for calculating average decay grade was unequal for wood species and treatments because some stakes failed as the test site who maintained and it was not possible to determine whether decay was the sole cause of failure.

Statistical Analyses

Analysis of variance (ANOVA) one-way, ANOVA multiple factors way and a post-hoc Tukey HSD multiple comparisons of means were carried out with open source R software.

The statistical analyses were on the following tests:

Laboratory test: chestnut and *Douglas fir* mass loss % of blocks exposed to *C. puteana*.

Field test: chestnut decay rating after six year; Douglas stakes rating of third (mid-term evaluation) and sixth year.

Results and Discussion

Laboratory Tests

Chestnut

Average uptake of oleothermic treatment on sapwood (n = 39) was 13% as (m/m) %.

The results obtained by natural and the conferred durability are reported in **Table 1**. Results due *G. trabeum* were omitted because the beech reference wood blocks had mass losses lower than minimum required by the standard.

The oleothermic treatment provided a good protection to sapwood but it was not fixed very well. Durability rating decreased from 1 to 3 when the wood was leached in accordance with ageing procedure EN 84. Durability of non-leached treated was similar to the inherent durability of non-leached heartwood. Weight losses of non-treated heartwood samples increasedwith leaching procedure, but the increase did not affect the durability class.

The one-way analysis of variance (Anova) of mass loss % of different treatment against *C. puteana*, indicated that there were significant differences between samples ($F_{3,16}$ = 2.2e−16***). The post hoc Tukey multiple comparisons of means, at a 95% family-wise confidence level, showed a significant difference in leached and non-leached treated chestnut underlined that the oleothermic gave a good durability (**Table 2**). In leached and non-leached heartwood the significant difference depended on leaching procedure that washed away the natural extractives. It was also interesting observe that only pair with not significant difference was untreated heartwood and treated sapwood where it was found a similar grade of effectiveness when exposed to *C. puteana*. This fact confirmed that sapwood treated with oleothermic treatment reached the same grade of durability of the heartwood.

Douglas fir

The average oleothermic uptake (n = 4) in sapwood and heartwood were 16% and 14% m/m % respectively.

The oleothermic treatment gave a good protection to *Douglas fir* sapwood, improving durability to class 2 (**Table 3**) in both leached and non-leached blocks.

In the untreated *Douglas fir* heartwood the leaching procedure determined a decreasing in durability from class 2 to 4. Probably this low durability was due to the fact that the leaching procedure washed away most part of the extractives.

Table 2.
Post hoc Tukey multiple comparisons of means at a 95% family-wise confidence level. T: treated, U untreated, L Leached, S sapwood, H heartwood.

Pair	Mean differ-ence	Lower bound	Upper bound	p value at a 95% confident level
TSL-TS	11.692	10.7594919	12.624508	0.0000000
UH-TS	0.116	−0.8165081	1.048508	0.9839834
UHL-TS	2.396	1.463491	3.328508	0.0000089
UH-TSL	−11.576	−12.5085081	−10.643492	0.0000000
UHL-TL	−9.296	−10.2285081	−8.363492	0.0000000
UHL-UH	2.280	1.3474919	3.212508	0.0000164

Table 3.
Median mass loss % of *Douglas fir* (H: heartwood, S: sapwood, T: Treated, L: leached, Y: yes, N:not). In the brackets the number of samples utilized for the determination of the medians are reported.D class: Durability class.

| H/S | T Y/N | L Y/N | Median Mass loss % | | | D class |
			C. puteana	*G. trabeum*	*P. placenta*	
S	Y	N	1.37 (2)	-	7.67 (2)	2
S	Y	Y	7.42 (5)	3.62 (3)	5.91 (4)	2
H	N	N	3.88 (3)	3.79 (2)	9.03 (4)	2
H	N	Y	12.37 (3)	1.45 (4)	19.67 (2)	4
S	N	N	11.26 (4)	5.02 (2)	13.89 (4)	3

The set of untreated sapwood gave reaching the durability class 3 (**Table 2**).

The one-way analysis of variance (Anova) of mass loss showed that there were significant difference among treatments with fungus *C. puteana*, gave significant differences between means (($F_{6,16}$ = 0.784800**). The post hoc Tukey multiple comparisons of means, 95% family-wise confidence level, evidenced significance in the mass loss in all combination with the exclusion of the pairs untreated heartwood-treated and leached sapwood, untreated heartwood-treated sapwood. The efficacy of treatment when applied into sapwood, and its fixation is evidenced by not significance between untreated heartwood-treated and leached sapwood.

In Field Test

Chestnut

The year-by-year results obtained by in field test of chestnut are reported in **Table 4**.

Only a slight difference was observed between treated and non-treated chestnut stakes. Four of eight treated stakes failed (decay grade 4) at the end of the test, and the mean life of these stakes was six years. Five of twelve untreated chestnut failed at sixth year with a mean life of 6 years, the number of ruptures was 6 on 12 and the mean life was, as well as treated ones, six years. The beech control stakes all failed after two years.

The analysis of variance of the decay grade of 68 months evidenced significant difference between treated and non-treated samples ($F_{2,27}$ = 0.03873*). The post hoc Tukey multiple comparisons of means at 95% family-wise confidence level, indicated that there are significant differences in decay grade were weather untreated chestnut and beech.

Douglas fir

In the **Table 5** are reported the overall results obtained from *Douglas fir* stakes.

In the treated heartwood, at the end of the test, six stakes on eight reached the rupture with a mean life 5.33 years. In the treated *Douglas fir* sapwood set at the end of the test 9 on 10 stakes reached the decay grade 4 with a mean life 5.11 years. The *Douglas fir* untreated set sapwood reached a mean life 4.60 years. In the untreated heartwood set only two stakes on 10 reached the rupture.

The decay grade of 68 months evidenced significance between samples ($F_{4,38}$ = 0.00064***) and the post hoc Tukey multiple comparisons of means, 95% family-wise confidencelevel, gave as significant the pairs untreated *Douglas fir*

Table 4.
Decay rating of average annual evaluations of Chestnut in field testing.

| Exposition time (month) | Average decay rating | | |
	Treated Chestnut, heartwood	Untreated Chestnut, heartwood	Beech
13	0.23	0.21	2.1
23	0.25	1.00	3.6
36	1.13	1.57	3.9
47	1.75	2.21	4.0
62	2.25	2.50	Failed
68	3.38	3.29	Failed

Evaluation of Durability Conferred by an Oleothermic Treatment on Chestnut and Douglas fir through Laboratory
and in Field Tests

95

Table 5.
Decay rating of field average annual evaluations of *Douglas fir.*

Exposition time (month)	Average decay rating				
	Treated *Douglas fir* sapwood	Treated *Douglas fir* heartwood	Untreated *Douglas fir* sapwood	Untreated *Douglas fir* heartwood	Scot pine sapwood
13	1.09	0.11	1.00	0.30	0.3
23	1.36	0.78	1.29	1.00	1.1
36	1.55	1.44	2.29	1.70	4
47	2.80	2.63	3.14	2.20	4
62	2.90	2.63	3.14	2.40	Failed
68	3.89	3.75	3.71	3.11	Failed

heart wood-treated *Douglas fir* sapwood, untreated *Douglas fir* heartwood-treated *Douglas fir* heartwood and Scot pine-untreated *Douglas fir* heartwood.

Conclusion

The results showed that oleothermic treatment improved the durability of chestnut and *Douglas fir*, both laboratory and fields exposition. Oleothermic treatment provided a good protection to chestnut sapwood as evidenced by laboratory experiments, but the treatment it was susceptible to the leaching during the field test.

With regarding *Douglas fir* field test is remarkable that the significant difference observed in the pair heartwood treated and not treated was due also to the presence of natural extractives inside the wood, responsible for the conferred durability. The protection given by oil was susceptible to leach and further studies are recommended to develop methods for fixing the treatment.

Acknowledgements

The author thanks the Regional Agency for Agriculture Developing and Innovation (ARSIA) that supported this research, CIRAD for theoleothermic treatment, and Mrs. Anna Maria Torniai from CNR Trees and Timber Institute who helped with fungi cultures.

REFERENCES

Borksholt, E., & Henriksen, H. K. (1992). Guideline for EN 252: Field test method for determining the relative protective effectiveness of wood preservatives in ground contact NWPC Information No. 23/90 ISSN 0358-707X.

CEN EN 113 (1996). *Wood preservatives. Test method for determining the protective effectiveness against wood destroying basidiomycetes. Determination of the toxic values.* Brussels: European Committee for Standardisation.

CEN EN 252 (1989). *Field test method for determining the relative protective effectiveness of a wood preservative in ground contact.* Brussels: European Committee for Standardisation.

CEN EN 84 (1996). *Wood preservatives. Accelerated ageing of treated wood prior to biological testing-Leaching procedure.* Brussels: European Committee for Standardisation.

CEN EN 335 (2006). *Durability of wood and wood-based products—Definition of use classes—Part 1: General.* Brussels: European Committee for Standardisation.

CEN EN ISO 3696 (1996). *Water for analytical laboratory use. Specification and test methods.* Brussels: European Committee for Standardisation.

CEN EN 350-1 (1996). *Durability of wood and wood-based products.*

Natural durability of solid wood. Guide to the principles of testing and classification of natural durability of wood. Brussels: European Committee for Standardisation.

CEN/TS 15083-1 (2005). *Durability of wood and wood-based products Determination of the natural durability of solid wood against wood-destroying fungi, test methods. Part 1: Basidiomycetes.* Brussels: European Committee for Standardisation.

Grenier, D., Bailleres, H., Meot J., Langbour, P., & Lanvin, J.-D. (2003). A study of water loss and oil adsorption during oleothermic treatment of wood. *Proceeding of the 1st European Conference on Wood Modification, Belgium,* 23-32.

Hughes, A. (2004). *The tools at our disposal.* Lisbon: Final Workshop Cost Action E22.

Militz, H., Krause, A., & Hof, C. (2004). Modified wood for window and cladding products. In A. Ceccotti (Ed.), *Proceeding of International Symposium on Advanced Timber and Timber Composite Elements for Building,* Florence: CNR-Ivalsa.

Palanti, S., & Susco, D. (2004). A new wood preservative based on heated oil treatment combined with triazole fungicides developed for above-ground conditions. *International Biodeterioration & Biodegradation, 54,* 337-342.

Pernak, J., Zabielska-Mateljuc, J., Kropacs, A., & Foksowicz-Flaczyk, J. (2004). Ionic liquid in wood preservation. *Holzforschung, 52,* 286-291.

Purlow, D.F. (1976). Result of field test on the natural durability of timber (1932-1975). Building Research Establishment Current Paper CP 6/76, Garston.

Sailer, M., & Rapp, A.O. (2001). Water repellency of some natural oil. Workshop and Meetings of European COST (Cooperation in the field of the Scientific and Technical Research). Action E22-Environmental Optimisation of Wood Protection. Reinbek.

Sanchez, D., Leinonen, A., Vesterinen P., Capaccioli, S., Gil, J., & Echeverria, I. (2005). Bio-south: Techno-economical assessment of production and use of biofuels for heating and cooling applications in south Europe. *14th European Biomass Conference,* 17-21 October 2005, Paris.

Thevenon, M. F. (2001). Oils and water repellents in wood protection research and development in France. Reinbek: Workshop on Optimising treatment levels and managing environmental risks, Cost E22.

Treu, A., Militz, H., & Breyne, S. (2001). Royal treatment-scientific background and practical application. Reinbek: Workshop on Optimising Treatment Levels and Managing Environmental Risks, Cost E22.

Van Eckeveld, A., Homan, W., & Militz, H. (2001). Water repellency of some natural oil. Workshop and Meetings of European COST (Cooperation in the field of the Scientific and Technical research). Reinbek: Action E22-Environmental Optimisation of Wood Protection.

Vernois, M. (2001). Heat treatment of wood in France—State of art in rewiew on heat treatments of wood. In A. O. Rapp (Ed.), *Proceeding of the Special Seminar COST E22,* Antibes.

16

New Methods to Quantify Canopy Structure of Leafless Boreal Birch Forest from Hemispherical Photographs

T. D. Reid[*], R. L. H. Essery

School of Geosciences, The University of Edinburgh, Edinburgh, UK

Hemispherical photography has been used for many years to measure the physical characteristics of forests, but most related image processing work has focused on leafy canopies or conifers. The boreal forest contains large areas of deciduous trees that remain leafless for over half the year, influencing surface albedo and snow dynamics. Hemispherical photographs of these sparse, twiggy canopies are difficult to acquire and analyze due to bright bark and reflections from snow. This Note presents new methods for producing binary images from hemispherical photographs of a leafless boreal birch forest. Firstly, a thresholding method based on differences between colour panes provides a quick way to remove bright sunlit patches on vegetation. Secondly, an algorithm for joining up fragmented pieces of tree after thresholding ensures a continuous canopy. These methods reduce the estimated hemispherical sky view fraction by up to 6% and 3%, respectively. Although the processing remains subjective to some degree, these tools help to standardize analysis and allow the use of some photographs that might have previously been considered unsuitable for scientific purposes.

Keywords: Hemispherical Photography; Image Processing; Leafless Canopies; Boreal Forest; Snow; Abisko

Introduction

Forest canopies strongly affect the radiation balance at the Earth's surface. One well-established technique for quantifying this effect is to take hemispherical photographs (hemiphotos) looking upwards from the forest floor. Hemiphotos can be analyzed to estimate forest parameters such as sky view fraction and plant area index van Gardingen 1999, but the first important step involves applying thresholds of brightness to produce a binary map of black (trees/horizon) and white (sky) pixels. The thresholding is usually done manually, and has been criticised for being subjective, prompting the development of automatic methods Nobis 2005, Ishida 2004. Sky conditions are also important, and overcast skies are preferable for their homogeneous quality. Zhang 2005 suggest measuring a reference exposure for the open sky before lengthening exposure time to increase contrast between sky and vegetation, while Pueschel 2012 suggest taking five photographs on different exposure settings. However, such complicated approaches are not always practical in remote, challenging field environments. It is beneficial to take photos quickly, with minimal equipment and simple protocols that can be performed by different operators, including students.

Moreover, most papers on hemiphoto analysis have focussed on leafed deciduous trees or needleleaf conifers. The motivation for this Note arose during fieldwork in March 2011 in an area of leafless mountain birch forest in Abisko National Park, Sweden. Hemiphotos taken in this region present considerable challenges; homogeneous overcast conditions are rare or accompanied by blizzard conditions, and the high albedo snow cover, low solar elevations and white birch bark cause considerable light scattering.

This Note describes new techniques that can aid processing of hemiphotos of sparse, leafless Arctic canopies. Section 3 describes a method of manipulating the colour panes of a hemiphoto to identify sunlit parts of the canopy that are too bright to be distinguished from the sky through a simple threshold. Section 4 describes a branch-joining algorithm that acts to repair hemiphotos in which branches with heterogeneous colour characteristics have become broken-up during the thresholding process. These new concepts are easy to use, and remove some subjectivity from the processing of hemiphotos. They would provide useful additions to popular existing software packages such as Gap Light Analyzer, or GLA Frazer 1999, which don't tend to include special settings or functions for leafless canopies.

Hemiphoto Acquisition and Thresholding

All photographs were taken using a Nikon Coolpix 4300 digital camera with a Nikon FC-E8 fisheye converter lens on an automatic exposure setting, under the mountain birch canopy around 3 km south of Abisko village (68.32° N, 18.83° E). Images were saved as JPEGs with resolution that gave a hemi-

---

spherical view of radius 1704 pixels. Algorithms were encoded using IDL (Interactive Data Language©, 2011 Exelis Visual Information Solutions). Canopy metrics-hemisphere-averaged sky view fraction v_s and effective plant area index P were calculated by first obtaining values of gap fraction, $v_{\theta\phi}$ as the ratio of sky pixels to total pixels in nine concentric 10 bands of elevation angle (θ) in the hemisphere. v_s was calculated by summing the weighted individual v_θ values according to Essery 2008, and P was calculated from logarithms of v_θ according to van Gardingen 1999.

The thresholding of images in this paper is all done manually. The automated threshold method of Nobis 2005 was tested on Abisko hemiphotos, but it resulted in binary images that looked far from the reality; large parts of trees were lost, perhaps because their software was developed for leafed canopies and has difficulty detecting sky-tree edges in the leafless canopy.

Colour Pane Manipulation

Previous studies have highlighted the benefit of selecting the blue colour pane of an RGB image to separate canopy and sky, because blue is enhanced in the sky and reduced on trees Frazer 1999, Nobis 2005. The method presented here expands on this concept, and that of Normalized Difference Vegetation Index (NDVI), which indicates the presence of vegetation via the difference in intensities at certain spectral ranges Kriegler 1969. **Figure 1** shows that a similar argument can be applied to mountain birch forest. A spectrophotometer (Analytical Spectral Devices, Inc. (ASD)) was used to measure reflectivity spectra of snow, white bark and darker copper-coloured bark in the area where hemiphotos were taken in 2011 (Richardson and Sandells, unpublished); these are similar to spectra measured by

Ovhed 1995. On applying these reflectivities to a typical sky spectrum for the Abisko region (replicated from Ovhed 1995), white bark produces a flatter spectrum in the visible region (400 to 700 nm) than the sky, which gives considerably higher irradiance at the blue end. This supports the argument for using the blue pane ($\approx$ 400 to 500 nm). However, the irradiance reflected from white bark is often higher than from blue sky and even some clouds. In this Note the shapes of the graphs are exploited, taking the red pane into account. This involves subtracting a proportion of the red (R) pixel values (with digital number (DN) between 0 and 255) from the blue (B) pixel values to produce a new grayscale image I_g, defined as:

$$I_g = B - f_r R \qquad (1)$$

where f_r is a number between 0 and 1. If I_g is then normalized back to the range 0:255, the difference between sky and tree is magnified. Finally, a suitable threshold t_b (=0 to 255) should be chosen and applied to I_g, producing a binary image.

Figure 2 shows an example hemiphoto in which the sunlit white bark produces many bright patches on trees. In the blue-pane-only image (upper row) these were too bright to remove through simple thresholding without starting to turn many sky pixels black as well. On applying the new method with a value of $f_r = 0.5$, a subjectively better binary image was produced. The trunks come out black at a lower threshold and the vegetation as a whole is filled out to more accurately represent

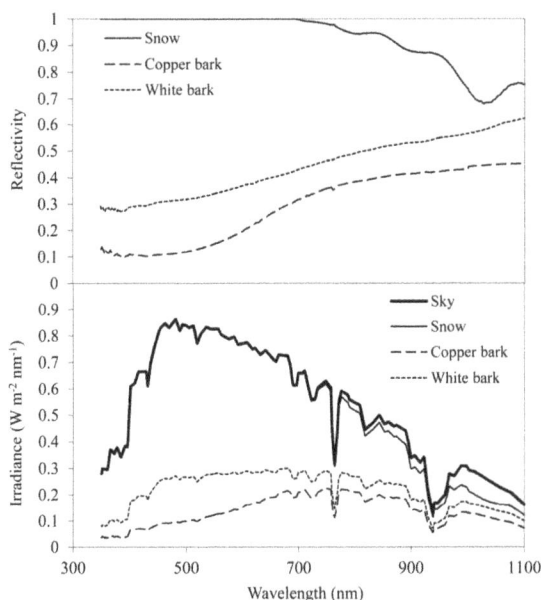

Figure 1.
Reflectivity spectra of snow, copper-coloured birch bark and white birch bark measured in Abisko in 2010 (upper panel). Lower panel shows a typical sky irradiance spectrum for Abisko replicated from Ovhed and Holmgren (1995), and the irradiance expected from snow, copper bark and white bark, where each is calculated by multiplying the sky irradiance by the relevant reflectivity.

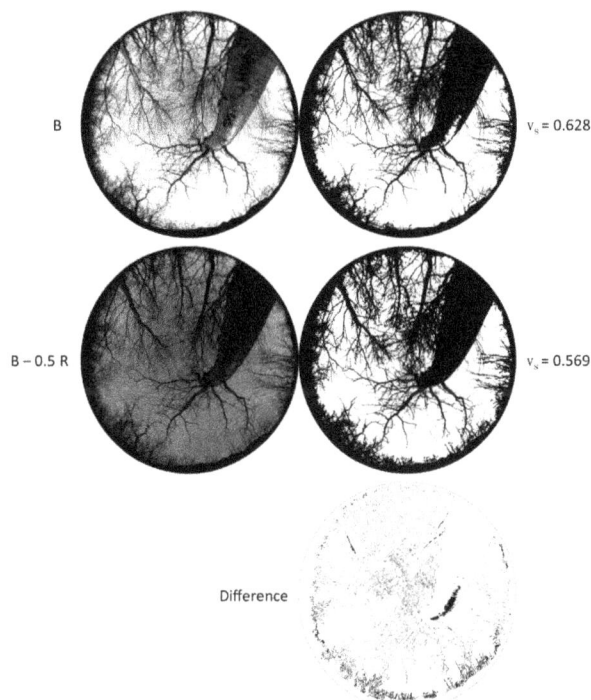

Figure 2.
Blue colour pane from an Abisko birch hemiphoto (top left) and the subjectively best binary image (top right) acquired by manually increasing the threshold until just before some sky pixels began turning black. By subtracting a fraction of the red pane (see Equation (1)) and renormalising (centre left), it is easier to correctly classify bright areas of bark without affecting the sky. This produces a different value of sky view fraction (v_s).

the real canopy. This causes a reduction in v_s from 0.628 to 0.569.

Incorporating this colour pane manipulation in a graphical user interface, using sliding bars to choose the best combinations of f_r and t_b, has proven useful for analyzing several hundred hemiphotos from the Abisko region. Such functionality is not included in packages such as GLA, which allows users to select an individual colour pane but doesn't allow panes to be combined in more complex ways. The implementation of Equation (1) or variations on it would be very straightforward and possibly beneficial for other forest situations.

This Note presents no automated method for removing snow-covered topography from hemiphotos. Snow could be removed by calculating horizon angles from a DEM Dozier 1990. Alternatively one could use a camera sensitive to near-infrared radiation (NIR) Chapman 2007, because as **Figure 1** shows there is a difference between visible and NIR reflectance for snow. For this Note, snowy horizons were shaded by hand, which can be done quite quickly and accurately.

Branch Joining Algorithm

It is reasonable to assume that there should be no flying twigs in a hemiphoto, and all dark areas of tree should be connected to the ground forming one continuous dark object in the image. When applying a brightness threshold to a gray colour pane, the priority is to remove all the sky pixels, but even with additional measures such as those described in Section 3, some brighter tree pixels will be saturated to the extent that they will be wrongly classified as sky. This means that the thresholded image has many twigs and branches appearing fragmented. This problem is enhanced for canopies with bright, reflective bark such as mountain birch.

For these reasons a branch joining algorithm (BJA) was developed to connect the disjointed parts of thresholded images. **Figure 3** illustrates this process. A connected-component algorithm was used to identify individual blobs in the image. For each blob, the BJA identifies the two furthest-apart pixels (pixels 1 and 2 in **Figure 3**), representing the extreme ends of a branch section, and performs searches in increasing radii around both those points to find out which of the two is closest to part of a different blob (pixel 3). The pixel is then joined to the other blob by a line; the width of this line is determined by the width of the original blob, defined as the number of pixels in the blob divided by its length in pixels, rounded to the nearest whole number. If it turns out that the blob is further than a specified maximum distance from the nearest other blob, the blob is deleted. For Abisko hemiphotos 30 pixels was chosen as the maximum separation, but this could be adjusted for different circumstances.

Figure 4 shows the BJA applied to an Abisko hemiphoto. On applying the BJA, the binary image is filled out considerably. For comparison, a small cropped part of the photo was processed by manually (laboriously!) shading in black all the branches seen in the original image. This manual shading is also subjective to some extent, but at least gives individual attention to every part of the image, so it is arguably the closest attainable representation of the truth. Counting the sky pixels in the three cropped images showed that the threshold-only approach provides a sky view fraction 3.5% higher than the hand-drawn method. On applying the BJA, the difference is only

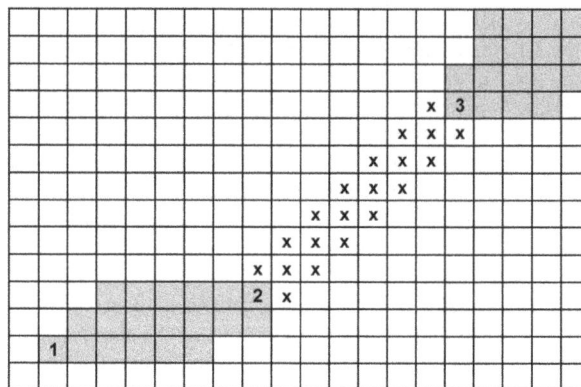

Figure 3.
Schematic illustration of the branch-joining algorithm. For each isolated blob (e.g. bottom left), the two furthest-apart pixels are identified (marked 1 and 2). The algorithm then finds out which of 1 or 2 are closest to a pixel of another blob-in this case it is pixel 2, which is joined to pixel 3 with a line (pixels marked with x) whose width is determined by the width of the original blob.

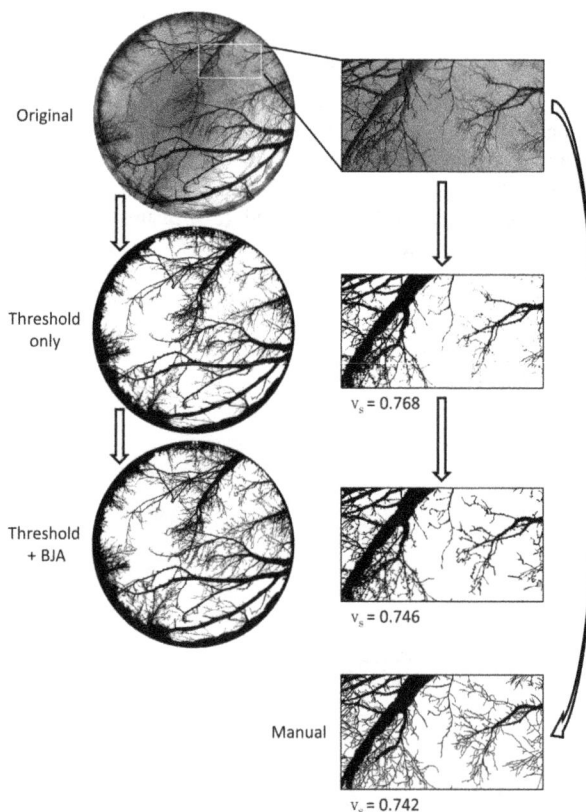

Figure 4.
Blue pane of an example hemiphoto (top row) processed by thresholding (second row) then by the BJA (third row). The close-up region in the right column was also subjected to a separate manual approach with twigs hand-drawn in black on the original image before thresholding (bottom right).

0.5%.

Figure 5 shows v_s and P calculated for 50 Abisko hemiphotos with just a threshold applied, and the same images after applying the BJA. The branch-joining algorithm decreases v_s and increases P, with maximum changes of −0.03 in v_s

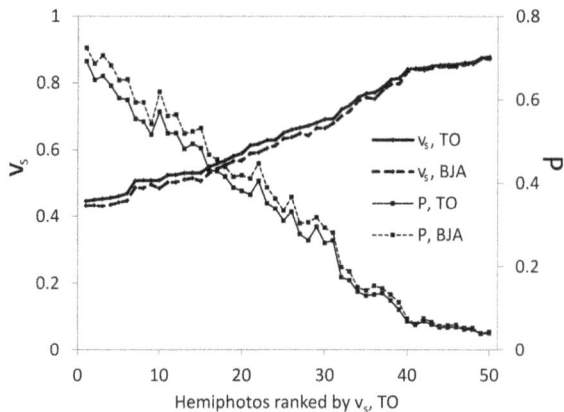

Figure 5.

v_s and p calculated for 50 hemiphotos from Abisko, numbered according to their sky view fraction. Solid lines show the results with image thresholding only (TO), and dotted lines are results after applying the branch-joining algorithm (BJA).

and +0.05 in P. As might be expected, the differences are larger when v_s is smaller, because such hemiphotos contain more vegetation to be processed by the BJA. A change in v_s of 3% may seem a small amount, but could have a significant effect on calculations of surface radiation balance, especially if conclusions were applied across large areas of forest.

It should be acknowledged that the BJA is rather ad-hoc in the way it accounts for canopy geometry. A more sophisticated BJA was trialled by preferentially attaching individual blobs to others that lay in the direction of the major axis of the original blob. This resulted in some twigs being drawn in strange places, because there are many small blobs of 5 pixels or less in which the major axis is not obvious. Overall, the simpler approach is usefully non-specific; further levels of sophistication might make the algorithm less generalizable to other canopies.

Conclusion

The techniques described in this Note address some of the difficulties encountered on analyzing hemiphotos of leafless Arctic canopies, and could provide useful additions to existing hemiphoto software. The manipulation of individual RGB colour panes has proven very useful for correctly classifying bright areas of canopy that would otherwise be missed out, and the BJA fills many of the gaps left by the thresholding process. Both techniques remain to be tested for different leafless forest types, but for at least the case of boreal birch, they help to standardize analysis across thresholded images.

The process of producing a binary image from a hemiphoto will always be subjective to some degree, but these new methods go some way to reducing that subjectivity. The colour pane manipulation provides a quick and easy way to correctly classify sunlit areas of bark, encouraging all users to pay special attention to those areas; meanwhile the repairs made by the BJA work to produce similar final results from binary images that may have had different levels of branch fragmentation resulting from thresholding by different users. The subjectivity could be further reduced in future by adapting existing automatic thresholding methods Nobis 2005, Ishida 2004 to work for leafless canopies.

Most usefully, these methods make it possible to use hemi-

photos taken quickly under automatic camera settings and non-ideal light conditions—often the only possible way to conduct fieldwork in the Arctic winter. To improve on this study without increased effort in the field, images should be saved in raw format to avoid loss of detail on conversion to JPEG and allow more sophisticated processing Lang 2010. Future work could compare hemiphoto-based radiation transfer modelling to sub-canopy radiometer measurements Link 2004 or terrestrial laser scanning Antonarakis 2009, Cote 2009 to assess the benefits of different image processing methods. Such independent verification would be highly beneficial for forest studies on the whole, because hemiphotos remain the fastest, cheapest and easiest way to record forest structure in the field.

Acknowledgements

This work was funded by the UK's Natural Environment Research Council grant number NE/H008187/1. The authors would like to thank all the support staff at Abisko Scientific Research Station (ANS), as well as Nick Rutter, Maya King, Cécile Ménard, Steve Hancock, Rob Holden, Michael Spencer and Megan Reid for assistance in the field, Mark Richardson and Melody Sandells for measuring and providing Abisko reflectivity spectra, and Casey Ryan for advice on processing hemiphotos.

REFERENCES

Antonarakis, A. S., Richards, K. S., Brasington, J., & Muller, E. (2010). Determining leaf area index and leafy tree roughness using terrestrial laser scanning. *Water Resources Research, 46,* W06510.

Chapman, L. (2007). Potential applications of near infrared hemispherical imagery in forest environments. *Agricultural and Forest Meteorology, 143,* 151-156.

Côté, J.-F., Widlowski, J.-L., Fournier, R. A., & Verstraete, M. M. (2009). The structural and radiative consistency of three-dimensional tree reconstructions from terrestrial lidar. *Remote Sensing of Environment, 113,* 1067-1081.

Dozier, J., & Frew, J. (1990). Rapid calculation of terrain parameters for radiation modeling from digital elevation data. *IEEE Transactions on Geoscience and Remote Sensing, 28,* 963-969.

Essery, R., Pomeroy, J., Ellis, C., & Link, T. (2008). Modelling long-wave radiation to snow beneath forest canopies using hemispherical photography or linear regression. *Hydrological Processes, 22,* 2788-2800.

Frazer, G., Canham, C., & Lertzman, K. (1999). *Gap light analyzer (GLA): Imaging software to extract canopy structure and gap light transmission indices from true-colour sheye photographs, users manual and program documentation.* Burnaby: Simon Fraser University.

Ishida, M. (2004). Automatic thresholding for digital hemispherical photography. *Canadian Journal of Forest Reseach, 34,* 2208-2216.

Kriegler, F., Malila, W., Nalepka, R., & Richardson, W. (1969). Pre-processing transformations and their effect on multispectral recognition. *Proceedings of the 6th International Symposium on Remote Sensing of Environment.* Ann Arbor, MI: University of Michigan, 97-131.

Lang, M., Kuusk, A., Mttus, M., Rautiainen, M., & Nilson, T. (2010). Canopy gap fraction estimation from digital hemispherical images using sky radiance models and a linear conversion method. *Agricultural and Forest Meteorology, 150,* 20-29.

Link, T., Marks, D., & Hardy, J. (2004). A deterministic method to characterize canopy radiative transfer properties. *Hydrological Processes, 18,* 3583-3594.

Nobis, M., & Hunziker, U. (2005). Automatic thresholding for hemisphericalcanopy photographs based on edge detection. *Agricultural and Forest Meteorology, 128,* 243-250.

Ovhed, M., & Holmgren, B. (1995). Spectral quality and absorption of solar radiation in a mountain birch forest, Abisko, Sweden. *Arctic and Alpine Research, 27,* 380-388.

Pueschel, P., Buddenbaum, H., & Hill, J. (2012). An efficient approach to standardizing the processing of hemispherical images for the estimation of forest structural attributes. *Agricultural and Forest Meteorology, 160,* 1-13.

van Gardingen, P., Jackson, G., Hernandez-Daumas, S., Russell, G., & Sharp, L. (1999). Leaf area index estimates obtained for clumped canopies using hemispherical photography. *Agricultural and Forest Meteorology, 94,* 243-257.

Zhang, Y., Chen, J. M., & Miller, J. R. (2005). Determining digital hemispherical photograph exposure for leaf area index estimation. *Agricultural and Forest Meteorology, 133,* 166-181.

17

Ponderosa Pine Seed Source Test in Nebraska in the Central Great Plains of the United States

Wayne A. Geyer[1]*, Keith D. Lynch[1], Peter Schaefer[2], William R. Lovette[3]

[1]Division of Forestry, Kansas State University, Manhattan, USA
[2]Department of Plant Science, South Dakota University, Vermillion, USA
[3]Nebraska Forest Service, Lincoln NE 68583 Nebraska Forest Service, Lincoln, USA

Ponderosa pine (*Pinus ponderosa* Laws) has been planted widely in the Great Plains for wind breaks. A 1968 study recommended using material from south-central South Dakota and north-central Nebraska. A second test to further delineate seed sources (families) from numerous collection sites in this region was established in 1986. This paper reports results for survival, height, diameter, and D^2H (diameter squared X height) measurements at 15 years. Results identify a wide range of suitable families within the region. A majority of the tested sources performed well, thus verifying the original recommendations.

Keywords: *Pinus ponderosa*; Provenance; Seed Source; Family; Tree Selection; Growth Characteristics

Introduction

Ponderosa pine (*Pinus ponderosa* Laws) is an important component of the windbreak agroforestry system in the Great Plains. Its drought tolerance, dense crown form, and tall growth habit make ponderosa pine excellent for windbreaks, sight barriers, and ornamental plantings (Flint, 1983). Ponderosa pine is one of the few tall trees that grow in the region. Because of its evergreen nature, it provides year-round protection to fields and farmsteads (Schaefer & Baer, 1985). The natural range of ponderosa pine extends from British Columbia, Canada, southward into northern Mexico, and from California eastward into the Great Plains, excluding for Kansas (Crichfield & Little, 1966). Although ponderosa pine has been widely planted in the plains region, it has performed inconsistently.

Western pine tip moth (*Rhyacioniabushnelli*) has caused widespread damage in the plains (Kopp et al., 1987), but outstanding performance of individual trees in the plains plantations suggests that proper selection could improve tree quality.

Early studies determined that trees grown from seed collected from the northeastern range of ponderosa pine performed best in most of the provenance test plantations (Deneke & Read, 1975; Baer & Collins, 1979; Read, 1983; Schaefer & Baer, 1985, 1992; Van Haverbeke, 1986). Six years data from a Kansas plantation showed that early growth appeared to be clinally related to elevation of seed provenances (Deneke & Read, 1975). Plains nurseries have focused much of their ponderosa pine production on seed collections near Ainsworth and Valentine, Nebraska, and Rosebud, South Dakota.

Trees from Jordan, Montana, also performed well in more than half of the early plantations (Read, 1983).

In 1986, the GP-13 Technical Committee of the Great Plains Agricultural Council initiated a second cooperative ponderosa pine study in cooperation with the North Central and Rocky Mountain Forest Experiment Stations. The intent of this study was to further sample recommended provenances identified in the 1968 study. Collection sites are shown in **Figure 1**. Nine progeny tests were established in Saskatchewan (Canada), Montana, North Dakota, South Dakota, Nebraska, Kansas, Oklahoma, Texas, and Minnesota. This paper reports data from the Nebraska test. A previous report (Geyer & Schaefer, 2011) indicated that 1968 recommendations were valid for Kansas and South Dakota.

Materials and Methods

The tree plantation discussed in this report here used seedlings representing 96 open-pollinated families (**Figure 1**, **Table 1**) that were planted in an individual tree factorial planting design with 8 replications and 3.7 × 3.7 m (12 × 12 ft) plots. Each replication was included 5 trees in single-tree, noncomtiguous plots. Two border rows surrounded the plantation. A total of 3840 trees were planted. Weeds were controlled by cultivation for the first 3 years. The plantation was near Republican City, Nebraska, on an alluvial site.

Superior height growth potential can be accurately identified at an early age (i.e., 5 to 15 years); (Lambeth, 1980; Read, 1983; Van Haverbeke, 1986; Schaefer & Baer, 1992). For analysis, we applied the GLM procedure of SAS (SAS Institute, 2003) for height, trunk diameter, and D^2H (a measure of trunk volume); Duncan's multiple range test for mean separation; and chi-square for survival, stem crook, sweep, and number of terminal buds. Correlations were determined among height, diameter, D^2H, and latitude and longitude of each major geographic (provenance) location. The 13 major geographic zones (provenances) and 96 seed sources (families) were compared.

*Corresponding author.

Table 1.
Collection zones (provenances) of ponderosa pine seed sources.

Geographic origin (#)	Tree additions (families) (#)	Local area (town/state)	Annual precipitation((mm)(in))	Annual mean temp. °C (°F)	Elevation m (ft)	Latitude (°N)	Longitude (°W)
720	02 - 11	Ainsworth NE	579 (22.8)	8.6 (47.6)	780 (2560)	42.59	100.00
721	01 - 11	Valentine NE	495 (19.5)	8.4 (47.2)	800 (2625)	42.88	100.55
757	01 - 40	Rosebud SD	955 (37.6)	8.5 (47.7)	850 (2789)	43.25	100.82
990	01 - 10	Springview NE	573 (22.7)	8.3 (47.0)	740 (2428)	42.82	99.75
991	01 - 05	Kilgore NE	516 (20.3)	8.3 (47.0)	800 (2625)	42.94	100.97
992	01 - 04	Drinkwater NE	495 (19.5)	8.3 (47.0)	859 (2800)	42.47	101.07
993	01 - 07	Nenzel NE	526 (20.3)	9.0(48.3)	950 (3117)	42.93	101.11
994	01 - 05	Bassett NE	635 (25.0)	9.0 (48.3)	710 (2329)	42.60	99.54
995	01 - 06	Snake River NE	495 (19.5)	8.3 (47.00)	866 (2840)	42.71	100.97
996	01 - 14	Sparks NE	495 (19.5)	8.3 (47.0)	800 (2625)	42.94	100.24

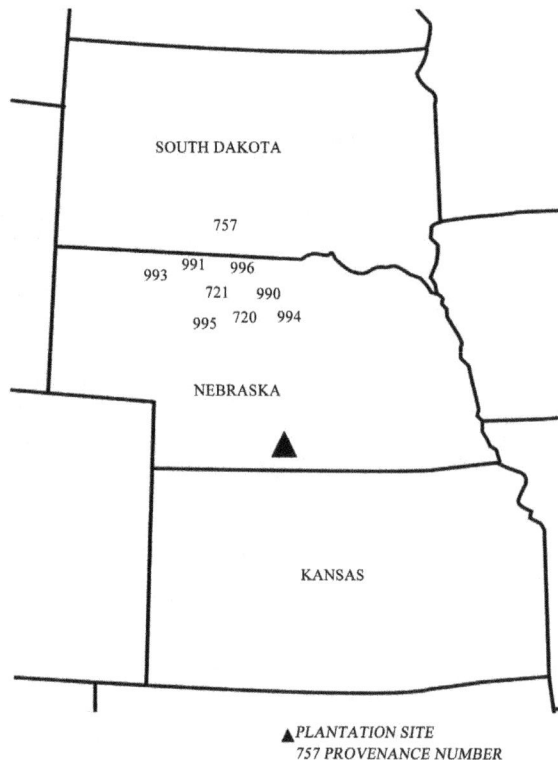

Figure 1.
Collection zones (provenances) of ponderosa pine seed sources.

Most of the sources were from north central Nebraska and southern South Dakota. An operational thinning was conducted at 15 years to remove insect-infested trees and trees of poor stature before an inventory.

Results and Discussion

Few differences existed in the size of most families because all trees were measured after thinning. Total family height varied by location. Mean height of all families was 6.8 m (22.4 ft) with a range from 5.9 to 7.8 m (19.0 to 25.5 ft). The top 10 families (**Table 2**) differed by 0.4 m (1.4 ft); and were significantly taller than only the shortest three families. The poorest 10 families ranged from 5.9 to 6.2 m (19.0 to 20.5 ft) and were significantly different (1% level) from only the tallest two families. The top 85 families did not differ significantly from each other, and the shortest family was 6.4 m (21.0 ft).

Mean diameter was 17.8 cm (7.0 in) with a range of 14.2 to 23.1 cm (5.6 to 9.1 in). The best 10 families differed by 3.6 cm (1.4 in) and the poorest 10 families differed by 3.0 cm (1.2 in). As with height, only families at either end of the ranking were significantly different. The top two families had significantly larger diameters than the bottom 84, whereas the bottom three families had significantly smaller diameters then the top 17. Although trees with the largest diameter tended to be taller ($r =$ 0.16179%, 1% level), only family 72,111 ranked in top 10 for both diameter and height. Survival was not good for the largest 10 families. Many families (49) had less than 20% survival. Survival of trees with the largest diameter, ranged from 3 to 20%.

Trunk volume was significantly different (1% level) among families (**Table 1**). The mean D^2H was 1126 units with a range from 651 to 1889. The largest 38 were not significantly different. The top six sources were significantly larger than the bottom 7.

Mean survival was surprisingly low, ranging from 3.3 to 56.7% with a mean of 21.3%. Four of the top 7 families were from Spring view, Nebraska (1990's), and exhibited significantly higher survival than the remaining 87 families (**Table 2**). As in the Kansas and South Dakota plantation (Geyer & Schaefer, 2011), the poorest families were from western provenances. The poorest 10 families had significantly poorer survival than the remaining sources.

Correlation Analysis

We found no strong relationship between elevation, latitude, or longitude and survival, diameter, height, and D^2H. The r values were significantly different, but very low (<0.15).

Table 2.
10 best and 10 worst families in the Nebraska ranking of plantation.

Survival		Diameter		Height		D^2H volume index	
Family[1]	(%)	Family	cm (in)	Family	m (ft)	Family	Volume
Top 10 families							
99,001	56.7a	99,203	23.1 (9.1a)	75,704	25.5 (7.8a)	75,704	1889a
99,006	53.3b	75,704	23.1 (9.1a)	75,726	25.5 (7.8a)	99,203	1724ab
72,002	53.3b	75,703	20.1 (7.9ab)	72,111	24.8 (7.6ab)	72,111	1547a-c
72,103	53.3b	72,111	19.8 (7.8a-c)	72,101	24.5 (7.5a-c)	75,703	1457a-d
99,008	53.3b	75,735	19.8 (7.8a-d)	75,716	24.3 (7.4a-d)	75,726	1444a-e
99,003	53.3b	75,710	19.8 (7.8a-d)	75,720	24.2 (7.4a-d)	75,735	1440a-e
99,606	47.7c	99,103	19.6 (7.7a-e)	75,717	24.2 (7.4a-d)	75,710	1408a-f
99,404	47.7c	75,706	19.6 (7.7a-e)	75,724	24.4 (a-d)	75,705	1404a-f
99,608	47.7c	72,004	19.6 (7.7a-e)	75,705	24.1 (7.3a-d)	99,611	1362b-g
99,405	47.7c	75,728	19.6 (7.7a-e)	99,611	24.1 (7.3a-d)	99,502	1354b-g
Bottom 10 families							
75,209	6.6h	99,204	16.3 (6.4b-h)	99,505	20.5 (6.2b-g)	99,201	861f-j
99,204	6.6h	99,007	16.3 (6.3b-h)	75,730	20.5 (6.2b-g)	99,614	850f-j
99,203	3.3h	99,101	16.3 (6.3c-h)	99,101	20.5 (6.2b-g)	75,708	832f-j
75,704	3.3h	75,709	15.7 (6.2d-h)	75,709	20.2 (6.2b-g)	99,204	818g-j
75,726	3.3h	99,614	15.7 (6.2d-h)	99,103	20.1 (6.0c-g)	99,101	812g-j
99,502	3.3h	72,101	15.5 (6.1e-h)	99,306	19.8 (6.0d-g)	72,106	801g-j
75,725	3.3h	72,106	15.5 (6.1e-h)	75,708	19.8 (6.0d-g)	75,709	780g-j
99,105	3.3h	72,105	14.7 (5.8f-h)	99,204	19.5 (5.9e-g)	72,105	754h-j
99,502	3.3h	99,306	14.7 (5.8gh)	99,303	19.2 (5.9f-g)	99,306	656i-j
99,505	3.3h	99,505	14.2 (5.6h)	99,201	19.0 (5.9g)	99,505	651j
All sources							
N	631		631		631		631
Mean	21.3		17.8 (7.0)		22.4 (6.8)		1126
Sign.	<1%		<1%		NS		<1%
	3.3 - 56.7		14.2 - 23.1 (5.6 - 9.1)		19.0 - 25.5 (5.9 - 7.8)		651 - 1126

[1]Note: First 3 numbers of the source code are collection zone and last 2 numbers are tree addition identifiers. Origin 811 was only planted in Kansas.

Summary and Conclusion

Fast-growing trees are desirable for windbreak establishment in the Great Plains. Ponderosa pine is often planted in homestead and field plantings in the central and northern plains.

Plains-wide studies conducted in the 1960s showed that south-central South Dakota and north-central Nebraska provided better planting materials. The present study was conducted to further refine selection areas for ponderosa pine sources. Materials from 13 collection areas and 76 individual trees were planted at Central Plains sites for evaluation. Within this relatively small area, analyses indicated that geographic origin (provenance) influenced growth of 15-year-old ponderosa pines.

The 5 best collection zones were from the eastern part of Nebraska (720—Ainsworth; 721—Valentine; 990—Spring view; 994—Bassett; and 996—Sparks). The poorest collection zoneswere found in the western areas. Growth of the top 10 individual tree families came from both Nebraska and South Dakota. Seventy percent grew well.

REFERENCES

Baer, N. W., & Collins, P. E. (1979). Ten-year performance of a ponderosa pine provenance study in eastern South Dakota. South Dakota State University Agricultural Experiment Station, Brookings, TB-52, 6 p.

Critchfield, W. B., & Little, E. L. Jr. (1966). *Geographic distribution of the pines of the world*. Washington, DC: USDA, Miscellaneous Publication 991, 97p.

Flint, H. L. (1983). *Landscape plants for eastern North America.* John Wiley & Sons, New York, 677 p.

Deneke, J. J., & Read, R. A. (1975). Early survival and growth of ponderosa pine provenances in East-Central Kansas. USDA Forest Service Research Note RM-297, Rocky Mountain Experiment Station, Ft. Collins, 4p.

Geyer, W. A. (2011). Evaluation of ponderosa pine seed sources for windbreaks in the Central Great Plains of the United States. *Arboriculture & Urban Forestry, 37,* 265-268.

Lambeth, C. (1980). Juvenile-mature correlations in Pineaceae and implications or early selection. *Forest Science, 26,* 571-580.

Koop, R. F., Geyer, W. A., Argent, R. M., & Lynch, K. D. (1987). Evaluation of ponderosa seed sources for the eastern Great Plains. *Journal Arboriculture, 13,* 139-144.

Read, R. (1983). Ten-year performance of ponderosa pine provenances in the Great Plains of North America. USDA Forest Service Research Paper,Rocky Mountain Experiment Station, Ft. Collins, 17p.

SAS Institute Inc. (2002-2003). *SAS/STAT user's guide* (Version 9.1). Cary, NC: SAS Institute Inc.

Schaefer, P. R., & Baer, N. W. (1985). Ponderosa pine provenance for windbreaks in eastern South Dakota. *Northern Journal of Forestry, 2,* 105-107.

Schaefer, P. R., & Baer, N. W. (1992). Stability of ponderosa pine provenance: results after 21 years in eastern South Dakota. *Northern Journal of Forestry, 9,* 102-107.

Van Haverbeke, D. F. (1986). Genetic variation in ponderosa pine: A 15-year test of provenances in the Great Plains. USDA Forest Service Rocky Mountain Forest Range Experiment Station Research Paper 80526, 16 p.

18

Local Communities Demand for Food Tree Species and the Potentialities of Their Landscapes in Two Ecological Zones of Burkina Faso

Daniabla Natacha Edwige Thiombiano[1*], Niéyidouba Lamien[2],
Ana M. Castro-Euler[3], Barbara Vinceti[4], Dolores Agundez[5], Issaka Joseph Boussim[1]

[1]Laboratoire de Biologie et d'Ecologie Végétales, Université de Ouagadougou, Ouagadougou, Burkina Faso
[2]Département de Production Forestière, Institut de l'Environnement et de Recherches Agricoles, Koudougou, Burkina Faso
[3]Instituto Estadual de Floresta, Embrapa Amapa Rodovia Juscelino Kubitschek, Macapa, Brazil
[4]Bioversity International, CGIAR, Headquaters: Via Dei TreDenari, Rome, Italy
[5]INIA-CIFOR, Dpto Sistemas y Recursos Forestales, Carretera de la Coruña, Madrid, Spain

We examined demand and supply of Food Tree Species (FTS) products in Burkina Faso. The hypotheses were: 1) demand for FTS products by local communities exceeds what can be sustainably extracted, and 2) local communities of the Sudanian zone have access to more diverse FTS products compared to Sahelian zone. Surveys were conducted in 300 households and 360 quadrats in landscapes surrounding 6 villages to determine the diversity, richness and availability of FTS. The results indicate that local communities tend to exploit FTS which are rare to find or absent in the landscape surrounding their village. While the range of FTS largely exploited tends to coincide across the two ecological zones, the diversity and density of the preferred FTS are discordant between the two zones. The results of the present study further support the need for conservation and restoration strategies to sustain the local communities demand for FTS products.

Keywords: Human Feeding; Food Security; Indigenous Knowledge; Quantitative Ethnobotany; Diversity, Threats

Introduction

Globally, 1.6 billion people strongly rely on forest resources for their livelihoods (Pimental et al., 1997). In most developing countries, forested landscapes play an important role for poor individuals and households (Langat & Cheboiwo, 2010). In sub-Saharan Africa, dry forest and woodlands surrounding rural settlements supply a vast array of wild natural resources (i.e., firewood, food, construction materials, medicine and fibers) for home consumption and sale (Shackleton & Gumbo, 2010).

In Burkina Faso, the basic diet in rural areas is not very diverse. The thick porridge processed from millet, sorghum or maize flour, locally referred to as "Tô", associated with sauce containing vegetables and condiments, constitutes the staple food. It is daily cooked five times per week and represents 83% of household meals (Soulama, 1990). Apart from the sauce, which can contain a variety of ingredients, starch is the main nutrient found in the porridge. With such dietary habits, what contributes to alleviating nutritional deficiencies in the diet of this country? Many studies have shown the pivotal role that Food Tree Species (FTS) play in balancing diets as additional ingredients and snacks, serving as staple foods during food

shortage periods, and generating income that enables to buy additional food products not available locally. Mertz et al. (2001) and Kristensen & Lykke (2003) in their studies in the south eastern and the south central parts of Burkina Faso respectively, have reported that about 17 trees species are used by local communities as ingredients for sauces or condiments. Products from a total of 21 fruit tree species were found to be consumed as snacks by children (Lamien et al., 2009). According to Shackleton et al. (2011) and Thiombiano et al. (2012), forests and woodlands supply rural dwellers with a wide range of foods, and contribute to food security and nutrition directly and indirectly by providing fruits, seeds and leaves. Most frequently, women are responsible for gathering, processing and selling non-timber forest products, generating income or using the products for home consumption; these activities indicate the key role played by women in food security strategies (Hasalkar & Jadhav, 2004; Shackleton et al., 2011).

Unfortunately, a sustainable provision of products and services by forest ecosystems is being threatened by their increasing degradation and accelerated conversion of forest land to alternative land uses (Langat & Cheboiwo, 2010). FAO (2001) statistics indicate that tropical regions lost 15.2 million hectares of forests per year during the 1990s. In Burkina Faso, Ouéd-

---

[*]Corresponding author.

raogo et al. (2010) found that conversion of forest land to crop-land proceeded at an annual rate of 0.96%, while human population density shifted from 17 to 30 inhabitants/km^2 in the 20 years between 1986 and 2006. Among the drivers of land use change, Paré (2008) identified demographic and economic factors as key ones, and detected a close correlation between biodiversity status or change in African ecosystems and the spatio-temporal variation in human population density, following the common pattern of a positive relationship between an increasing population density and magnitude of the threats to biodiversity.

Ethnobotanical research, in combination with quantitative ecological sampling methods, based on the use of plots or transects, has gained considerable importance over last decades (Cotton, 1996; Thomas et al., 2009). Despite the many ethnobotanical studies carried out in Burkina Faso (Mertz et al., 2001; Kristensen & Lykke, 2003; Belem et al., 2008; Lamien et al., 2009), up to now, little work has been undertaken using quantitative ethnobotany approaches, which enable to assess the sustainability in exploitation of non-timber forest products by local communities, and whether demand for these products goes beyond the potential supply. Knowledge of these elements is of paramount importance to gear policy-making towards a more sustainable use of the diversity of woody species, in a context of expanding cultivated lands due to a rapid population growth and of increasingly unpredictable precipitation patterns. The aim of the present study is to establish evidence that local communities demand for FTS products is exceeding the poten-

tial of their surrounding landscapes, thereby stimulating conservation and restoration actions. We formulated the following hypotheses: 1) the demand for FTS products by local communities exceeds what can be sustainable extracted within the surrounding landscape in both ecological zones, and 2) due to the more favorable environmental conditions (i.e., higher rainfall), the local communities of the Sudanian zone have access to a more diverse and richer range of FTS compared to those within the Sahelian zone in Burkina Faso. Therefore, surveys were conducted at households and landscape levels, to determine what FTS products are frequently used by households to meet their food and income needs, and to understand what is their availability in the surrounding landscapes.

Material and Methods

Study Area

The study was conducted at 6 sites located in the Sahelian and Sudanian ecological zones of Burkina Faso as they are described by Fontes and Guinko (1995). In each ecological zone (**Figure 1**), 3 villages in the Sudanian zone (namely Barcé, Péni and Sara) and 3 in the Sahelian zone (Barsalgho, Bourgou, & Pobé-Mengao) were randomly selected for the study from Burkina Faso's INSD (2010) most recent census database. The characteristics of the selected villages are summarized in **Table 1**

The sites in the Sahelian zone, the vegetation before human

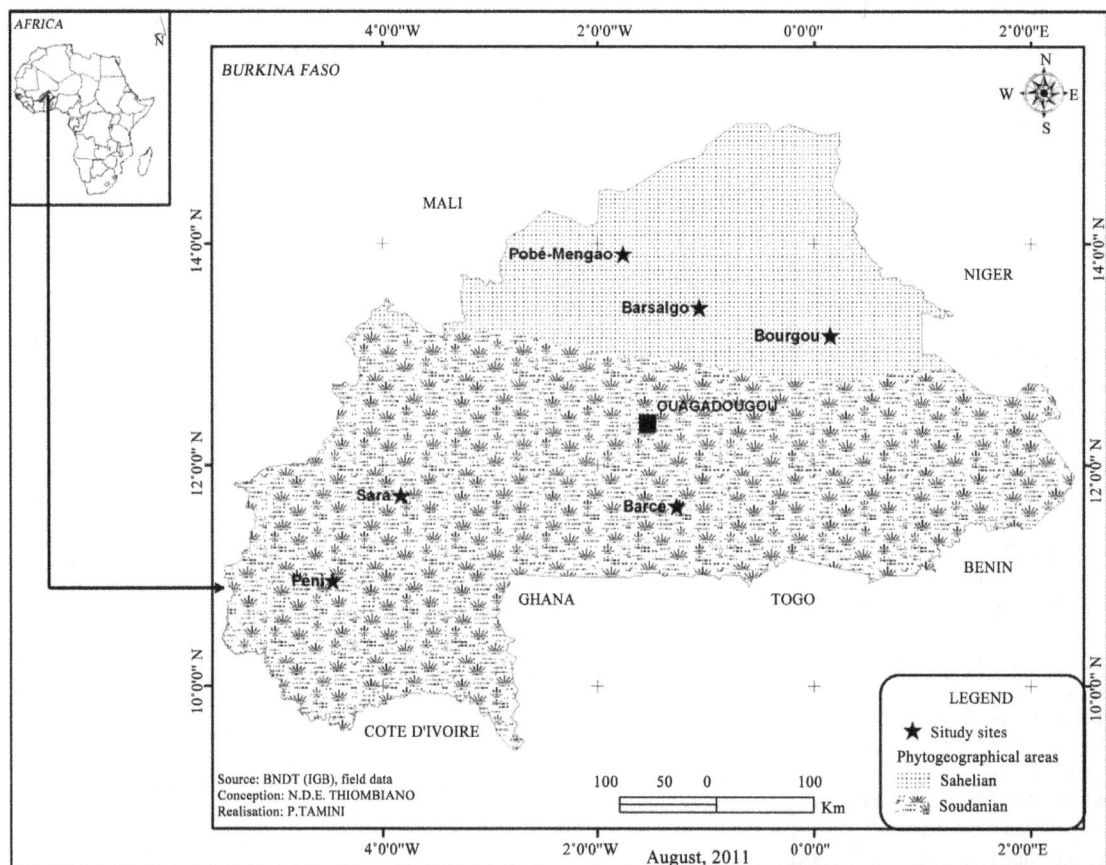

Figure 1.
Location of the different study sites in Burkina Faso.

Table 1.
Characteristics of the study sites.

Ecological zones	Villages	No households	% households surveyed	Population densities (Inh·km^{-2})	Coordinates	Annual rainfall (mm)
Sahelian	Barsalgho	1571	3	64	13°25'N, 01°04'W	400 - 600
	Bourgou	353	14	48	13°09'N, 00°09'W	
	Pobé-Mengao	578	9	28	13°53'N, 01°46'W	
Sudanian	Barcé	124	40	68	11°36'N, 01°16'W	700 - 1000
	Péni	659	8	82	10°56'N, 04°29'W	
	Sara	608	8	40	10°57'N, 04°25'W	

impact was semi-desert grasslands and thorny shrub lands to wooded grasslands and bush land. The traditional parkland systems (integrated crop-tree-livestock systems), are the main source of food, income and environmental services, but are rapidly deteriorating; biodiversity and cover of woody species is being lost, soil fertility is declining from already low levels due to exhaustive cropping practices and soil erosion (Bationo et al., 2003).

The sites in the Sudanian zone, the original vegetation is woodland and dry forest. At present, the main land uses type are: fallow, agro-forestry parklands with food crops, pockets of gallery forests without cultivation, some sacred hills. Clear-cutting systems are used in areas under both short and long term fallow.

Ethnobotanical Survey

In each ecological zone, 50 households per village were randomly selected using the village households list from the most recent population census database (INSD, 2010). A total of 300 informants from the selected 300 households were interviewed. The informants were all female as it is known in the region that women have a more significant role in the collection, trade and processing of the majority of non-timber forest products (Gausset et al., 2005). Using a semi-structured questionnaire, informants were asked to list down all the tree species which are sources of edible products, directly consumed as staple foods, snacks, vegetables or as supplements to diets during food shortage periods. Respondents were interviewed in isolation to avoid influencing each other's answers. Each interview was carried out in the informant's local language, in order to facilitate understanding of the questions and to obtain precise information on local dietary habits.

Ecological Sampling

The FTS inventory was carried out in the landscapes surrounding the 6 villages listed in **Table 1**. With regard to sampling design, among the various methods for abundance estimation described in Krebs (1999), line transect, based on individual quadrats as sampling units, was adopted, as recommended by Mahamane and Saadou (2008) for the type of vegetation found in the study sites. Four transects of 5 km each were established north, south, west and east from each village, in order to cover the FTS products harvesting areas around each village. The results of surveys showed that women, who generally cover long distances to harvest FTS products in the areas surrounding the village, cannot go beyond 5 km from their houses.

Along each transect, a total of 15 quadrats of 50×50 m^2 and 20×20 m^2 were defined respectively for the inventory of adult trees and seedlings (Mahamane & Saadou, 2008). A total of 360 quadrats were established in both Sahelian and Sudanian zones. The distance considered between two quadrats was 333 m to cover the 5 km radius. The smaller quadrats to record seedlings were sub-plots established within the larger quadrats of 50×50 m^2. FTS stems with girth ≥ 10 cm at 20 cm from the soil were considered as adult trees while stems with girth < 10 cm were considered as seedlings (Zida, 2007). The list of FTS to be inventoried was determined by the preferences expressed in the households' survey, in the same village. The number of stems of adult trees and seedlings was recorded.

Data Analysis

To establish the list of FTS and their relative importance in the households, the frequency of use per species was estimated by taking into account the proportion of times a plant was mentioned out of the total number of interviews. This technique reveals the relative importance of each cited species. The adult stems and seedlings densities of each species was estimated by extrapolating the number of individuals found in the 50×50 m^2 and 20×20 m^2 quadrats to a hectare. Descriptive statistics were computed from these derived variables. To compare adult trees and seedlings densities within and between ecological zones, the dataset was submitted to normality test. The Kolmogrov-Smirnov statistic was significant, indicating a violation of the assumption of normality. The comparison was therefore, performed using the Mann-Whitney U test, alternative to the independent-samples t-test (Pallant, 2010).

Among the various indices of diversity found in the literature, the reciprocal of Simpson's (D) index was used to estimate the FTS richness, as it provide a good estimate of diversity for a relatively small sample size (Magurran, 2004). The index was derived from households surveys and landscape-scale FTS inventories. We assumed the household similar to the quadrat in terms of number of species cited or found in the different units. The Richness Shared Index (RSI) suggested by Ladio and Lozada (2004) was adopted to analyze the richness cited per informant or found per quadrat with respect to the total richness mentioned by the total number of people interviewed or found in the total number of quadrats. An independent sample t-test was used to compare the Simpson reciprocal diversity index ($1/D$) and the Richness Share Index (RSI) between ecological zones and between households' survey and landscapes FTS data, which were normally distributed. The guidelines for Cohen (1988) partial eta squared statistic (η_p^2), reported in Pallant (2010), were used for effect size interpretation.

In order to determine how the species used in the households or found in the quadrats are similar among the different plant communities, we computed the Jaccard (*JSI*) and Morisita-Horn (C_{MH}) similarity indices (Magurran, 2004). The different secondary variables were calculated using the Microsoft Office Excel 2007 and all the statistical analyses were performed with IBM SPSS Statistic 19.

Results

Importance of FTS in Households and Landscapes

The proportion of households which have indicated a particular species as source of food, and the proportion of quadrats

where this species was found, according to the ecological zone, are presented in **Table 2**. A total of 30 FTS from 16 families were recorded in the household surveys and quadrat inventories. The inventory data showed that 13 species were common to the two ecological zones, 5 were specific to the Sahelian zone and 12 were specific to the Sudanian zone. The households survey resulted in 10 common species, 5 restricted to the Sahelian and 10 to the Sudanian zone.

In both ecological zones, species such as *Adansonia digitata*, *Lannea microcarpa*, *Parkia biglobosa* and *Vitellaria paradoxa* were mentioned by more than 70% of the respondents as sources of food, but were seldom found in the quadrats of the landscapes surrounding the villages, with the exception of *Vitellaria paradoxa* and *Parkia biglobosa* which were respec-

Table 2.

Proportions (%) of households (n = 300) and quadrats (n = 360) where FTS were recorded in the two ecological zones of Burkina Faso.

Species	Sahelian zone		Sudanian zone	
	Households	Quadrats	Households	Quadrats
Adansonia digitata	98	8	95	2
Afzelia africana	0	0	14	3
Annona senegalensis	0	0	5	15
Balanites aegyptiaca	1	66	1	7
Bombax costatum	74	1	28	10
Boscia senegalensis	6	9	0	0
Cadaba farinosa	3	4	0	0
Capparis corymbosa	0	0	2	1
Ceiba pentandra	0	0	5	0
Detarium microcarpum	1	0	11	13
Diospyros mespiliformis	4	4	0	16
Ficus sycomorus	0	1	1	1
Gardenia erubescens	0	0	13	2
Grewia bicolor	0	3	0	4
Landolphia heudelotii	0	0	10	1
Lannea acida	0	0	0	23
Lannea microcarpa	91	4	86	15
Leptadenia hastata	57	0	5	0
Maerua crassifolia	5	4	0	0
Parkia biglobosa	81	0	97	32
Piliostigma reticulatum	0	35	0	6
Saba senegalensis	39	1	73	1
Sclerocarya birrea	0	13	16	6
Securidaca longepedunculata	0	0	0	8
Strychnos spinosa	0	0	11	11
Tamarindus indica	1	7	0	2
Vitellaria paradoxa	71	3	97	77
Vitex doniana	1	0	17	6
Ximenia americana	0	2	19	5
Ziziphus mauritiana	0	12	0	0

tively encountered in 77% and 32% of the quadrats in the Sudanian zone. Some species such as *Afzelia africana*, *Gardenia erubescens*, *Landolphia heudelotii*, *Vitex doniana* and *Ximenia americana* were restricted to the Sudanian zone. On the opposite, very few informants from the Sahelian zone mentioned *Balanites aegyptica* and *Piliostigma reticulatum* as sources of food although they were encountered in 66% and 35% of the quadrats respectively. In the Sudanian zone, important species with similar use and presence patterns were *Lannea acida* (cf. 23% and 0% from household surveys and quadrat inventories, respectively) and *Diospyros mespiliformis* (cf. 16% and 0% from household surveys and quadrat inventories, respectively). *Piliostigma reticulatum* and *Securidaca longepedunculata* were not recorded in household surveys as sources of food but these

two species were considered in the quadrat inventories because it is well known from other sources that these species are used as food ingredients. *Leptadenia hastata*, a liana, was reported by half of the households in the Sahelian zone and few households in the Sudanian zone but the FTS inventory did not include it.

Landscapes' Potentialities of Food Tree Species

The potentialities of FTS in the landscapes are expressed in terms of list of the species that occur in the area, and the density of their adult trees and seedlings in the two ecological zones (**Table 3**). A Mann-Whitney U test revealed greater seedlings densities than adult trees, both in the Sahelian zone (seedlings

Table 3.
Adult trees and seedlings of FTS densities (Stems/ha ± sd) in the two ecological zones of Burkina Faso.

Species	Sahelian zone		Sudanian zone	
	Adult trees	Seedlings	Adult trees	Seedlings
Acacia macrostachya	0 ± 0	9 ± 6	0 ± 0	15 ± 10
Adansonia digitata	5 ± 3	4 ± 0	8 ± 7	4 ± 0
Afzelia africana	0 ± 0	0 ± 0	7 ± 6	32 ± 17
Annona senegalensis	0 ± 0	0 ± 0	13 ± 11	76 ± 72
Balanites aegyptiaca	15 ± 12	54 ± 65	10 ± 6	19 ± 16
Bombax costatum	8 ± 6	8 ± 0	5 ± 2	14 ± 10
Boscia senegalensis	11 ± 10	36 ± 31	0 ± 0	0 ± 0
Cadaba farinosa	5 ± 2	11 ± 11	0 ± 0	0 ± 0
Capparis corymbosa	0 ± 0	9 ± 4	4 ± 0	4 ± 0
Detarium microcarpum	0 ± 0	0 ± 0	24 ± 28	83 ± 100
Diospyros mespiliformis	10 ± 6	11 ± 8	7 ± 5	32 ± 35
Ficus ingens	0 ± 0	0 ± 0	0 ± 0	9 ± 2
Ficus sycomorus	4 ± 0	0 ± 0	10 ± 8	4 ± 0
Gardenia erubescens	0 ± 0	0 ± 0	9 ± 5	35 ± 12
Grewia bicolor	5 ± 2	8 ± 0	4 ± 0	13 ± 6
Landolphia heudelotii	0 ± 0	0 ± 0	4 ± 0	11 ± 9
Lannea acida	0 ± 0	0 ± 0	8 ± 6	23 ± 14
Lannea microcarpa	5 ± 2	4 ± 0	10 ± 7	13 ± 12
Maerua angolensis	0 ± 0	5 ± 2	0 ± 0	4 ± 0
Maerua crassifolia	5 ± 2	15 ± 13	0 ± 0	0 ± 0
Parkia biglobosa	0 ± 0	0 ± 0	10 ± 8	17 ± 17
Piliostigma reticulatum	10 ± 10	66 ± 96	9 ± 7	21 ± 26
Saba senegalensis	4 ± 0	0 ± 0	6 ± 3	6 ± 4
Sclerocarya birrea	6 ± 3	9 ± 6	9 ± 7	17 ± 18
Securidaca longepedunculata	0 ± 0	0 ± 0	8 ± 6	62 ± 119
Strychnos spinosa	0 ± 0	0 ± 0	9 ± 9	69 ± 58
Tamarindus indica	6 ± 5	8 ± 0	5 ± 2	10 ± 8
Vitellaria paradoxa	4 ± 0	0 ± 0	25 ± 23	110 ± 163
Vitex doniana	0 ± 0	0 ± 0	4 ± 1	15 ± 12
Ximenia americana	4 ± 0	9 ± 2	5 ± 2	21 ± 19
Ziziphus mauritiana	7 ± 5	18 ± 17	0 ± 0	8 ± 6

$Md = 16$, $N = 373$ and adult trees $Md = 8$, $N = 322$; $U = 31322$, $z = -11.08$, $p = 0.000$, $r = 0.42$) and the Sudanian zone (seedlings $Md = 24$, $N = 603$ and adult trees $Md = 8$, $N = 481$; $U = 72,503$, $z = -14.31$, $p = 0.000$, $r = 0.43$). The eta squared values ($\eta_p^2 = 0.42$ in Sahelian zone and 0.43 in Sudanian zone) indicate a large size effect within ecological zones.

Comparison between ecological zones revealed greater densities in the Sudanian than the Sahelian zone both for seedlings (Sudanian zone $Md = 24$, $N = 603$ and Sahelian zone $Md = 16$, $N = 373$; $U = 94,737$, $z = -4.16$, $p = 0$ .000, $r = 0.13$) and adult trees (Sudanian zone $Md = 8$, $N = 481$ and Sahelian zone $Md = 8$, $N = 322$; $U = 69,024$, $z = -2.74$, $p = 0$ .006, $r = 0.10$). The eta squared values ($\eta_p^2 = 0.13$ for seedlings and 0.10 for adult trees) also indicates a large effect between ecological zones.

Richness and diversity of FTS in households and landscapes.

The species richness and diversity were measured through the reciprocal of Simpson diversity index ($1/D$), the mean species richness per households and quadrat, and the Richness Share Index (RSI). These are presented in **Table 4**.

To determine the difference between Sahelian and Sudanian zones in the $1/D$ estimated from households' surveys and landscape FTS inventories, an independent-samples t-test was conducted. There was no significant difference in $1/D$ both for households (Sahelian $M = 6.726$, sd = 0.952 and Sudanian $M = 6.145$, sd = 0.576; $t(4) = 0.905$, $p = 0.43$ two-tailed) and landscapes (Sahelian $M = 3.283$, sd = 1.265 and Sudanian $M = 3.330$, sd = 0.276; $t(4) = -0.063$, $p = 0.95$ two-tailed). The magnitude of the $1/D$ differences (means difference = 0.58, 95% $CI = -1.202$ to 2.365) was larger ($\eta_p^2 = 0.16$) for households survey data than for landscapes inventory data (means difference = -0.047, 95% $CI = -2.123$ to 2.029; $\eta_p^2 = 0.0009$).

An independent-samples t-test was also conducted to compare the reciprocal of Simpson indices ($1/D$) between households' survey and landscape FTS inventory data sets within each of the two ecological zones. There was a significant difference in $1/D$ both for the Sahelian zone (Households $M = 6.726$, sd = 0.952 and Landscapes $M = 3.283$, sd = 1.265; $t(4) = 3.766$, $p = 0.020$ two-tailed) and the Sudanian zone (Households $M = 6.145$, sd = 0.576 and Landscapes $M = 3.330$, sd = 0.276; $t(4) = 7.627$, $p = 0.002$ two-tailed). The magnitude of the $1/D$ differences (households means difference = 3.44, 95% $CI = 0.905$ to 5.981; $\eta_p^2 = 0.78$-landscapes means difference = 2.81, 95% $CI = 1.790$ to 3.839; $\eta_p^2 = 0.93$) was considerable both for

Sahelian and Sudanian zones.

To compare the Richness Share Indices (RSI) between households' survey and landscape FTS inventory data sets in the Sahelian and Sudanian ecological zones, an independent-samples t-test was performed. There were significant differences in RSI both for Sahelian (Households $M = 0.511$, sd = 0.049 and Landscapes $M = 0.019$ sd = 0.00; $t(2) = 17.494$, $p = 0.003$ two-tailed) and Sudanian (Households $M = 0.470$, sd = 0.045 and Landscapes $M = 0.018$, sd = 0.000; $t(2) = 17.53$, $p = 0.003$ two-tailed). The magnitude of the RSI differences (households means difference = 0.491, 95% $CI = 0.028$ to 0.370; $\eta_p^2 = 0.99$) and landscapes means difference = 0.452, 95% $CI = 0.341$ to 0.563; $\eta_p^2 = 0.99$) were large for both ecological zones.

To compare the Richness Share Indices estimated from households' survey and landscape FTS inventory between Sahelian and Sudanian ecological zones, an independent-samples t-test was conducted. There was no significant difference in RSI for households data set (Sahelian $M = 0.511$, sd = 0.049 and Sudanian $M = 0.470$, sd = 0.045; $t(4) = 1.065$ $p = 0.35$ two-tailed), but significant difference for landscapes data set (Sahelian $M = 0.019$ sd = 0.00 and Sudanian $M = 0.018$, sd = 0.000; $t(4) = 5.682$, $p = 0.005$ two-tailed). The magnitude of the RSI differences (means difference = 0.041, 95% $CI = -0.065$ to 0.146) was large (eta squared = 0.22) and not significant for households data set, but greater and significant for landscapes data set (means difference = 0.002, 95% $CI = 0.001$ to 0.003; eta squared = 0.89).

Similarity of FTS between Ecological Zones and Informants Data Sets

To measure FTS similarity between the Sahelian and Sudanian zone, the Jaccard (JSI) and Morisita-Horn (C_{MH}) similarity indices were computed for households and landscapes data sets (**Table 5**). The JSI, which is based on the presence or absence of a species in a data set, indicates about 50% of similarity between FTS occurring in the Sahelian and the Sudanian zone, while the Morisita-Horn similarity index, which is based on the number of individual of a species recorded in a data set, indicates 86% of similarity between the two ecological zones for household surveys data. At the landscapes scale, 45% of similarity in FTS was detected between in the Sahelian and the Sudanian landscapes, according to Jaccard index, while the

Table 4.
Species richness.

Species richness Indices	Sahelian zone		Sudanian zone	
	Households	Landscape	Households	Landscape
Nb of families	11	12	14	16
Nb Genus	15	17	20	23
Nb of Species (S)	15	18	20	25
Reciprocal of Simpson index ($1/D$)	6.726 ± 0.952	3.28 ± 1.27	6.145 ± 0.576	3.33 ± 0.28
Mean species richness per household and quadrat	5 ± 0.1	2 ± 1	6 ± 0.1	3 ± 1
RSI	0.51 ± 0.03	0.02 ± 00	0.47 ± 0.03	0.02 ± 00

Table 5.
Recorded FTS similarity between the Sahelian and Sudanian zones in households survey and landscapes inventory data sets in Burkina Faso.

Ecological zones	Origin of data sets	Sahelian		Sudanian	
		Household	Landscape	Household	Landscape
Jaccard Similarity Index (*JSI*)					
Sahelian	Household	1	0.55	0.50	0.43
	Landscape		1	0.29	0.45
Soudanian	Household			1	0.69
	Landscape				1
Morisita-Horn's Index of Similarity (*C_{MH}*)					
Sahelian	Household	1	0.04	0.86	0.39
	Landscape		1	0.03	0.11
Soudanian	Household			1	0.74
	Landscape				1

Morisita index indicated only 11% of similarity.

Discussion

This study focuses on FTS products that rural communities use and the potential of the landscape surrounding their villages, to supply these products. Data were collected at households and landscapes levels to measure whether the current demand of FTS products by some selected rural communities in Burkina Faso, matches the supplying capacity of the environment around the villages, and assess whether exploitation patterns are sustainable.

Our results indicate that rural communities are still using tree species as source of food in Burkina Faso. A total of 30 species were recorded in the two main ecological zones (Sahelian and Sudanian). With regard to what species are considered most important, the results of this study are supported by those obtained from previous investigations: 15 tree species whose products are used to prepare sauces and relishes were reported also by Mertz et al. (2001); furthermore, 17 tree species which are source of edible fruits and ingredients for sauces in the Sudanian zone, were reported also by Kristensen and Lykke (2003). Finally, 14 of the tree species recorded in this study as supplements to diets in the Sahelian region were reported also in a study by Ganaba et al. (2002). These results seem to indicate a general consensus on what tree species constitute a notable source of foods, with some specificity according to ecological regions.

Comparing frequencies of species occurrence in households and landscape data sets, considerable differences emerged between ecological zones. In accordance with our first hypothesis, some important FTS (according to household surveys) occurred at low density in the landscapes surrounding the villages studied, suggesting limited potentialities of these species to match the demands of FTS products. This situation was further supported by the low density of adults, except for *Vitellaria paradoxa* in the Sudanian zone. For example, *Parkia biglobosa* was recorded as important in 81% of the households surveyed in the Sahelian zone but was totally absent in the landscapes surrounding the study sites, both in the stage of seedlings and adults. This indicates that its products were brought in the area from sudanian areas and sold in the market place. Cunningham (2000) reported that the first typical response to resource scarcity is to increase the harvest range.

For most species, density of seedlings appeared greater than that of adults, in each ecological zone, suggesting that conditions are favorable to recruitment. Nevertheless, seedling densities seem low if compared to those reported by Bognounou et al. (2009) for a tree species giving non edible products, *Pteleopsis suberosa* (Combretaceae), with 21630 to 26120 individuals ha^{-1} and those reported by Koadima (2008) for a FTS, *Securidaca longepeduncula* (Polygalaceae), with 7600 individuals ha^{-1} in landscapes surrounding villages of the Sudanian zone. Authors such as Boffa (2000) and Kozlowski (2002) pointed out the difficulties encountered by most tree species to regenerate and reach the adult stage, notably on annual cropland, due to various factors (failure of seeds to germinate, predators, pathogens, seedlings mortality and anthropogenic disturbances).

With regard to our second hypothesis about a predicted more diverse and richer wealth of FTS in the Sudanian zone, compared to the Sahelian zone, due to its higher rainfall, the comparison of the reciprocal of Simpson's diversity indices between the two zones did not exhibit any difference both for household surveys and landscape inventories datasets. The similarities in landscapes data could be explained by the fact that each zone has specific FTS but a large number of FTS are common. Similarities in households data can be explained by the fact that even if species such *Parkia biglobosa* and *Detarium microcarpa* were absent from the Sahelian landscape inventoried, these species appeared in the associated household surveys because traders import their products from the Sudanian zone. The trade of non-timber forest products across regions in Burkina Faso is considerable. Kernels of *Sclerocarya birrea* seeds and dry leaves of *Adansonia digitata* are largely exported from the Sahelian to the Sudanian zone (Lamien & Traoré, 2002), while flour and seeds of *Parkia biglobosa*, and dry fruits of *Detarium microcarpa*, are exported from the Sudanian to the Sahelian zone (Lamien N., personal observations in 2008). Fresh fruits of *Saba senegalensis* and *Lannea micro-*

carpa are also exported from production areas to main cities (Lamien et al., 2010). The comparison of the $1/D$ index between households and landscapes data sets indicates greater values of $1/D$ in households than landscapes data both in the Sahelian and Sudanian zones. The explanation could be linked to convergent customary habits, across households within the same ecological zone, of exploitation of a well-defined set of FTS, since ancestral times. Household surveys tend to produce homogeneous results for what concerns the range of most used species, with very limited differentiation. This is also supported by Kristensen and Lykke (2003) who found that all their informants had given the same answer to 28% of their questions. This suggests a consolidate transmission of traditional knowledge from a generation to the next, generating a cultural inertia in the use of natural resources found in the surrounding environment (Ladio & Lozada, 2004; Cotton, 1996). This finding is further supported by other results in this study, such as the mean species richness per household and the RSI, whose values also are greater compared to those found from the landscapes data set; however, for these values, no significant difference between the Sahelian and Sudanian zone was found, suggesting convergent patterns between households of the two ecological zones. Nevertheless, differences in landscapes' RSI values between the two ecological zones, although very low, were significant, indicating a certain variability of species richness. The high Jaccard Similarity Index did not support this situation which seems more related to the species individuals as indicated by the low Morisita-Horn's similarity index.

Conclusion

In accordance with the first hypothesis formulated, this study reveals that the demand of FTS products by local communities, in some selected sites from two ecological zones of Burkina Faso, is exceeding the supply generated within in the areas surrounding the rural communities investigated. There is evidence that without adequate forest conservation and restoration initiatives, the areas surrounding villages will no longer meet the demand for FTS products. Lack of availability in a particular region is compensated by import of products from other areas. However, even those regions where climatic conditions are more favorable to support a more diverse range of FTS (e.g., Sudanian ecological zone), the most important FTS occur at low frequencies, both as seedlings and adults. The diversity and richness of FTS used by local communities are similar across the two regions, showing a tendency to utilize similar FTS. This seems linked to customary habits. On the contrary to the second hypothesis formulated in this paper, there is no evidence that the range of FTS in the Sudanian zone is more diverse and richer compared to that of the Sahelian zone of Burkina Faso, although densities of individual species are significantly different between the two zones. Our results call for conservation strategies to sustain the local communities' food tree species demand.

Acknowledgements

The authors are grateful to Africa-Brazil agricultural innovation marketplace and the National Institute for Agricultural and Food Scientific Research and Technology, (INIA, Spain), through a collaboration between Bioversity International, the Forest Research Center (CIFOR) of INIA, and several national partners from African countries, under the umbrella of SA-FORGEN, the sub-Saharan African Forest Genetic Resources programme, which supported financially this study. Thanks to Dr Patrice Savadogo for his valuable comments on the manuscript.

REFERENCES

Bruce Bationo, A., Traore, Z., Kimetu, J., Bagayoko, M., Kihara, J., Bado, V., Lompo, M., Tabo, R., & Koala, S. (2003). Cropping systems in the Sudano-sahelian zone: Implications on soil fertility management. http://www.syngentafoundation.org/db/1/432.pdf

Belem, B., Smith-Olsen, C., Theilade, I., Bellefontaine, R., Guinko, S., Lykke, A. M. et al. (2008). Identification des arbres hors forêt préférés des populations du Sanmatenga (Burkina Faso). *Bois et-Forêts des Tropiques, 298,* 53-60.

Boffa, J. M. (2000). West African agroforestry parklands: Keys to conservation and sustainable management. *Unasylva, 51,* 11-17.

Bognounou, F., Savadogo, P. Thiombiano, A., Tigabu, M., Boussim, I. J., Oden, P. C., & Guinko, S. (2009). Impact of disturbance from roadworks on *Pteleopsis suberosa* regeneration in roadside environments in Burkina Faso, West Africa. *Journal of Forestry Research, 20,* 355-361.

Cotton, C. M. (1996). *Ethnobotany: Principles and applications.* Chichester, England: John Wiley and Sons.

Cunningham A. B. (2000). *Applied ethnobotany: People, wild plant use and conservation* (300 p). London: Earth Scan.

FAO (2001). *Global forest resource assessment.* Rome: Food and Agriculture Organization of the United Nations.

Fontes, J., & Guinko, S. (1995). *Carte de végétation et de l'occupation d'un sol du Burkina Faso. Notice explicative. Ministère de la coopération française, Projet campus* (8813101). Toulouse: Université Paul Sabatier.

Ganaba, S., Ouadba, J. M., & Bognounou, O. (2002). Utilisation des ressources végétales spontanées comme complément alimentaire en région sahélienne du Burkina Faso. *Annales de Botanique de l'Afrique de l'Ouest, 2,* 101-112.

Gausset, Q., Yago-Ouattara, E. L., & Belem, B. (2005). Gender and trees in Péni, South-Western Burkina Faso. Women's needs, strategies and challenges. *Danish Journal of Geography, 105,* 67-76.

Hasalkar, S., & Jadhav, V. (2004). Role of Women in the use of Non-Timber. Forest produce: A Review. *Journal Social Science, 8,* 203-206.

INSD (2010). *Résultats préliminaires du recensement général de la population et de l'habitat de 2009.* Ouagadougou: Institut National des Statistiques et de la Démographie (INSD).

Koadima, M. (2008). *Inventaire des espèces ligneuses utilitaires du parc W et terroirs riverains du Burkina Faso et état des populations de trois espèces à grande valeur économics* (56 p). Ouagadougou: Université de Ouagadougou.

Kozlowski, T. T. (2002). Physiological ecology of natural regeneration of harvested and disturbed forest stands: Implication for forest management. *Forest Ecology and Management, 158,* 195-221.

Krebs, J. C. (1999). *Ecological methodology.* New York: Addison-Wesley Educational Publishers Inc.

Kristensen, M., & Lykke, A. M. (2003). Informant-based valuation of use and conservation preferences of savanna trees in Burkina Faso. *Economic Botany, 57,* 203-217.

Ladio, A. H., & Lozada, M. (2004). Patterns of use and knowledge of wild edible plants in distinct ecological environments: A case study of a Mapuche community from north western Patagonia. *Biodiversity and Conservation, 13,* 1153-1173.

Lamien, N., & Traoré, S. (2002). Commercialisation des produits non ligneux des arbres de la zone semi aride du Burkina Faso: Cas des feuilles sèches de Baobab (*Adansonia digitata* L.). In *Nouer des liens entre la recherche et le développement en agroforesterie dans les basses terres semi arides d'Afrique de l'ouest, 2ème atelier régional sur les aspects socio-économiques de l'agroforesterie au sahel* (pp. 1-9), Bamako: ICRAF.

Lamien, N., Lingani-Coulibaly, P., & Traore-Gue, J. (2009). Importance of local fruits consumption in diet balance in Burkina Faso, West Africa. *ActaHorticulturae, 806*, 203-208.

Lamien, N., Traore, S., & Kini, F. (2010). Potentialités productive, nutritive et économique de la liane goïne (*Saba senegalensis* A. DC. Pichon) dans le sahel burkinabè. *Sahelian Studies and Research, 15*, 115-124.

Langat, D., & Cheboiwo, J. (2010). To conserve or not to conserve: A case study of forest valuation in Kenya. *Journal of Tropical Forest Science, 22*, 5-12.

Magurran, A. E. (2004). *Measuring biological diversity* (234 p). Oxford: Black-Well Science Ltd.

Mahamane, A., & Saadou, M. (2008). Méthodes d'étude et d'analyse de la flore et de la végétation tropicales (82 p). Actes de l'atelier sur l'harmonisation des méthodes, tenu à Niamey du 4 au 9 août 2008.

Mertz, O., Lykke, A. M., & Reenberg, A. (2001). Importance and seasonality of vegetable consumption and marketing in Burkina Faso. *Economic Botany, 55*, 276-289.

Ouedraogo, I., Tigabu, M., Savadogo, P., Compaore, H., Oden, P. C., & Ouadba, J. M. (2010). Land cover change and its relation with population dynamics in Burkina Faso, West Africa. *Land Degradation and Development, 21*, 1-10.

Pallant, J. (2010). *SPSS survival manual: A step by step guide to data analysis using the SPSS program* (4th ed.). England: McGraw-Hill Companies.

Paré, S. (2008). *Land use dynamics, tree diversity and local perception of dry forest decline in Southern Burkina Faso, West Africa*. Ph.D. Thesis, Uppsala: Swedish University of Agricultural Sciences.

Pimental, D., Nair, M. Mc., Buck, L., Pimental, M., & Kamil, J. (1997). The value of forests to world food security. *Human Ecology, 25*, 91-120.

Shackleton, S., & Gumbo, D. (2010). Contribution of non-wood forest products to livelihoods and poverty alleviation. In E. Chidumayo, & D. A. Gumbo (Eds.), *The dry forests and woodlands of Africa: Managing for products and services* (pp. 63-91). The Earthscan Forestry Library.

Shackleton, S., Charlie, I., Shackleton, C., & Shanley, P. (2011). *Non-timber forest products in the global context* (285 p). Heidelberg, Dordrecht, London, New York: Springer,

Soulama, S. (1990). Analyse économique des systèmes et structures alimentaires en zones à déficit céréaliers au Burkina Faso. In *Stratégies et politiques alimentaires au Sahel. De la recherche à la prise de décision* (pp. 175-198). UO/CILSS/Centre Sahel UL.

Thiombiano, D. N. E., Lamien, N., Dibong, D. S., Boussim, I. J., & Belem, B. (2012). Le rôle des espèces ligneuses dans la gestion de la soudure alimentaire au Burkina Faso. *Sécheresse, 23*, 86-93.

Thomas, E., Vandebroek, I., Van Dammea, P., Goetghebeur, P., Douterlungne, D., Sanca, S., & Arrazola, S. (2009). The relation between accessibility, diversity and indigenous valuation of vegetation in the Bolivian Andes. *Journal of Arid Environments, 73*, 854-861.

Zida, D. (2007). *Impact of forest management regimes on ligneous regeneration in Sudanian Savanna of Burkina Faso*. Doctoral Thesis, Uppsala: Swedish University of Agricultural Sciences.

19

The Relative Importance of Natural Disturbances and Local Site Factors on Woody Vegetation Regeneration Diversity across a Large, Contiguous Forest Region

Gerardo P. Reyes[1,2*], Daniel Kneeshaw[1], Louis de Grandpré[1,3]

[1]Department of Biological Sciences, Centre for Forest Research, University of Quebec in Montreal, Montreal, Canada
[2]Faculty of Interdisciplinary Studies, Lakehead University, Orillia, Canada
[3]Natural Resources Canada, Canadian Forest Service, Laurentian Forestry Centre, Ste-Foy, Canada

Stand-level diversity after natural disturbance can potentially differ across a large, contiguous forest region despite being dominated by the same canopy species throughout as differences in disturbance types and local site conditions can regulate species distribution. Our main objective was to examine the relative importance of natural disturbances (spruce budworm (*Choristoneura fumiferana*) outbreak, windthrow, and their interaction) and local site factors (climate, physiography, and stand structure and composition variables) on woody vegetation diversity among three, physiographically distinct locations across a large, contiguous forest region. Seventy-six *Abies balsamea-Betula* spp. stands affected by natural disturbance were compared and analysed using canonical ordination methods, diversity indices, and ANOVA. Different combinations of factors were important for vegetation re-establishment at each location. Differences in alpha (α), beta (β), gamma (γ), Shannon's *H'*, and evenness (*J*) diversity indices were observed among locations across the study region. Our findings indicate that while certain processes are important for maintaining canopy dominance by *Abies balsamea* and *Betula* spp. throughout the region, different combinations of factors were important for creating variation in woody species diversity among locations that resulted in greater woody species diversity at the regional scale.

Keywords: Regeneration; Natural Disturbance; Environmental Factors; Species Diversity; Eastern Boreal Mixedwood Region

Introduction

Climate determines the regional species pool (Dansereau, 1954; Neilson et al., 1992; Messaoud et al., 2007). However, woody species distribution and abundance across the landscape is largely determined by the disturbance regime (Gromtsev, 2002; Belote et al., 2009). Consequently, a distinct spatio-temporal arrangement of woody species is often observed at local scales due to variation in disturbance type, severity, and frequency across the landscape (Chen & Popadiouk, 2002; Reyes et al., 2010). Apart from natural disturbances, local site factors also play key roles in forest development, as they are capable of influencing the disturbance regime itself, and can directly affect species diversity after disturbance (Peterson, 2000; McGuire et al., 2002; Wang et al., 2006).

Different types of disturbance influence regeneration patterns in distinctive ways depending on the severity of effects on forest vegetation and stand structure, and the variation in the temporal release of newly available resources. Windthrow and spruce budworm (*Choristoneura fumiferana*) outbreak, for example, have been shown to result in alternative successional

trajectories in boreal ecosystems due to differences in the degree of soil disturbance and in the timing of tree mortality (Krasny & Whitmore, 1992; Nagel & Diaci, 2006). In the case of windthrow, this disturbance type can substantially alter canopy species dominance whereas budworm damage can simply delay or promote development towards late-succession conditions (Bergeron, 2000; Chen & Popadiouk, 2002). However, variation in forest regeneration among stands experiencing the same disturbance type can also occur because of differences in local site conditions (Turner et al., 1997; Reyes & Kneeshaw, 2008). For example, where advance regeneration is abundant, windthrow may cause only structural *versus* compositional changes in the tree canopy (Peterson, 2000; Reyes et al., 2010). Moreover, the sub-canopy tree and shrub community can display variation in responses to natural disturbances depending on light availability and level of soil disturbance; and depending on the relative speed of development and growth after disturbance, can ultimately impact the composition of the regenerating tree canopy (Gray et al., 2005; Hart & Chen, 2006). Other local site factors such as pre-disturbance stand composition, physiography, and stand structural characteristics have been shown to affect woody vegetation diversity after natural disturbance at the stand scale (Dupont et al., 1991; Osumi et al., 2003;

---

*Corresponding author.

Rodriguez-Garcia et al., 2011). However, it is unclear if the same factors play key roles in determining re-vegetation patterns throughout an entire forest region. Factors can differ in importance among locations across a region, operate at local or regional scales, and can work synergistically to affect species diversity (Wimberly & Spies, 2001; Hillebrand & Blenckner, 2002; Huebner & Vankat, 2003). Examining woody vegetation distribution after natural disturbances at various locations across a large, contiguous forest region dominated by the same canopy species can thus help to resolve potential contrasts by determining the different factors responsible for any observed variation across the region, for differences in responses to the same factors, or conversely, potentially link fine and broad-scale vegetation recovery patterns to a few important factors that have consistent effects on woody species diversity across the region as a whole.

The purpose of our study is to document the relative importance of natural disturbances and local site factors on woody vegetation regeneration diversity among three, physiographically distinct locations across a large, contiguous forest ecosystem dominated by the same canopy species throughout. We specifically addressed the following questions: is the canopy composition maintained after disturbance or are different successional trajectories occurring? Does the type of natural disturbance determine woody species diversity or do local site factors more strongly influence regeneration distribution? Alternatively, do natural disturbances work synergistically with local site factors in determining woody species diversity? Lastly, do different combinations of disturbances and local site factors influence woody species diversity for specific locations or can a few common factors explain species distribution across the region? We hypothesized that given the physiographical differences among locations across the region, that different combinations of factors would be responsible for woody species diversity after natural disturbance at the local level; and consequently, that if different factors influence regeneration dynamics at local scales, it is likely that species diversity will differ across the region. Thus, although woody species diversity should be the highest at the regional scale, diversity should also reflect the intra-regional differences in disturbance history and the unique local site characteristics of these forests.

Study Area

The *Abies balsamea-Betula* spp. boreal mixedwood zone spans east to west across southern Quebec, Canada from 46° to 50°N and 64° to 80°W (**Figure 1**). It encompasses 23.8 million ha and represents 18.6% of the forested land of the province (MRNQ, 2003). The physiography of the region is highly variable. The western portion is continental, and is relatively flat to rolling. Topography becomes increasingly hilly to montane towards the east, as the eastern portion of the region includes the northeastern limit of the Appalachian mountain chain that runs southwest into the United States. The eastern portion also borders the Gulf of St. Lawrence, and thus contrasts with the western portion of the region by having a maritime influence. Elevations of forested areas range from 80 to 400 m in the west, and from sea level to 900 m in the east. Surface deposits are mostly glacial till or of lacustrine origin (Robitaille & Saucier, 1998). The soil moisture regime is classified as xeric-mesic to mesic, while soil drainage ranges from imperfect to rapid.

Numerous disturbances such as fire, insect pests, and windthrow occur at varying frequency and severity in eastern boreal mixedwoods. The current return interval length for catastrophic

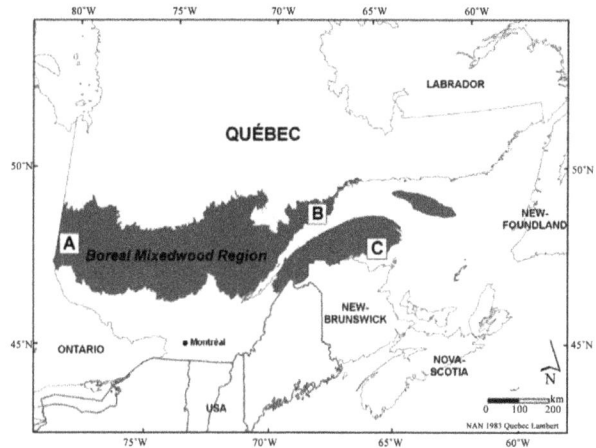

Figure 1.
Study locations within the boreal mixedwood region of Quebec, Canada. (A) continental—Abitibi-Temiskaming; (B) northern coastal—the North Shore; (C) southern coastal—the Gaspé Peninsula.

fires across the region ranges from 170 to 645 years (Bergeron et al., 2006), suggesting that less severe, but more frequent disturbances now assert greater influence on post-disturbance boreal mixedwood dynamics than in the recent past. Thus, we focused our attention on three types of non-fire disturbances that are common to boreal mixedwood forests: spruce budworm (*Choristoneura fumiferana*) outbreak, windthrow, and their interaction (nb., where a stand affected by spruce budworm outbreak is then subjected to windthrow prior to canopy recovery).

We sampled within three physiographically distinct, widely dispersed locations (between 200 and 500 km) across the boreal mixedwood region of Quebec, Canada (**Figure 1**). The western sites are located in the Abitibi and Temiskaming municipalities (77°30' to 79°10'W and 47°30' to 48°20'N). These sites have a flat to rolling topography and are influenced by continental climate conditions. The two eastern coastal locations both have hilly to rolling topography but differ mainly by latitude and elevation. The northern coastal sites are situated along the north shore of the St. Lawrence River, within the Haute Côte Nord and Manicouagan municipalities, and between 48°30' to 50°00' N and 68°00' to 69°50'W. Stands in this location are near the northern limit of the boreal mixedwood region. The southern coastal sites are situated within the southern part of the Gaspé Peninsula, along the Chaleur Bay area of the Atlantic Ocean (between 48°10' to 48°35'N and 65°45' to 66°15'W). Other climate, physiographic, and forest structure and composition characteristics are compared among geographic locations in **Table 1**.

Relative species composition of the forest canopy can be quite variable within this forest type. We limited sampling to stands having two of the more common relative species compositions: stands were either conifer-dominated (≥75% density of conifers in canopy) or mixed coniferous-deciduous (50% - 74% density of conifers in canopy) prior to disturbance. In all cases, *Abies balsamea*, *Betula papyrifera*, and *Betula alleghaniensis* dominated the forest canopy prior to disturbance, and represented at least 60% of the coniferous and deciduous components within each stand, respectively. *Picea glauca*, *Picea mariana*, and *Acer rubrum* were also abundant in some sites while *Thuja occidentalis*, *Pinus resinosa*, *Pinus strobus*, *Pinus*

Table 1.
Comparison of climate, physiographic, and forest structure and composition variables at each location using factorial ANOVA. Values in parentheses indicate ±1 Standard Error of the mean. Unlike letters along a row indicate values are significantly different among locations (p < 0.05).

Variable	Location		
	Continental	Northern coastal	Southern coastal
Annual rain (mm)	691.3 (5.2) **a**	691.0 (7.5) **a**	724.3 (2.0) **b**
Annual snow (mm)	274.3 (6.2) **a**	303.4 (4.5) **b**	238.6 (0.6) **c**
Annual precipitation (mm)	965.6 (11.4) **ab**	988.9 (10.3) **b**	963.1 (1.4) **a**
Annual temperature (°C)	2.1 (0.1) **a**	2.2 (0.1) **a**	3.6 (0.1) **b**
Mean monthly temperature-summer (°C)	16.3 (0.1) **a**	15.3 (0.1) **a**	16.6 (0.1) **a**
Mean monthly temperature-winter (°C)	−10.4 (0.1) **a**	−8.8 (0.1) **b**	−7.3 (0.1) **c**
Mean windspeed (km·h^{-1})	12.9 (0.1) **a**	15.6 (0.0) **b**	18.7 (0.1) **c**
Maximum windspeed (km·h^{-1})	52.8 (0.7) **a**	75.6 (0.1) **b**	76.0 (0.2) **b**
Maximum gust speed (km·h^{-1})[†]	98.4	105.2	no data
Physiography			
Elevation (m)	311.2 (11.8) **a**	274.2 (25.4) **a**	421.4 (9.7) **b**
Slope[*] (%)	1.0 (0.0) **a**	1.2 (0.2) **a**	2.1 (0.1) **b**
Latitude (degrees)	47.3 (0.1) **a**	49.0 (0.1) **b**	48.5 (2.3) **b**
Longitude (degrees)	78.4 (0.1) **a**	68.8 (0.1) **b**	66.1 (2.8) **c**
Soil drainage[†]	3.5 (0.2) **a**	3.1 (0.2) **a**	2.3 (0.1) **b**
Forest structure and composition			
Mean disturbance area (ha)	11.6 (1.6) **ab**	8.9 (1.7) **a**	13.2 (1.0) **b**
Snag density (ha^{-1})	138.2 (17.8) **a**	294.1 (33.3) **b**	499.3 (29.4) **c**
Course woody debris density (ha^{-1})	1176.4 (10.9) **a**	1729.1 (82.2) **b**	2159.9 (105.2) **c**
Percentage of trees uprooted	28.0 (3.5) **a**	31.9 (3.4) **a**	37.2 (1.9) **a**
Decay class average [‡]	11.2 (0.1) **a**	11.2 (0.1) **a**	10.4 (0.1) **b**
Tree regeneration density (ha^{-1})	126 + 907.9 (1936.6) **a**	36825.6 (2593.6) **b**	15284.5 (1525.3) **a**
Shrub regeneration density (ha^{-1})	11982.9 (2288.1) **a**	33,969.6 (3407.6) **b**	7 454.0 (732.1) **a**
Total regeneration density (ha^{-1})	24890.7 (2043.0) **a**	70,895.2 (3279.9) **b**	22 738.6 (1463.2) **a**
Coniferous legacy tree density (ha^{-1})	1.3 (.3) **a**	8.9 (1.2) **b**	0.1 (0.2) **c**
Deciduous legacy tree density (ha^{-1})	0.1 (0.03) **a**	2.7 (0.6) **b**	0.2 (0.2) **a**
Crown cover of coniferous legacy trees (%)	3.5 (0.8) **a**	3.5 (0.9) **a**	4.2 (1.0) **a**
Crown cover of deciduous legacy trees (%)	1.6 (0.7) **a**	1.7 (0.6) **a**	6.8 (1.2) **b**
Stand density prior to disturbance (ha^{-1})	926.6 (89.3) **a**	1474.0 (83.8) **b**	1718.7 (80.7) **b**

Note: [†]Insufficient data to include in analyses. [*]1: 0° to 10°, 2: >10° to 20°, 3: >20° to 30°, 4: >30° to 40°, 5: >40°; [†]2: good, 3: moderate; 4: imperfect; [‡]See Imbeau & Desrochers (2002).

banksiana, Larix laricina, Acer saccharum, Fraxinus americana, Fraxinus nigra, Populus balsamifera, and *Populus tremuloides* were occasionally present.

Sampling Methods

Disturbance type, climate, and physiographical data were derived from digitised aerial photos, eco-forestry maps, or sourced from various provincial and federal government agencies (MNRQ, 2003; Environment Canada, 2004), with the exception of slope and elevation, which were determined on site using a clinometer and GPS unit, respectively. A total of 76 sites affected by spruce budworm outbreak, windthrow, or their interaction were examined: 24, 22, and 30 sites for the continental (Abitibi-Temiskaming), northern coastal (North Shore), and southern coastal (Gaspé Peninsula) locations, respectively.

Up to six 20 × 20 m quadrats were sampled within each disturbed site. We limited our sampling to mature stands (≥80 years old-verified within each site by counting annual rings of remnant trees or old stumps) having at least 0.2 ha of contiguous canopy mortality to reduce variation resulting from differences in stand age, disturbance severity, and the spatial extent of disturbance. Further, all quadrats were at least 40 m from the nearest intact forest and 30 m from the nearest logging road to avoid edge effects.

Density of tree and shrub regeneration was quantified within each quadrat for three size classes: (1) 1 to 2 m tall, (2) >2 m tall and <4 cm dbh (1.37 cm), and (3) between 4 and 8 cm dbh using a nested plot design. Different height classes were used to determine which individuals were more or less likely to be recruited into the canopy. Some species in the region such as *Abies balsamea* can produce 1000 s of regenerating seedlings per hectare, many of which remain suppressed even after canopy disturbance (Reyes & Kneeshaw, 2008). Class 1 regeneration was sampled in a 2 × 10 m area, class 2 in a 5 × 10 m area, and class 3 within the entire quadrat. Density of snags (standing dead trees >10 cm dbh) and coarse woody debris (downed trees >10 cm dbh) were determined and classified using a modified decay classification scale developed by Imbeau & Desrochers (2002). Coarse woody debris was also categorised as uprooted or snapped when possible; n.b., past research has shown that certain boreal species establish better in exposed mineral soils *versus* other substrates (Kuuluvainen & Juntunen, 1998). Thus, if more coarse woody debris is uprooted *versus* snapped, then one could expect a greater proportion of these types of species as well. Crown cover (m^2) of legacy trees (mature overstory trees that survived the natural disturbance) within each quadrat, and density of deciduous and coniferous legacy trees within a 35 m radius of the quadrat centre were also determined. This was done to provide some indication of the amount of shade available to the regeneration, which can thus potentially influence what species regenerate in close proximity, and provide information on potential sources of seed rain, respectively.

Analyses

Various analyses were used to compare and contrast woody vegetation diversity after natural disturbance between the continental, northern coastal, and southern coastal locations. Comparisons of the climate, physiographic, and forest structure and composition characteristics that were quantified for each location were made using analysis of variance (ANOVA) (SPSS 10.0, 1999) followed by the Student-Newman-Keuls multiple range test when significant differences were observed (at $p < 0.05$) (**Table 1**).

Direct Gradient Analyses

We used a series of redundancy analyses (RDA) (van den Wollenberg, 1977) to examine the relationships between local site factors (i.e., the various disturbance types, climate, physiographic, and stand structure and composition variables) and woody vegetation species distribution for each of the three study locations. For each analysis, the forward selection option was implemented to both rank the importance of each site factor variable and to exclude redundant and non-significant variables from the model. The significance of the explanatory effect of a site factor variable was determined using a Monte Carlo permutation test (200 permutations, $p < 0.05$) prior to the addition of the next best fitting variable. CANOCO 4.0 software (ter Braak & Smilauer, 1998) was used to run the analyses. Variables were centred and standardized as the site factor variables were measured using different units.

Species Diversity Estimates

Five measures of species diversity were calculated at two spatial scales: for both the entire study region and for each of the three study locations (continental, northern coastal, southern coastal). Thus, comparisons between regional and local diversity levels, as well as among the three study locations could be made. Calculations were also made for the entire woody vegetation community and for each of the woody vegetation layers separately; i.e., canopy tree regeneration *versus* sub-canopy tree & shrub regeneration. This was done as different disturbance types can affect the understory community differently (Veblen & Ashton, 1978; Hart & Chen, 2006). Thus, diversity of the canopy tree regeneration may be influenced by the sub-canopy tree and shrub community that survived the disturbance or by sub-canopy species that can quickly establish soon after (Gray et al., 2005).

Regional level species diversity estimates were the following: alpha diversity (α) represented mean site level richness, beta diversity (β) represented differences in richness among sites across the study region; i.e., the differences in species composition among spatial units, while gamma diversity (γ) represented the total richness across the study region (Whittaker, 1960; Novotny & Weiblen, 2005). Beta diversity (Whittaker, 1960) was computed as:

$$\beta = (\gamma/\alpha) - 1 \quad . \tag{1}$$

Shannon's diversity index (Shannon & Weaver, 1949):

$$H' = -\sum p_i \ln p_i \tag{2}$$

where p_i = proportion of the total sample belonging to species i (in our case, the relative density of a species), and evenness (J):

$$J = H'/\ln \alpha \tag{3}$$

where J is an index of how relative abundances of species are distributed (Pielou, 1966), were determined for each regeneration layer for each site. Analysis of variance followed by the Student-Newman-Keuls multiple range tests were used to compare the various diversity estimates among study locations and vegetation layers ($p < 0.05$).

Species richness is partly a function of spatial scale (Palmer & White 1994). We acknowledge that differences in richness could be an artifact of sampling effort among locations. The total disturbed area examined for the continental, northern coastal, and southern coastal sites were 405.2, 505.6, and 1238.3 ha, respectively. We sampled the three locations using 35, 57, and 95 quadrats, respectively, to make sampling effort more equitable among locations. However, bias in species richness may also occur due to the unequal number of quadrats. Sample rarefaction (Krebs, 1989) was used to compute a species accumulation curve as a function of the number of quadrats examined for each location. PAST version 1.94b (Hammer et al., 2001) was used to run analysis. Results show that differences in species richness due to differences in the number of quadrats examined in each location was negligible (**Figure 2**). Thus, we felt that making comparisons of diversity estimates among the three locations using our sampling protocol was a valid undertaking.

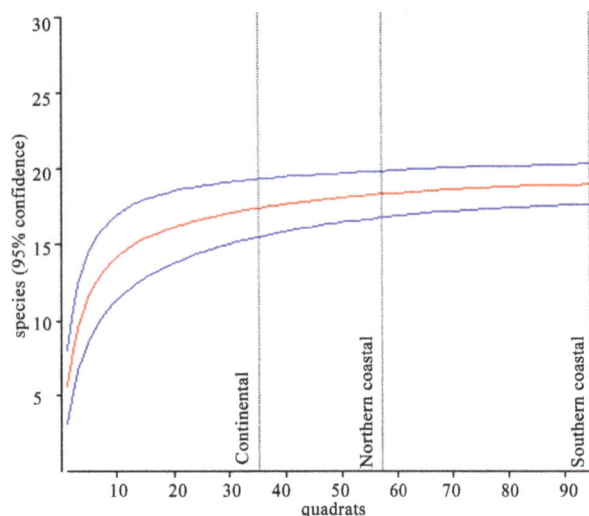

Figure 2.
Sample rarefaction analysis to determine if differences in species richness estimates were associated with variation in sampling effort. A total of 35, 57, and 95 quadrats (20 × 20 m) were used in the continental, northern coastal, and southern coastal locations, respectively. Curves (in blue) above and below mean species values represent 95% confidence intervals.

Results

Factors Influencing Vegetation Distribution across the Forest Region

Disturbance type was not the primary driver of species distribution patterns in all locations across the boreal mixedwood region. The different disturbance types had only a minor influence on regeneration distribution in the coastal sites, whereas windthrow produced distinctive woody vegetation distribution patterns in the continental sites (**Table 2**). There was no common group of variables that influenced regeneration distribution at all locations. Different combinations of factors strongly influenced woody vegetation regeneration at each location (**Table 2**). In decreasing order of importance, woody vegetation distribution in the continental sites was primarily driven by windthrow, annual rainfall, and coarse woody debris density. Species distribution for the northern coastal sites were most strongly associated with latitude, elevation, annual rainfall, coarse woody debris density, and windthrow, while species distribution in the southern coastal sites were associated mostly by stand composition prior to disturbance, coarse woody debris density, decay class, spruce budworm outbreak, and elevation (**Table 2**, **Figure 3**). Coarse woody debris density was the sole factor affecting distribution patterns in all locations, accounting for a large proportion of the variation throughout the boreal

Table 2.
Canonical correlation coefficients between significant environmental variables and the first four ordination axes for redundancy analysis examining species-environment relationships for each study location ($p < 0.05$; ns indicates non-significance).

Location & Environmental Variable	Axis			
	1	2	3	4
Continental				
Windthrow	−0.70	−0.33	ns	ns
Annual rainfall (mm)	0.75	−0.39	ns	ns
Coarse woody debris density (ha^{-1})	0.27	ns	−0.66	ns
Northern coastal				
Latitude	0.57	ns	ns	ns
Elevation (m)	−0.18	ns	0.24	0.39
Annual rainfall (mm)	−0.19	0.35	ns	ns
Coarse woody debris density (ha^{-1})	ns	0.24	−0.25	0.39
Windthrow	−0.10	−0.01	−0.43	ns
Southern coastal				
Conifer-dominated stand prior to disturbance	0.54	ns	ns	ns
Coarse woody debris density (ha^{-1})	00.51	ns	−0.30	ns
Decay class	−0.29	−0.34	−0.31	ns
Spruce budworm outbreak	00.28	ns	ns	0.30
Elevation (m)	ns	−0.44	ns	ns

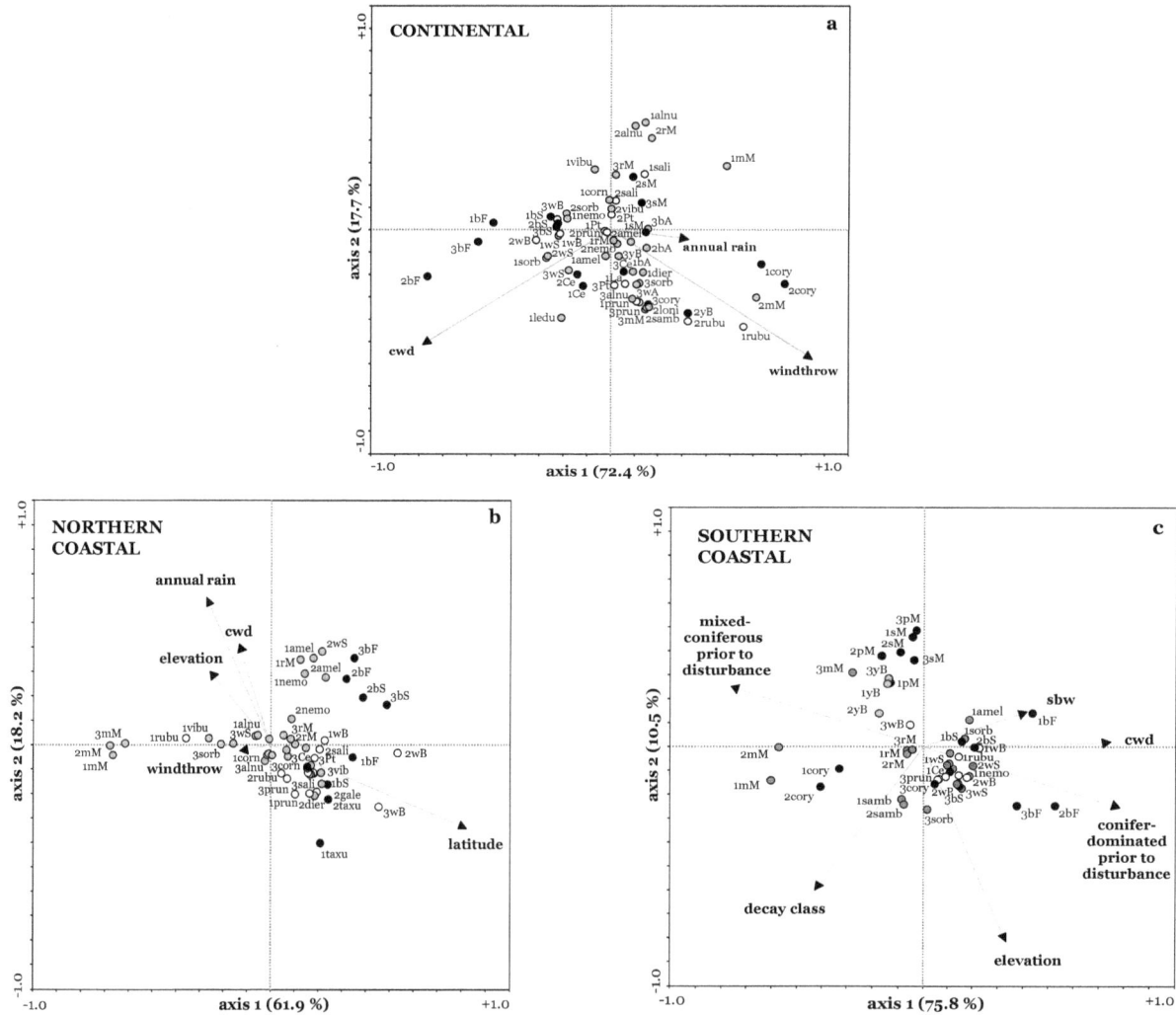

Figure 3.
Examination of woody vegetation distribution patterns for three locations within the boreal mixedwood region of Quebec, Canada in relation to disturbance type, climate, physiography, and stand structure and composition variables using Redundancy analysis (RDA). The first two canonical axes in RDA explained 27.2% and 6.6%, 15.3% and 4.5%, and 18.2% and 2.6% of the cumulative variance in the species data for the continental, northern coastal, and southern coastal sites, respectively. Length and position of vectors and points from the origin in relation to axes 1 and 2 indicate strength of relationships among variables in ordination space, where greater distances from the origin in conjunction with closer positions to either axis 1 or 2 indicate stronger associations. Only the environmental variables having significant relationships with axes 1 or 2 are shown. Some species names near the origin were removed to reduce clutter. Numbers preceding species codes indicate the following regeneration size classes: (1) 1 to 2 m tall, (2) >2 m tall and <4 cm dbh (1.37 m), (3) 4 to 8 cm dbh. Species and environmental variable codes are as follows: wA = *Fraxinus americanus*, bA = *Fraxinus nigra*, wB = *Betula papyrifera*, yB = *Betula alleghaniensis*, Ce = *Thuja occidentalis*, bF = *Abies balsamea*, La = *Larix laricina*, mM = *Acer spicatum*, pM = *Acer pensylvanicum*, rM = *Acer rubrum*, sM = *Acer saccharum*, Pt = *Populus tremuloides*, wP = *Pinus strobus*, bS = *Picea mariana*, wS = *Picea glauca*, alnu = *Alnus* spp., amel = *Amelanchier* spp., corn = *Cornus stolonifera*, cory = *Corylus cornuta*, dier = *Diervella lonicera*, gale = *Myrica gale*, kalm = *Kalmia angustifolia*, ledu = *Ledum groenlandicum*, loni = *Lonicera* spp., nemo = *Nemopanthus mucronata*, prun = *Prunus* spp., rubu = *Rubus* spp., sali = *Salix* spp., samb = *Sambucus* spp., sorb = *Sorbus* spp., vibe = *Viburnum edule*, vibu = *Viburnum cassinoides*, taxu = *Taxus canadensis*, BF-WB = *Abies balsamea-Betula papyrifera* Ecozone, BF-YB = *A. balsamea-B. alleghaniensis* Ecozone, and cwd = coarse woody debris. Shade tolerance: ● high, ◐ mid, ○ low.

mixedwood region. Further, all height classes of each species generally responded in the same manner, indicated by the clustering of intra-specific species points in the biplots (**Figure 3**).

Species Diversity across the Region

A total of 33 woody species were observed in the system (12 canopy trees, 21 sub-canopy tree and shrubs) (**Table 3**). No species were considered rare or endangered. Relative abundances of *Abies balsamea*, *Betula papyrifera*, and *Betula al-*

leghaniensis, the dominant coniferous and deciduous species in the system, were generally maintained, although the northern coastal sites had increases in the coniferous component relative to pre-disturbance conditions in mixed-coniferous stands (**Figure 4**). Cyclical regeneration patterns occurred in stands that were conifer-dominated prior to disturbance as *Abies balsamea*, a shade-tolerant conifer species, dominated the regeneration layer (**Figure 4**, **Table 3**). Stands that were mixed-coniferous prior to disturbance maintained their deciduous canopy tree components, while deciduous sub-canopy tree and shrub com-

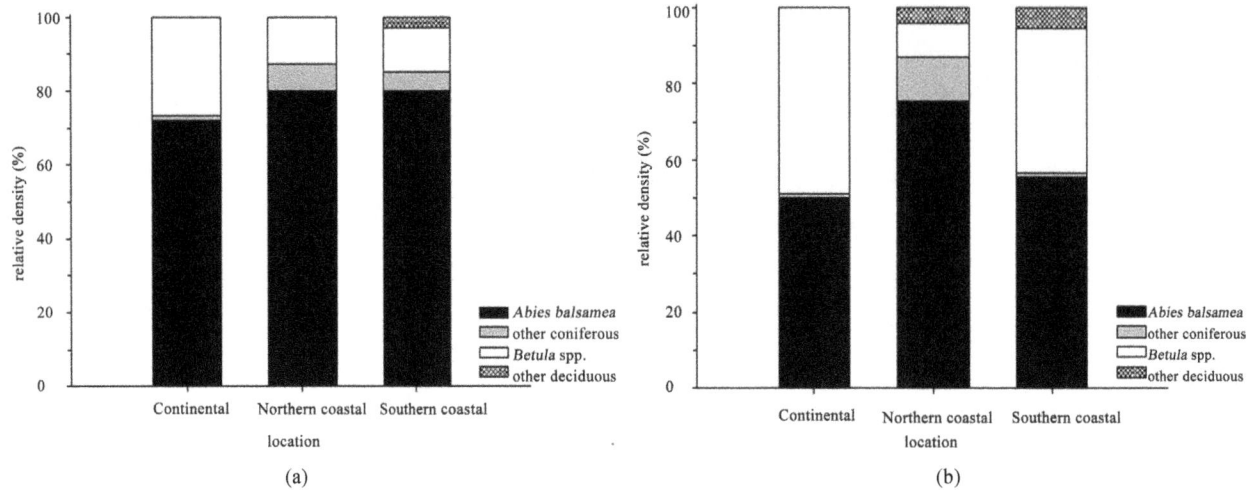

(a)   (b)

Figure 4.
Relative densities of principal canopy tree species regeneration for each location within the boreal mixedwood study region for (a) conifer-dominated stands (≥75% conifer canopy prior to disturbance) and (b) mixed-coniferous stands (50% - 74% conifer canopy) prior to disturbance.

petitors such as *Acer spicatum* and *Corylus cornuta* were more abundant there. Shade-tolerant conifer tree species were strongly associated with high densities of coarse woody debris. However, *Betula papyrifera*, a shade-intolerant deciduous species, was able to maintain its relative abundance in stands that were conifer-dominated prior to disturbance, and was able to establish in areas with high coarse woody debris densities. Closer examination showed that most of the *Betula papyrifera* regeneration was restricted to microsites with exposed mineral soils resulting from tree uprooting (G. Reyes, personal observation).

Comparing Species Diversity among Locations

Alpha (α) diversity for canopy tree species was similar at all locations throughout the region ($p > 0.05$) (**Table 4**). Alpha diversity for total woody vegetation regeneration and for sub-canopy tree and shrub regeneration was higher in the northern coastal *versus* the southern coastal sites ($p < 0.05$) while α diversity in the continental sites was intermediate between the other two locations and did not significantly differ from either coastal location ($p > 0.05$). Beta (β) diversity for all vegetation layers was generally higher in the continental sites, indicating greater variation in species diversity among sites there. This coincided with the continental sites having the largest species pool among the three locations (n.b., 28 woody species observed **Table 4**), as gamma (γ) diversity was the highest there for all vegetation layers. Shannon's diversity index (H') was similar among locations for total woody vegetation and for the canopy tree layer ($p > 0.05$). Sub-canopy tree and shrub H' was greater in the northern coastal *versus* the southern coastal locations ($p < 0.05$). Dominance by certain canopy tree species was more prevalent in the northern *versus* southern coastal locations, as indicated by the lower evenness (J) estimates ($p < 0.05$). Indeed, the majority of the canopy tree regeneration in the northern coastal sites consisted of either *Abies balsamea*, *Picea mariana*, or *Betula papyrifera* (**Table 3**). The continental sites had intermediate J values and did not differ from either coastal location. Alpha, β, γ, and H' values were greater for the sub-canopy tree and shrub layer relative to the canopy tree regeneration within and among all geographical locations and for

across the region as a whole.

Discussion

Factors Influencing Vegetation Distribution across the Forest Region

While different factors influenced post-disturbance community composition among locations across the region, relative species compositions of the tree canopy were either maintained or were developing towards dominance by the shade-tolerant conifer, *Abies balsamea*. The likely mechanisms by which this is occurring are via the competitive advantage provided by an abundant advance regeneration layer and the protection provided by coarse woody debris.

Because the disturbances examined here did not cause extensive damage to the understory vegetation present prior to disturbance, stand composition after natural disturbance was dominated by *Abies balsamea*, a species that develops an abundant advance regeneration layer over time (Morin, 1994). The presence of advance regeneration reduced the ability of shade-intolerant deciduous species to establish after natural disturbance by limiting available growing space, light, and other resources (Morin et al., 2008).

Density of coarse woody debris was the only environmental factor significantly affecting regeneration patterns across the entire region. Its irregular spatial distribution and variation in level of decay throughout the landscape was important for the creation of stand heterogeneity, to which the various species responded. For example, shade-tolerant conifer tree species such as *Abies balsamea* and *Picea mariana* dominated areas where the density of coarse woody debris was high. Coarse woody debris was beneficial for the survival of the advance regeneration perhaps by mitigating the adverse changes in microsite conditions after disturbance by providing cover and shade (Gray & Spies, 1997; Elliott et al., 2002). Consequently, cyclical regeneration patterns (Baskerville, 1975) were observed in conifer-dominated stands. Conversely, areas where coarse woody debris densities were low or where it was absent altogether allowed for the establishment of more shade-intolerant deciduous species such as *Populus tremuloides*, *Betula*

Table 3.
Mean regeneration density (ha^{-1}) for woody species establishing within the boreal mixedwood region of Quebec, Canada. Shade tolerance: high, ●
◐ mid, ○ low.

Species	Location		
	Continental	Northern coastal	Southern coastal
Canopy trees			
Abies balsamea ●	2576.1	7427.8	2714.7
Picea mariana ●	194.3	459.8	22.5
Thuja occidentalis ●	22.9	1.0	0.8
Acer saccharum ●	21.4	-	-
Picea glauca ◐	73.7	211.7	25.1
Betula alleghaniensis ◐	7.3	-	183.1
Pinus strobes ◐	-	-	2.1
Fraxinus americanus ◐	1.4	-	-
Fraxinus nigra ◐	8.6	-	-
Larix laricina ○	1.4	-	-
Betula papyrifera ○	268.9	1103.4	856.7
Populus tremuloides ○	50.3	2.8	16.2
Sub-canopy trees & shrubs			
Acer pensylvanicum ●	-	-	38.0
Corylus cornuta ●	1004.5	-	169.0
Taxus canadensis ●	-	1170.6	-
Acer rubrum ◐	44.3	90.4	98.3
Acer spicatum ◐	491.8	3011.0	884.0
Alnus spp. ◐	437.9	525.1	-
Amelanchier spp. ◐	21.4	686.5	204.2
Cornus stolonifera ◐	15.7	104.4	2.1
Diervella lonicera ◐	1.4	58.3	-
Kalmia angustifolium ◐	-	318.9	-
Ledum groenlandicum ◐	5.7	26.3	-
Lonicera spp. ◐	1.4	-	-
Myrica gale ◐	-	31.6	-
Nemopanthus mucronata ◐	90	30.3	3.7
Sambucus spp. ◐	1.4	-	18.9
Sorbus spp. ◐	197.1	641.7	348.3
Viburnum cassinoides ◐	40.0	-	-
Viburnum edule ◐	236.4	486.8	-
Prunus spp. ○	66.6	72.5	83.3
Rubus spp. ○	302.8	1236.8	13.7
Salix spp. ○	37.1	1.3	-

Table 4.
Estimates of alpha (α), beta (β), gamma (γ), Shannon's (*H*) and evenness (*J*) diversity indices according to location and regeneration layer. Testing for differences in α, *H*, and J among locations for each vegetation layer was made using ANOVA. Within a column, values with post-script indicate significant differences were observed for the particular vegetation layer-where unlike letters among locations for the vegetation layer in question indicate significant differences at *p* < 0.05. Values in parentheses indicate ±1 Standard Error of the mean.

Location & vegetation layer	Indice				
	α	β	γ	*H*	*J*
Continental					
All woody vegetation	8.5 (0.5) **ab**	2.3	28	1.3 (0.1)	0.5 (0.1)
Canopy trees	3.7 (0.2)	1.9	11	0.6 (0.1)	0.5 (0.1) **ab**
Sub-canopy tree & shrubs	4.7 (0.4) **ab**	2.6	17	1.0 (0.1) **ab**	0.7 (0.1)
Northern coastal					
All woody vegetation	9.2 (0.5) **a**	1.4	22	1.4 (0.1)	0.5 (0.1)
Canopy trees	3.7 (0.2)	0.6	6	0.7 (0.1)	0.4 (0.1) **a**
Sub-canopy tree & shrubs	5.8 (0.4) **a**	1.8	16	1.1 (0.1) **a**	0.6 (0.1)
Southern coastal					
All woody vegetation	7.5 (0.4) **b**	1.5	19	1.2 (0.1)	0.5 (0.1)
Canopy trees	3.4 (0.2)	1.4	8	0.7 (0.1)	0.6 (0.1) **b**
Sub-canopy tree & shrubs	4.1 (0.4) **b**	1.7	11	0.8 (0.1) **b**	0.6 (0.1)
Regional					
All woody vegetation	8.3	3.0	33	1.3	0.5
Canopy trees	3.6	2.3	12	0.6	0.6
Sub-canopy tree & shrubs	4.8	3.4	21	1.0	0.6

spp., and *Acer spicatum*. Retaining a deciduous tree canopy component in areas with a high density of coarse woody debris was also observed, but was mostly restricted to microsites with exposed mineral soils resulting from tree uprooting. Uprooting occurred in approximately 1/3 of all tree mortality throughout the region, and accordingly, may be an important mechanism for maintaining the proportion of *Betula* spp. and other shade-intolerant species across the landscape.

Mixed-coniferous stands either maintained their relative co-niferous-deciduous species ratios or increased in the deciduous component. However, much of this increase was related to greater densities of deciduous sub-canopy tree and shrub species. The influx of sub-canopy tree and shrub species can temporarily alter species composition ratios relative to pre-disturbance conditions and/or delay canopy development for a number of years (Hart & Chen, 2006), but will have little effect on canopy species composition once the tree canopy grows beyond a few metres in height. In fact, when considering only the tree canopy regeneration, mixed-coniferous stands generally maintained their pre-disturbance relative coniferous-deciduous species ratios or displayed increases in the conifer component.

Species Diversity across the Region

Differences in species diversity among locations was largely due greater gamma (γ) diversity in the tree canopy layer in continental sites relative to coastal sites, and to variation in the sub-canopy tree and shrub layer γ diversity among coastal locations. The greater degree of variation in the sub-canopy and shrub component in the coastal locations suggests that these species were more sensitive to differences in local site conditions while canopy tree species had a wider range of habitat tolerances. This is shown via the lower beta (β) diversity values for canopy trees *versus* sub-canopy trees and shrubs in all locations.

Comparing regional γ diversity *versus* γ diversity at each location suggests that the continental sites were more limited by the regional species pool, whereas local processes and conditions were important limiting factors in the coastal sites. Only five of the 33 species observed throughout the region were not observed in continental sites whereas 11 and 14 species were absent in the northern and southern coastal sites, respectively. Thus, it appears that a broader depth of habitat types is available in the continental sites relative to coastal locales. This is corroborated by the highest β diversity being observed in the continental sites, followed by the northern coastal then southern coastal sites.

Previous studies have reported a β diversity-latitude gradient of decreasing values from south to north (Qian & Ricklefs, 2007; Lenoir et al., 2010). Similar patterns were observed here for the canopy tree layer and for total woody vegetation as β diversity was greatest in the continental areas (lowest latitudes) and lowest in the northern coastal locations (the highest latitudes). The sub-canopy tree & shrub layer did not follow this pattern as β diversity was higher in the northern *versus* southern coastal sites. Further, α and γ diversity for sub-canopy tree & shrubs did not follow this latitudinal pattern along the coast either, as these diversity components were also higher in northern coastal sites. Other site-level factors, such as density of coarse woody debris may have been important in influencing regeneration distribution. Differences in β diversity among locations coincided with coarse woody debris and snag densities being lowest in the continental sites and the highest in the southern coastal sites, which would suggest greater availability of growing space in the continental areas and lowest available space in the southern coastal sites.

Conclusion

The effects of windthrow, spruce budworm, and interaction natural disturbances on local site conditions were not severe enough to alter canopy species dominance. Dominance by the main canopy species, *Abies balsamea* and *Betula* spp. was maintained throughout the region, suggesting that these species had a wider tolerance range and wider niche overlap relative to other woody species, and/or that the severity of these natural disturbances wasn't sufficient to alter succession trajectories. Moreover, differences in the importance of factors influencing responses of the woody vegetation among locations across the boreal mixedwood region reinforces the concept that species distribution is not controlled by natural disturbance alone, but that local environmental characteristics and constraints on plant biology also dictate re-establishment success. Each location across the region had different and unique site characteristics, and accordingly, species composition and abundance varied from one location to the next. Thus, while the specific disturbance types did not considerably alter canopy tree composition, as development towards old-growth, conifer-dominated conditions was maintained, these natural disturbances were important for creating a mosaic of structural legacies that promote stand heterogeneity, which can ultimately help to maintain biodiversity in the region.

Acknowledgements

Many thanks to all my field assistants (J.-F. Mnudles-Gagnon, Maude Beauregard, David Saucier, Julie Messierer, and Isabelle Nault). Your various personalities, quirks, and habits that were amplified during long, buggy, hot and/or rainy days in the field made data collection quite memorable and fun! Additionally, this study would not have been possible without financial and/or technical support from TEMREX, NSERC-CFS and the SFMN.

REFERENCES

Baskerville, G. L. (1975). Spruce budworm: Super silviculturist. *The Forestry Chronicle, 51,* 138-140.

Belote, R. T., Sanders, N. J., & Jones, R. H. (2009). Disturbance alters local-regional richness relationships in Appalachian forests. *Ecology, 90,* 2940-2947.

Bergeron, Y. (2000). Species and stand dynamics in the mixed woods of Quebec's southern boreal forest. *Ecology, 81,* 1500-1516.

Bergeron, Y., Cyr, D., Drever, R., Flannigan, M., Gauthier, S., Kneeshaw, D., Lauzon, E., Leduc, A., Le Goff, H., Lesieur, D., & Logan, K. (2006). Past, current, and future fire frequencies in Quebec's commercial forests: Implications for the cumulative effects of har-

vesting and fire on age-class structure and natural disturbance-based management. *Canadian Journal of Forest Research, 36,* 2737-2744.

Chen, H. Y. H., & Popadiouk, R. V. (2002). Dynamics of North American boreal mixedwoods. *Environmental Reviews, 10,* 137-166.

Dansereau, P. (1954). Climax vegetation and the regional shift of controls. *Ecology, 35,* 575-579.

Dupont, A., Belanger, L., & Bousquet, J. (1991). Relationships between balsam fir vulnerability to spruce budworm and ecological site conditions of fir stands in central Quebec. *Canadian Journal of Forest Research, 21,* 1752-1759.

Elliott, K. J., Hitchcock, H. L., & Krueger, L. (2002). Vegetation response to large scale disturbance in a southern Appalachian forest: Hurricane Opal and salvage logging. *Journal of the Torrey Botanical Society, 129,* 48-59.

Environment Canada (2004). Canadian climate normals or averages 1971-2000. URL (last checked 24 July 2006).
http://climate.weatheroffice.ec.gc.ca/climate_normals/index_e.html

Gray, A. N., & Spies, T. A. (1997). Microsite controls on tree seedling establishment in conifer forest canopy gaps. *Ecology, 78,* 2458-2473.

Gray, A. N., Zald, H. S. J., Kern, R. A., & North, M. (2005). Stand conditions associated with tree regeneration in sierran mixed-conifer forests. *Forest Science, 51,* 198-210.

Gromtsev, A. (2002). Natural disturbance dynamics in the boreal forests of European Russia: A review. *Silva Fennica, 36,* 41-55.

Hammer, O., Harper, D. A. T., & Ryan, P. D. (2001). PAST: Paleontological statistics software package for education and data analysis. *Palaeontologia Electronica, 4,* 1-9.

Hart, S. A., & Chen, H. Y. H. (2006). Understory vegetation dynamics of North American boreal forests. *Critical Reviews in Plant Sciences, 25,* 381-397.

Hillebrand, H., & Blenckner, T. (2002). Regional and local impact on species diversity—From pattern to process. *Oecologia, 132,* 479-491.

Huebner, C. D., & Vankat, J. L. (2003). The importance of environment vs. disturbance in the vegetation mosaic of Central Arizona. *Journal of Vegetation Science, 14,* 25-34.

Imbeau, L., & Desrochers, A. (2002). Foraging ecology and use of drumming trees by three-toed woodpeckers. *Journal of Wildlife Management, 66,* 222-231.

Krasny, M. E., & Whitmore, M. C. (1992). Gradual and sudden forest canopy gaps in Allegheny northern hardwood forests. *Canadian Journal of Forest Research, 22,* 139-143.

Krebs, C. J. (1989). *Ecological methodology.* New York, NY: Harper and Row Publishers Inc., 654 p.

Kuuluvainen, T., & Juntunen, P. (1998). Seedling establishment in relation to microhabitat variation in a windthrow gap in a boreal *Pinus sylvestris* forest. *Journal of Vegetation Science, 9,* 551-562.

Lenoir, J., Gégout, J.-C., Guisan, A., Vittoz, P., Wohlgemuth, T., Zimmermann, N. E., Dullinger, S., Pauli, H., Willner, W., Grytnes, J.-A., Virtanen, R., & Svenning, J.-C. (2010) Crossscale analysis of the region effect on vascular plant species diversity in southern and northern European mountain ranges. *PLoS ONE, 5,* e15734.

McGuire, A. D., Wirth, C., Apps, M., Beringer, J., Clein, J., Epstein, H., Kicklighter, D. W., Bhatti, J., Chapin III, F. S., de Groot, B., Efremov, D., Eugster, W., Fukuda, M., Gower, T., Hinzman, L., Huntley, B., Jia, G. J., Kasischke, E., Melillo, J., Romanovsky, V., Shvidenko, A., Vaganov, E., & Walker, D. (2002). Environmental variation, vegetation distribution, carbon dynamics and water/energy exchange at high latitudes. *Journal of Vegetation Science, 13,* 301-314.

Messaoud, Y., Bergeron, Y., & Leduc, A. (2007). Ecological factors explaining the location of the boundary between the mixedwood and coniferous bioclimatic zones in the boreal biome of eastern North America. *Global Ecology and Biogeography, 16,* 90-102.

Morin, H. (1994). Dynamics of balsam fir forests in relation to spruce budworm outbreaks in the boreal zone of Quebec. *Canadian Journal of Forest Research, 24,* 730-741.

Morin, H., Laprise, D., Simard, A.-A., & Amouch, S. (2008). Régime des épidémies de la Tordeuse des bourgeons de l'épinette dans l'Est de l'Amérique du Nord. In *Aménagement écosystemique en forêt boréale* (pp. 165-192). Quebec: University of Quebec Press.

MRNQ. (2003). Ministère des resources naturelles du Quebec. URL (last checked 24 July 2006). http://www.mrn.gouv.qc.ca

Nagel, T. A., & Diaci, J. (2006). Intermediate wind disturbance in an old-growth beech-fir forest in southwestern Slovenia. *Canadian Journal of Forest Research, 36,* 629-638.

Neilson, R. P., King, G. A., De Velice, R. L., & Lenihan, J. M. (1992). Regional and local vegetation patterns: The responses of vegetation diversity to subcontinental air masses. In *Ecological Studies 92: Landscape boundaries: consequences for biotic diversity and ecological flows* (pp. 129-149). New York: Springer-Verlag.

Novotny, V., & Weiblen, D. B. (2005). From communities to continents: beta diversity of herbivorous insects. *Annales Zoologici Fennici, 42,* 463-475.

Osumi, K., Ikeda, S., & Okamoto, T. (2003). Vegetation patterns and their dependency on site conditions in the pre-industrial landscape on north-eastern Japan. *Ecological Research, 18,* 753-765.

Palmer, M., & White, P. S. (1994). Scale dependence and the species-area relationship. *American Naturalist, 144,* 717-740.

Peterson, C. J. (2000). Catastrophic wind damage to North American forests and the potential impact of climate change. *The Science of the Total Environment, 262,* 287-311.

Pielou, E. C. (1966). The measurement of diversity in different types of biological collections. *Journal of Theoretical Biology, 13,* 131-144.

Qian, H., & Ricklefs, R. E. (2007). A latitudinal gradient in large-scale beta diversity for vascular plants in North America. *Ecology Letters, 10,* 737-744.

Reyes, G., & Kneeshaw, D. (2008). Moderate-severity disturbance dynamics in *Abies balsamea-Betula* spp. forests: The relative importance of disturbance type and local stand and site characteristics on woody vegetation response. *Ecoscience, 15,* 241-249.

Reyes, G., Kneeshaw, D., De Grandpré, L., & Leduc, A. (2010). Changes in woody vegetation abundance and diversity after natural disturbances causing different levels of mortality. *Journal of Vegetation Science, 21,* 406-417.

Robitaille, A., & Saucier, J. P. (1998). Paysages régionaux du Québec méridional. Québec: Les Publications du Québec.

Rodriguez-Garcia, E., Gratzer, G., & Bravo, F. (2011). Climatic variability and other site factor influences on natural regeneration of *Pinus pinaster* Ait. in Mediterranean forests. *Annals of Forest Science, 68,* 811-823.

Shannon, C. E., & Weaver, W. (1949). *The mathematical theory of communication.* Urbana, IL: University of Illinois Press.

SPSS Inc. (1999). *Professional base system software for statistical analysis* (v.10.0). Chicago, Illinois: SPSS Inc.

terBraak, C. J. F. & Smilauer, P. (1998). *CANOCO reference manual and user's guide to CANOCO for windows: Software for canonical community ordination* (Version 4.02). New York: Microcomputer Power, Ithaca.

Turner, M. G., Dale, V. H., & Everham III, E. H. (1997). Crown fires, hurricanes, and volcanoes: A comparison among large-scale disturbances. *BioScience, 47,* 758-768.

van den Wollenberg, A. L. (1977). Redundancy analysis: An alternative for canonical correlation analysis. *Psychometrika, 42,* 207-219.

Veblen, T. T., & Ashton, D. H. (1978). Catastrophic influences on the vegetation of the Valdivian Andes, Chile. *Vegetatio, 36,* 149-167.

Wang, X.-P., Tang, Z.-Y., & Fang, J. Y. (2006). Climatic control on forests and tree species distribution in the forest region of northeast China. *Journal of Integrative Plant Biology, 48,* 778-789.

Whittaker, R. H. (1960). Vegetation of the Siskiyou Mountains, Oregon and California. *Ecological Monographs, 30,* 279-338.

Wimberly, M. C., & Spies, T. A. (2001). Influences of environment and disturbance on forest patterns in coastal Oregon watersheds. *Ecology, 82,* 1443-1459.

20

Family Forest Owners' Motivation to Control Understory Vegetation: Implications for Consulting Forestry

Alex C. Londeau, Thomas J. Straka
School of Agricultural, Forest, and Environmental Sciences, Clemson University, Clemson, USA

Forest vegetation management has evolved as a recognized component of intensive forest management practice. It involves the management of competing vegetation necessary to obtain the high yields expected in modern forest plantations via control of interfering plants that influence regeneration outcome, impact timber stand development, and limit native plant and wildlife diversity. It includes cultural control, fire control, mechanical control, biological control, and chemical control. The public perception of forest vegetation management, especially chemical control, is sometimes negative due to health and environmental concerns. It is an important tool in the forest management alternatives available to consulting foresters managing family forest lands (the vast majority of private forest land in the United States). We report on a study that addresses the motivations of family forest owners that implement forest vegetation management practices and the motivation of those who chose not to implement after forester recommendations to do so. For those who do implement forest vegetation management, improvement of wildlife habitat and increased timber growth was the main motivation. For those who did not, cost was the main concern. Size of forest holding plays a major role in determining who will practice intensive forestry.

Keywords: Forest Vegetation Management; Chemical Control; Family Forest Owners; Consulting Forestry

Introduction

The evolution of forest vegetation management (FVM) as a recognized component of intensive forest management practice was described by Wagner et al. (2006). FVM involves the management of competing vegetation necessary to obtain the high yields expected in modern forest plantations. It operates by controlling interfering plants that influence regeneration outcome, impact timber stand development, and limit native plant and wildlife diversity (Stout & Finley, 2001). Interfering plants mainly suppress natural plants by shading the understory, but also compete for water, space, and nutrients. By impacting forest stand development, interfering vegetation tends to create a suboptimal future forest stand, one with a lower future timber value (Jackson et al., 2009).

FVM is not just one practice, but an integrated series of practices (Jackson & Finley, 2011; Wiensczyk et al., 2011). It can include cultural control (maintenance of a "healthy" mix of desirable trees species that supports a robust forest with an unopportunistic environment for interfering plants), fire control (use of prescribed burning to reduce undesirable vegetation), mechanical control (removal of interfering plants by hand tools or machinery by cutting or pulling, commonly called weeding), biological control (introducing an insect or disease that is detrimental to the interfering plant), and chemical control (use of herbicides to reduce competition between undesirable interfering plants and the trees in the productive forest).

Integrated vegetation management (IVM) involves the use of the proper range of control methods. IVM uses the concept of a pyramid to delineate the treatments: cultural control is at the bottom and chemical control at the top of the pyramid. As a forester moves from cultural to fire to mechanical to biological to chemical control, the practices become more complex and costly. IVM involves starting at the bottom of the pyramid and moving up to more complicated and costly treatments (Nowak & Ballard, 2005; Miller, 2006; Smallidge, 2009).

In both recently-established forest plantations and naturally-regenerated forests, tree survival and growth is controlled by succession, a major process impacted by competition between trees and other natural vegetation. This competition for resources takes place in a battle for nutrients, water, and light, and by the temporal and spatial segregation of roots and shoots of neighboring plants that allows for completion for space and time (Balandier et al., 2006). The foundation of FVM is developed around this competition to employ techniques that optimize timber production while ensuring plant diversity. It is an intrinsic constituent of silviculture. The primary decision on competitive factors impacting tree survival and growth is silvicultural system: even-aged (clear-cutting, seed tree, or shelterwood) or uneven-aged (individual or group selection), with intermediate treatments of competition control and thinning (Nunamaker & Valachovic, 2007).

The first textbook on FVM defines the term as "the practice of efficiently channeling limited site resources into usable forest products rather than into noncommercial plant species" (Walstad & Kuch, 1987). Wagner (1994) defined it as "managing the course and rate of forest vegetation succession to achieve silvicultural objectives." In North American silviculture textbooks FVM generally includes release and site preparation treatments (Wagner et al., 2006). Timber volume gains from FVM are well-documented for the various forest-growing regions of the world and FVM has become associated with intensive forest management, primarily from chemical control (use of herbicides), especially in the American South (Wagner et al., 2006). Herbaceous and woody vegetation control in southern forest plantations has become a recognized practice when intensive forest management is practiced (Stringer et al., 2010; Osiecka & Minoque, 2011). Herbaceous weed control with herbicides has become closely associated with FVM. It is the top of the pyramid; it may be very costly, but it is also every effective (Vasic et al., 2012). The general public has also come to associate FVM with chemical control and this has resulted in public perceptions of FVM sometimes not enhancing sustainable forestry practices.

Public Perceptions of FVM

The use of FVM is a common component of silvicultural practice around the world; however, preferred methods differ by continent. For example, in Europe most herbaceous weeds are controlled by site preparation and in North America chemical vegetation control is most common (Ammer et al., 2011). Within North America, the southern US has attempted to develop competitive advantages in the international timber market with a focus on intensive high-yield forestry and strong support for the use of herbicides as necessary for effective FVM (McCormack, 1994). North American forestry studies have shown chemical control in timber production produces low risks to humans, soil, water and wildlife, while Europeans feel herbicides are a "serious threat for the maintenance of the set of multiple functions that forests provide" (Ammer et al., 2011). European-based forest certification systems, like the Forest Stewardship Council, take a much harsher view of chemical control than North American-based forest certification systems, like the Sustainable Forestry Initiative.

Not all forestry studies support the North American conclusion that FVM is a necessary part of intensive forest management. Some find results from research studies to be inconsistent in terms of the permanency of the growth and yield gains from FVM, the effects of overstory and gap size, the effects on ground vegetation control, the opposite effects of woody and herbaceous vegetation management, and the effect on species richness (Ammer et al., 2011).

Even in North America, some perceive FVM to be detrimental to forest sustainability goals. Research has tended to focus on the high yields that result from FVM and not on contribution to ecosystem management goals. Changes in cultural patterns may be necessary to gain a focus on ecosystem management concerns (Newton, 2006). Wagner et al. (1998) surveyed public perceptions of risk and acceptability of FVM alternatives in Ontario. Participants were given nine FVM alternatives and asked to rank them from riskiest to least risky. In order of riskiness, the alternatives were: aerially-applied herbicides, biological control, ground-applied herbicides, mulches, prescribed fire, site preparation, cover cropping, grazing animals, and manual cutting. Public acceptance was lowest for aerially-applied herbicides (18%) and second lowest for ground-applied herbicides (37%). There results suggest that strong public support can be obtained for FVM programs that do not include herbicides

(Wagner et al., 1998).

Chemical control using herbicides has gained favor in North America due to its effective and low cost relative to results. International forest certification systems certainly discourage broad use of forest herbicides. Herbaceous weed control methods have developed somewhat independently and the use of non-herbicide methods (mechanical, manual, thermal, and biological) seem to be almost discouraged. Greater public concern, changing perceptions of risk, and international forest certification systems may combine to encourage a wider use of control alternatives (Little et al., 2006). The changing demographics of America's family forest owners (FFOs) certainly will affect public perception of these alternatives. What impact could this have on consulting foresters who have these owners as their main clients?

FFOs and Undesirable Vegetation

Family forest owners (FFOs), the main clientele of forestry consultants are surveyed nationally by the USDA Forest Service. The last survey was in 2006 (Butler, 2008). That survey showed FFOs had a strong concern over the issue of undesirable vegetation, or the problem solved by FVM. Statistics in that survey were stated in terms of area owned by FFOs and number or population of FFOs. This was an important distinction. At the extremes of forest tract size, say owners of small tracts (less than 5 ha) and owners of large tracts (greater than 2000 ha), there are many, many owners of small tracts and few owners of large tracts. Thus, statistic alone, area or number of owners will skew an interpretation of impact on the total forest.

Data from that survey, the National Woodland Owners Survey, is available for analysis using table making software developed by the USDA Forest Service (Butler et al., 2013). **Tables 1** and **2** were developed using that software. The survey included key health concerns of FFOs, including the issue of undesirable vegetation. Nearly one-third of FFOs (32%) were concerned with undesirable vegetation and these owners controlled one-third of family forest area.

Table 1 shows the distribution of FFOs by size of forest holding by both area and owners. **Table 1** also shows this same distribution for only the one-third concerned with undesirable vegetation. The two distributions are surprisingly similar. Notice

Table 1.

FFOs concerned with issue of undesirable vegetation by size of forest holding by total FFO population and area and the one-third of FFOs concerned with the issue, 2006 (Butler, 2008).

Size of forest holding (ha)	Total family forest		One-third concerned	
	%Area	%Owners	%Area	%Owners
1 - 3	7	6.4	7	60.6
4 - 20	22	30.8	23	27.9
21 - 40	16	6.8	15	6.3
41 - 200	32	5.3	30	4.7
201 - 400	8	0.4	8	0.3
401 - 2000	10	0.2	11	0.2
2000+	5	0.1	6	-

Table 2.
Characteristics of one-third FFOs expressing concern over issue of undesirable vegetation, 2006 (Butler et al., 2013).

Characteristic	%Area	%Owners
Owns over 20 ha of forest area	70	2.8
Over 55 years of age	70	65.1
College graduate (B.S. or higher)	45	34.5
Income greater than $50,000	64	52.4
Had a forest management plan	22	5.4
Harvested timber with forester advice	47	25.4
Received management advice	43	17.1
Source of management advice:		
State forestry agency	27	9.4
Federal agency	15	3.9
Extension	10	4.1
Forestry consultant	22	4.9
Forest industry	8	1.3
Logger	10	2.7
Non-profit organization	2	0.6
Another landowner	10	3.3

while it may seem that not a great a proportion of FFOs are concerned with the issue, when forest area owned by concerned owners is considered, then the issue becomes one of major importance. Tracts greater than 40 ha included only 6% of FFOs in the concerned third, but that 6% of owners controlled 55% of family forest area. Area, and not just number of owners, must be part of any analysis that involves FFOs.

The National Woodland Owners Survey Table Maker also allowed for a detailed analysis of that nearly one-third of FFOs who had a concern over undesirable vegetation (Butler et al., 2013). **Table 2** shows the proportions of these owners in terms of key characteristics. Only 13% of them own tracts larger than 20 ha in size, but 70% of the area owned by the group is in tracts larger than 20 ha. There may be a few of them, but they own very large holdings.

Older, more educated, FFOs, with higher incomes, tend to be more concerned with the issue (**Table 2**). As you'd expect, more active managers (those with management plans or who recently harvested timber or consulted a forester for timber harvesting advice) were more greatly concerned with the issue. The forestry advice question is one of the more interesting questions on the survey, at least in relation to consulting forestry. Those FFOs who sought forest management advice were much more likely to be concerned with the issue. But where did they get that advice? State forestry and other agencies was the top listed source. However, forestry consultants were a close second. Federal agencies and extension have a primary responsibility to provide advice, but both had lower percentages than consultants. Other sources were also much lower than consultants in terms of advice. Apparently, consulting foresters are one

of the largest sources of forest management advice, at least in terms of undesirable vegetation.

The source for forest management advice is such a surprising result it seems necessary to ask the question: how does this relationship affect other issues in the survey? The survey had seven biophysical issues (like water pollution or fire) and ten sociopolitical issues (like endangered species and property taxes). The relationship of consulting forestry being a very strong secondary source of forest management advice held across both biophysical and sociopolitical issues consistently in terms of area and nearly always in terms of owners. In fact, the survey identified state forestry agencies and other state agencies separately. They are combined into one category for **Tables 1** and **2**. If state agencies are separated out, forestry consultants are the number one forest management advice source in dealing with endangered species, lawsuits, regulations for timber harvesting, and timber theft issues.

Forestry Consultants' Clients and FVM

FVM is one of the costliest practices clients of consulting foresters confront. Site preparation and planting are also costly, but they are usually considered mandatory to the type of FFOs that invest in a forestry consultant. FVM is not as mandatory; the results are not as predictable and the increased yields not guaranteed. There are many factors that influence the decision to use FVM or not; it boils down to benefit/cost analysis by the forest owner who will make an ultimate decision on cost-effectiveness (Howle, Straka, & Nespeca, 2010).

FFOs that used FVM (or chose not to use it in situations where it was recommended) and were clients of a consulting forester were surveyed to determine their attitudes towards and motivations concerning FVM. This study was confined to FFOs in the southeastern United States. A survey questionnaire was designed to elicit basic information about two central questions 1) does the landowner utilize FVM and 2) what motivation(s) drove their decision, along with a few demographic questions. A large southern forestry consulting firm provided a list of clients that had used or rejected FVM options.

The clients were allowed to suggest motivations for implementing or not implementing FVM practices. If possible, they were grouped into predetermined categories on the "questionnaire" used to lead the conversation towards specific questions. Seven categories of motivations for FVM were eventually developed from the discussion: 1) eliminating competition for improved timber growth, 2) improvements to wildlife habitat, 3) most cost effective, 4) aesthetics, 5) fuel reduction/safety, 6) forester recommendation, and 7) proximity to residential areas. Six motivations for not implementing FVM were also developed: 1) not cost effective, 2) pollution, 3) too early in rotation, 4) never implemented control before, 5) short holding, and 6) not intensely managed.

Results and Discussion

The survey of FFOs in the Southeast resulted in 53 usable responses out of the 132 clients that were contacted and resulting in a response rate of 40%. The 53 clients that responded to the survey give a total of 64 motivations for either implementing (34 responders) or not implementing (19 responders) FVM. The leading motivation for FFOs to implement FVM was to improve wildlife habitat (**Table 3**). This is somewhat surprising

Table 3.
Motivations of FFOs to implement FVM, percentage of responses, n = 34.

Motivation	Percent of respondents
Wildlife habitat	38%
Timber growth	24%
Aesthetics	18%
Safety or fuel reduction	18%
Cost-effective	6%
Residential area	6%
Site preparation	6%
Tradition	6%
Demonstration	3%
Forester recommendation	3%
Pine straw production	3%

Table 4.
Motivations of FFOs not to implement FVM, percentage of responses, n = 19.

Motivation	Percent of respondents
Cost	58%
Never have/no need	21%
Too early in rotation	11%
Pollution	5%
Lack of help	5%
Short holding	5%
Tract too small	5%

Table 5.
FFOs implementing FVM by tract size (n = 54).

Tract size	Percent implementing FVM
<200 ha	50%
200 - 400 ha	60%
>400 h	86%

Table 6.
Treatment type by FFOs who implemented FVM (n = 34).

Treatment type	Percent using type
Fire control	63%
Mechanical control	19%
Chemical and fire control	15%
Chemical control	3%

as costly treatments are usually justified by increased wood yields. Reducing competition for improved timber growth was the second most cited motivation by 24% of respondents. The small percentage is also surprising for the same reason. FVM is a very costly treatment and increased wood yield would seem to be the driving force. However, secondary motivations like aesthetics and fuel reduction were major motivations also. One landowner said, "Fuel reduction and the threat of wildfire is a major concern for me." The range of motivations also downplays the importance of wood yield, as one landowner put it, "I just like the way my forest looks without all that underbrush." It is obvious that FFOs are not driven primarily by cost-effectiveness when considering the FVM decision and that a wide array of motivations control the use of chemical control in forest management.

Cost was the leading motivation for not implementing FVM at 58% (**Table 4**). A strong secondary reason was disagreement over need. FFOs who use consulting foresters pay for advice and tend to be more business-minded that the average owners. They are much more likely to be concerned with cost-effectiveness issues. Pollution (health concerns) was a minor issue, although one landowner was quoted saying, "I'm worried but runoff of herbicides from my property, as well as adding CO_2 to the atmosphere from burning."

Some demographics were obtained: how many forested area owned, age, highest level of education, and primary management goal. Educational level did not seem to be a major factor in a landowner's decision to implement or not implement FVM, while respondents older than 55 seemed to be involved in more intensive management regimes. This is a little different from the earlier study cited (Butler, 2008), but considering the major variable of size of forest holding, the results were identical with the earlier study (**Table 5**). As size of forest holding increases (tract size increases), so does the percent of FFOs that chose to use FVM; this can be answered by economies of scale (Cubbage, 1983).

The types of treatments and their frequency are shown in **Table 6**. Other studies have found fire control to be a minor reason for FVM (Ammer et al., 2011), but our results show fire control was the most popular form of FVM by FFOs in the Southeast. Mechanical treatments included chopping, mowing, and cutting, and occurred on tracts closer to residential areas.

Implications and Conclusion

The decision by FFOs to engage in FVM is significantly affected by the size of the landholding. Like any producer, the typical landowner seeks to maximize the benefits from their timberlands. The fact that some landowners were willing to use multiple, more expensive, and/or more labor intensive treatment types represents their willingness to invest in their family forests. FVM by FFOs is an important aspect of the intensely managed forest regimes of the south eastern US.

The most prevalent type of FVM found in this study was fire control, which most closely mimics the natural disturbance regime for the region (Gilliam & Platt, 1999). The most common motivation for using fire control was improving wildlife habitat; indicating that most landowners are more concerned with the annual benefits associated with wildlife, whether by revenues from hunting leases or by maximizing their utility function, rather than the discounted values of future timber harvesting. Cost being the most frequently cited motivation for not implementing FVM signifies that other landowners are reluctant to invest in their family forest. More importantly, the findings of this study show that the size of landholdings correlates directly with the level and intensity of forest management.

Acknowledgements

This research was supported by the Practicing Foresters Institute.

REFERENCES

Ammer, C., Balandier, P., Bentsen, N. S., Coll, L., & Löf, M. (2011). Forest vegetation management under debate: An introduction. *European Journal of Forest Research, 130,* 1-5.

Balandier, P., Collet, C., Miller, J. H., Reynolds, P. E., & Zedaker, S. M. (2006). Designing forest vegetation management strategies based on the mechanism and dynamics of crop tree competition by neighbouring vegetation. *Forestry, 79,* 3-27.

Butler, B. J. (2008). *Family forest owners of the United States, 2006.* General Technical Report NRS-27. Newtown Square, PA: USDA Forest Service, Northern Research Station.

Butler, B. J., Miles, P. D., & Hansen, M. H. (2013). National woodland owner survey table maker web application, version 1.0. Amherst, MA: USDA Forest Service, Northern Research Station. http://apps.fs.fed.us/fia/nwos/tablemaker.jsp

Cubbage, F. W. (1983). *Economics of forest tract size: Theory and literature.* General Technical Report SO-41. New Orleans, LA: USDA Forest Service, Southern Forest Experiment Station.

Gilliam, F. S., &. Platt, W. J. (1999). Effects of long-term fire exclusion on tree species composition and stand structure in an old-growth Pinus palustris (Longleaf pine) forest. *Plant Ecology, 140,* 5-26.

Howle, M. B., Straka, T. J., & Nespeca, M. C. (2010). Family forest owners' perceptions of chemical methods for invasive species control. *Invasive Plant Science and Management, 3,* 253-261.

Jackson, D. R., & Finley, J. C. (2011). *Herbicides and forest vegetation management: Controlling unwanted trees, brush, and other competing forest vegetation.* University Park, PA: Pennsylvanian State University.

Jackson, D. R., Wolf, M. T., & Finley, J. C. (2009). *Regenerating hardwood forests: Managing competing plants, deer, and light.* Forest Stewardship No. 15. University Park, PA: Pennsylvania State University, College of Agricultural Sciences.

Little, K. M., Willoughby, I., Wagner, R. G., Frochot, H., Gava, J., Gous, S., Lautenshlager, R. A., Örlander, G., Sankaran, K. V., & Wei, R. P. (2006). Towards reduced herbicide use in forest vegetation management. *South African Forestry Journal, 207,* 63-79.

McCormack, M. L. (1994). Reductions in herbicide use in forest vegetation management. *Weed Technology, 8,* 344-349.

Miller, J. H. (2006). Forest vegetation management: Development in the science and practices. *Forestry, 79,* 1-2.

Newton, M. (2006). Taking charge in forest vegetation management. *Canadian Journal of Forest Research, 36,* 2357-2363.

Nowak, C. A., & Ballard, B. D. (2005). A framework for applying integrated vegetation management on rights-of-way. *Journal of Arboriculture, 31,* 28-37.

Nunamaker, C., & Valachovic, Y. (2007). *Forest vegetation management.* Forest Stewardship Guide 6, Publication 8236. Oakland, CA: University of California.

Osiecka, A., & Minoque, P. J. (2011). *Considerations for developing herbicide prescriptions for forest vegetation management (FOR273).* Gainesville, FL: University of Florida.

Smallidge, P. J. (2009). Strategies to control undesirable and interfering vegetation in your forest. *Forest Landowner, 68,* 38-39.

Stout, S., & Finley, J. (2001). What are interfering plants, and when are they a problem? *Sustainable Forestry Initiative of Pennsylvania, Summer,* 8-10.

Stringer, J. W., Clatterbuck, W., & Seifert, J. (2010). *Site preparation and chemical control guidelines for hardwood tree plantings.* Athens, GA: Southern Regional Extension Forestry.

Vasic, V., Konstantinovic, B., & Orlovic, S. (2012). Weeds in forestry and possibilities of their control (Chapter 8). In: A. J. Price (Ed.), *Weed control* (pp. 147-170). Rijeka, Croatia: InTech.

Wagner, R. G. (1994). Toward integrated forest vegetation management. *Journal of Forestry, 92,* 26-30.

Wagner, R. G., Flynn, J., & Gregory, R. (1998). Public perception of risk and acceptability of forest vegetation management alternatives in Ontario. *The Forestry Chronicle, 75,* 720-727.

Wagner, R. G., Little, K. M., Richardson, B., & McNabb, K. (2006). The role of vegetation management for enhancing the world's forests. *Forestry, 79,* 57-79.

Walstad, J. D., & Kuch, P. J. (1987). *Vegetation management for conifer production.* New York, NY: John Wiley & Sons.

Wiensczyk, A. Swift, K., Momeaulf, A., Thiffault, N., Szuba, F., & Bell, W. (2011). A review of the efficacy of vegetation management alternatives for conifer regeneration in boreal forests. *Forestry Chronicle, 87,* 175-200.

21

Selection of Landscape Tree Species of Tolerant to Sulfur Dioxide Pollution in Subtropical China[*]

Xizi Zhang[1], Ping Zhou[2], Weiqiang Zhang[2], Weihua Zhang[3], Yongfeng Wang[2]

[1]International Department, The Affiliated High School of South China Normal University, Guangzhou, China
[2]Department of Forest Ecology, Guangdong Academy of Forestry, Guangzhou, China
[3]Department of Forest Breeding and Silviculture, Guangdong Academy of Forestry, Guangzhou, China

Sulfur dioxide (SO_2) is a major air pollutant, especially in developing countries. Many trees are seriously impaired by SO_2, while other species can mitigate air pollution by absorbing this gas. Planting appropriate tree species near industrial complexes is critical for aesthetic value and pollution mitigation. In this study, six landscape tree species typical of a subtropical area were investigated for their tolerance of SO_2: *Cinnamomum camphora* (L.) J. Presl., *Ilex rotunda* Thunb., *Lysidice rhodostegia* Hance, *Ceiba insignis* (Kunth) P. E. Gibbs & Semir, *Cassia surattensis* Burm. f., and *Michelia chapensis* Dandy. We measured net photosynthesis rate, stomatal conductance, leaf sulfur content, relative water content, relative proline content, and other parameters under 1.31 $mg \cdot m^{-3}$ SO_2 fumigation for eight days. The results revealed that the six species differed in their biochemical characteristics under SO_2 stress. Based on these data, the most appropriate species for planting in SO_2 polluted areas was *I. rotunda*, because it grew normally under SO_2 stress and could absorb SO_2.

Keywords: Sulfur Dioxide; Fumigation; Landscape Trees; Air Pollutant Tolerance; Sulfur Content; Net Photosynthesis Rate

Introduction

As one of the six major atmospheric pollutants, sulfur dioxide (SO_2) levels are currently a health concern. Sulfur dioxide can cause asthma and other respiratory health problems in people, forming acid rain that can damage forests and crops, and erode buildings. Because of industrialization and urbanization, developing countries, especially China, suffer from increasing concentrations of SO_2 in the air. Since 1990, SO_2 generated in China has been responsible for about one-fourth of the global emissions and more than 90% of the East Asian emissions. From 21.7 Tg (1 Tg = 10^{12} g) in 2000, SO_2 emissions increased by 53% to 33.2 Tg in 2006, at an annual growth rate of 7.3%. In 2007, Guangdong, a province in the Pearl River Delta Industrial district, emitted a total of 1177 Gg of SO_2, about 97% of which was emitted by power plants and industries (Lu, 2010).

Damage to plants is an important consequence of atmospheric SO_2. Gaseous pollutants, particularly SO_2, enter plants through the stomata by the process of photosynthesis and respiration. Nitrogen dioxide (NO_2) and SO_2 react with water on the cell walls inside leaves; by transfer and assimilation, the resulting sulfurous, sulfuric, nitrous, and nitric acids, react with other compounds and are transported to various parts of plants. If plants are exposed to air pollutants for a long time or the pollutant concentrations exceed a critical threshold, plants may be injured (Jim, 2007). Plant injury is usually cumulative in nature, reducing growth and yield and accelerating senescence.

The injury often has no overt visible symptoms aside from some degree of chlorosis (WHO, 2000). Because of the harmful effects of SO_2, plants cannot grow robustly and some also die in severely polluted industrial districts, creating "dead zones" without greenery in these areas. Many studies have investigated various aspects of the damage caused by SO_2 to plants, including photosynthesis (Swanepoel, 2007), stomatal density and function (Haworth, 2012), and carbon fixation efficiency (Chung, 2010).

Each plant is a living entity, and individuals vary in their adaptations to the environment and abilities to absorb pollutants. Suitable plants must be carefully selected for cultivation; otherwise they may not thrive or may die in adverse conditions of environmental pollution (Chung, 2010). In 2000, about 42.62 Mg of SO_2 was removed from the atmosphere by urban trees in Guangzhou. Because it costs less to remove SO_2 in the air in China compared to other developed countries, the monetary value of this service is low (Jim, 2007). Some studies have not only investigated the effects of air pollutants on plants, but also evaluated suitable air pollutant-tolerant plants, for example near a lignite-based thermal power station (Govindaraju, 2011), industrial complexes (Lee, 2004), and a coal-fired power plant (Sharma, 2008). The other studies assessed SO_2-tolerant plants, e.g., among wetland plants (Sha, 2010). Nevertheless, few investigations have continuously observed the responses of landscape tree species under high SO_2 concentrations in the subtropical areas of southern China. This study aims to understand how trees adapt to the stress of SO_2 and to facilitate the identi-

---

[*]Selection of high SO_2 tolerance species.

fication of species that can assimilate atmospheric SO_2 while growing normally in this area.

Materials and Methods

Tree Seedlings and Growing Conditions

Six popular landscape tree species in Southern China were selected for this experiment: *Cinnamomum camphora* (L.) J. Presl, *Ilex rotunda* Thunb., *Lysidice rhodostegia* Hance, *Ceiba insignis* (Kunth) P. E. Gibbs & Semir, *Cassia surattensis* Burm. f. and *Michelia chapensis* Dandy. For each species, eighteen healthy 1-year-old seedlings of approximately the same size were potted into 2 kg bags (height: 12 cm, radius: 4 cm) with loess containing 0.302 $g·kg^{-1}$ N, 0.3 $g·kg^{-1}$ P, 9.761 $g·kg^{-1}$ K, 21.21 $mg·kg^{-1}$ hydrolysable N, 4.3 $mg·kg^{-1}$ rapidly available P, and 28.47 $mg·kg^{-1}$ rapidly available K.

The average tree height, root collar diameter, and canopy of the trees are listed in **Table 1**. These seedlings were grown under natural conditions for 1 month with regular watering to allow their physiology to stabilize before the experiment began.

Controlled Environmental Conditions

In the experiment, three seedlings of each species (18 seedlings total) were placed under natural conditions with daily watering as a control group for normal growth without treatment. The other 15 samples of each species were placed together as an experimental group in a 2.0 m × 1.2 m × 1.8 m phytotron at Guangdong Academy of Forestry, China. From January 28 to February 5, 2013, these seedlings experienced 8 days of fumigation with 1.31 $mg·m^{-3}$ (= 0.5 ppm) SO_2 (MIC-SO_2) with the following conditions: temperature, 15°C - 25°C; relative humidity (RH), 50% - 60%; concentration of carbon dioxide, 380 - 400 ppm; and light intensity, 600 $μmol·m^{-2}·s^{-1}$. According to the result of some studies, SO_2 can impact the growth and yield of plants while reducing its foliar starch and protein contents, pigmentation, and WUE at concentrations as low as 0.06 - 0.15 ppm (Swanepoel, 2007). We used an unnaturally high concentration of SO_2 (1.31 $mg·m^{-3}$) to determine the relative sensitivities of species for which this information was almost unknown. According to the Pearl River Delta regional air quality monitoring reports (2006, 2007, 2008, 2009, 2010), the average of the monthly maxima of hourly averages of SO_2 in Huijingcheng (Foshan), one of the most severely polluted areas, is 0.394 $mg·m^{-3}$. The SO_2 concentration in this study was about three times that value. During the fumigation, plants were watered daily.

Measurements of Biochemical Characteristics

Leaf parameters were measured at regular time intervals during the SO_2 fumigation treatment. To observe changes in different parts of the seedlings, the first round of tests were conducted the day before the SO_2 treatment to serve as a baseline. Then, every 2 days (on Jan 30, Feb 1, Feb 3, Feb 5), three seedlings (replicates) of each species were removed from the phytotron and three to four of their leaves were picked off to measure relative water content, relative electrolytic leakage and proline content. Five rounds of tests were done.

Relative chlorophyll content was measured with a portable chlorophyll content meter (CCM-200 plus, OptiSciences, Hudson, NH, USA) on six young fully expanded leaves for each seedling. Relative water content was determined by the following equation:

$$R_{WC} = (W_f - W_d)/(W_s - W_d) \times 100\% \qquad (1)$$

The fresh weight (W_f), saturated fresh weight (W_s), which was the weight after soaking the leaves in distilled water for 24 hours, and dry weight (W_d), which was got by drying the fresh leaves in an oven of 80°C overnight, were measured by a electronic scale of 0.01 g (JJ500, G & G GmbH, Neuss, Germany). Relative electrolytic leakage, P, was evaluated using a conductivity meter (DSSJ-308A, China) and calculated by the following equation:

$$P = ((C_1 - C_0)/(C_2 - C_0)) \times 100\% \qquad (2)$$

where the conductivity of distilled water (C_0) and of the sample solute before boiling (C_1) and after boiling (C_2) were known. Proline content in the leaves was determined as described by Chen & Wang (2006). To measure photosynthesis, a portable photosynthesis system (Li-6400, LI-COR, Lincoln, NE, USA) was used to test three healthy leaves near the top of each seedling. To ensure the consistency of incident light intensity and leaf surface temperature, we tested the following parameters in both the control and treatment groups from 9:00 - 11:00 A.M. for 2 days after fumigation (on Feb 6 and 7): net photo synthetic rate (P_n, $μmol·m^{-2}·s^{-1}$), stomatal conductance (G_s, $mol·m^{-2}·s^{-1}$), transpiration rate (T_r, $mmol·m^{-2}·s^{-1}$), intercellular carbon dioxide concentration (C_i, $μmol·mol^{-1}$), photosynthetic

Table 1.
Growth status of the plants before sulfur dioxide fumigation.

Species	Code	Growth parameters			Family
		Height (cm)	Root collar diameter (cm)	Canopy (cm²)	
Cinnamomum camphora	A	51.20 ± 0.70	0.47 ± 0.01	242.20 ± 17.16	Lauraceae
Ilex rotunda	B	67.13 ± 0.80	0.66 ± 0.03	139.07 ± 11.78	Aquifoliaceae
Lysidice rhodostegia	C	54.47 ± 1.06	0.59 ± 0.02	377.60 ± 46.51	Caesalpinioideae
Ceiba insignis	D	87.53 ± 1.37	1.60 ± 0.05	977.60 ± 88.73	Malvaceae
Cassia surattensis	E	46.73 ± 0.76	0.55 ± 0.02	225.53 ± 12.74	Caesalpinioideae
Michelia chapensis	F	64.93 ± 1.07	0.70 ± 0.01	281.80 ± 24.36	Magnoliaceae

cally active radiation (PAR, $\mu mol \cdot m^{-2} \cdot s^{-1}$), atmospheric carbon dioxide concentration (C_a, $\mu mol \cdot mol^{-1}$), atmospheric temperature (T_a, °C), leaf temperature (T_l, °C), RH (%) and water use efficiency (WUE, $\mu mol \cdot mmol^{-1}$). WUE was calculated as:

$$WUE = P_n / T_r \qquad (3)$$

The sulfur content was tested by barium sulfate turbidimetry with various parts of the leaves selected. The ratio of leaf injury was estimated based on the percentage of visible leaf damages.

Statistical Analysis

Statistical analyses were conducted with Microsoft Office Excel 2007 (Redmond, WA, USA) and SPSS 16.0 (IBM, Chicago, IL, USA). ANOVAs and multiple comparisons were used to analyze significant difference of the relative proline content, sulfur content in leaves and the net photosynthesis rate. The tests of homogeneity were checked before multiple comparisons. The change of P_n (R, %), which was calculated by the equation:

$$R = \frac{\left(P_n\right)_{\text{after}} - \left(P_n\right)_{\text{before}}}{\left(P_n\right)_{\text{before}}} \times 100\% \qquad (4)$$

and the absolute differences between the value of sulfur content on 0 hour and 192 hour are calculated to compare the ability of species of tolerance to the SO_2 fumigation.

Results

In general, the sulfur content in the leaves of all six species increased significantly before and after the treatment ($P < 0.05$) *Cassia surattensis* had both the highest original sulfur content in the leaves (4.49 ± 1.035 $\mu g \cdot g^{-1}$) and the greatest increase (to 9.345 ± 1.172 $\mu g \cdot g^{-1}$), which demonstrated its strong ability to absorb SO_2. It was followed by *I. rotunda*. The sulfur content in leaves of *I. rotunda* was higher under SO_2 fumigation than in the control (5.155 ± 0.411 versus 2.273 ± 0.123 $g \cdot kg^{-1}$). It showed higher ability to absorb SO_2 gas than *C. camphora*, *L. rhodostegia*, *C. insignis*, or *M. chapensis* (**Figure 1**). The sulfur content in the leaves of *C. camphora* remained low at around 0.795 ± 0.236 and only increased to 2.616 ± 0.385 $g \cdot kg^{-1}$ during the treatment. The sulfur content of *L. rhodostegia* was not initially high nor did it increase much during treatment. The low sulfur contents in *C. insignis* and *M. chapensis* showed their weak ability to absorb SO_2.

In this study, three species showed significantly decline in the P_n after fumigation: *L. rhodostegia*, *C. insignis*, *Cassia surattensis* ($P < 0.05$) (**Figure 2**). *Ceiba insignis* declined the greatest amount in P_n, from 5.77 ± 1.33 to 0.75 ± 0.08 $\mu mol \cdot m^{-1} \cdot s^{-1}$, in G_s, from 0.09 ± 0.03 $mol \cdot m^{-2} \cdot s^{-1}$ to 0.02 ± 0.003 $mol \cdot m^{-2} \cdot s^{-1}$, and in T_r, from 1.23 ± 0.38 to 0.27 ± 0.06 $mmol \cdot m^{-2} \cdot s^{-1}$. In addition, WUE was reduced by fumigation from 4.91 ± 0.35 to 2.92 ± 0.47 $\mu mol \cdot mmol^{-1}$. *Lysidice rhodostegia* had a relatively large decrease in P_n, from 3.28 ± 0.48 to 0.95 ± 0.09 $\mu mol \cdot m^{-1} \cdot s^{-1}$. Also, both G_s and T_r decreased during fumigation from 0.04 ± 0.006 to 0.02 ± 0.001 $mol \cdot m^{-2} \cdot s^{-1}$ and from 0.64 ± 0.07 to 0.35 ± 0.02 $mmol \cdot m^{-2} \cdot s^{-1}$, respectively. WUE declined the most of the six species, from 5.13 ± 0.25 $\mu mol \cdot mmol^{-1}$ before fumigation to 2.71 ± 0.12 $\mu mol \cdot mmol^{-1}$ afterwards. C_i increased from 262.7 ± 4.4 to 337 ± 2.5 $\mu mol \cdot mol^{-1}$. The P_n of *C. surattensis* declined from 4.22 ± 0.07 to 2.90 ± 0.29 $\mu mol \cdot m^{-1} \cdot s^{-1}$, while its C_i increased from 299.4 ± 3.9 to 338.5 ± 1.2 $\mu mol \cdot mol^{-1}$, and its T_r reduced markedly from 1.24 ± 0.05 to 0.77 ± 0.04 $mmol \cdot m^{-2} \cdot s^{-1}$. Moreover, its electrolytic leakage increased distinctly from 12.54 ± 0.97 to $29.62\% \pm 4.94\%$, indicating that part of the plasma membrane was damaged by SO_2 in fumigation, thus affecting its normal growth.

The other species did not show significant changes in Pn before and after fumigation: *C. camphora*, *I. rotunda*, and *M. chapensis* ($P > 0.05$) (**Figure 2**). The P_n of *I. rotunda* remained between 3.66 ± 0.51 and 4.01 ± 0.39 $\mu mol \cdot m^{-1} \cdot s^{-1}$. *Cinnamomum camphora* changed little in P_n, which ranged between 2.75 ± 0.18 and 3.16 ± 0.09 $\mu mol \cdot m^{-1} \cdot s^{-1}$. No visible damage was observed on the surfaces of its leaves. However, there was a decrease in *WUE*, from 5.25 ± 0.59 to 4.39 ± 0.59 $\mu mol \cdot mmol^{-1}$. Similarly, *M. chapensis* did not have a significant change in P_n, which was 3.81 ± 0.25 before and 3.34 ± 0.73 $\mu mol \cdot m^{-1} \cdot s^{-1}$ after fumigation.

The proline could prevent folded proteins from denaturing, interact with phospholipids to stabilize cell membranes, scavenge hydroxyl radicals, and function as an energy and nitrogen source (Claussen, 2004). An increase in proline content indicated that the plant was stressed (**Figure 3**). The proline was vital in adjusting osmotic pressure in *M. chapensis*, *C. insignis* and *C. surattensi*. The proline content of *M. chapensis* was the highest among the species during fumigation, with an initial value of 224.76 ± 50.84 $\mu g \cdot g^{-1}$ and the highest value of 350.46 ± 43.97 $\mu g \cdot g^{-1}$ during the process of fumigation. The proline content of *C. surattensis* was also high (207.25 ± 5.05 $\mu g \cdot g^{-1}$) and increased during fumigation to 327.00 ± 21.82 $\mu g \cdot g^{-1}$. The

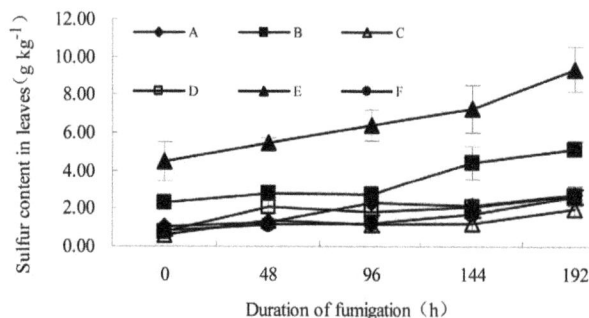

Figure 1.
Changes in the sulfur content of leaves during sulfur dioxide fumigation. A, *Cinnamomum camphora*; B, *Ilex rotunda*; C, *Lysidice rhodostegia*; D, *Ceiba insignis*; E, *Cassia surattensis*; F, *Michelia chapensis*. The error bar on each point indicated standard error.

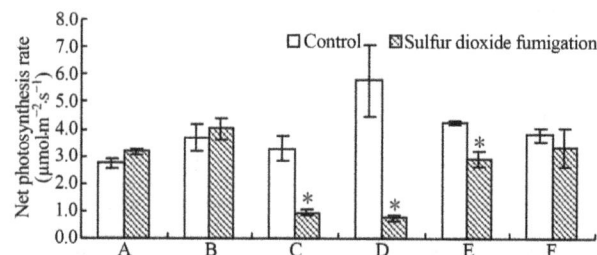

Figure 2.
Net photosynthesis rate before and after sulfur dioxide fumigation. A, *Cinnamomum camphora*; B, *Ilex rotunda*; C, *Lysidice rhodostegia*; D, *Ceiba insignis*; E, *Cassia surattensis*; F, *Michelia chapensis*. The error bar on each point indicated standard error. The asterisks indicated significant differences at 0.05.

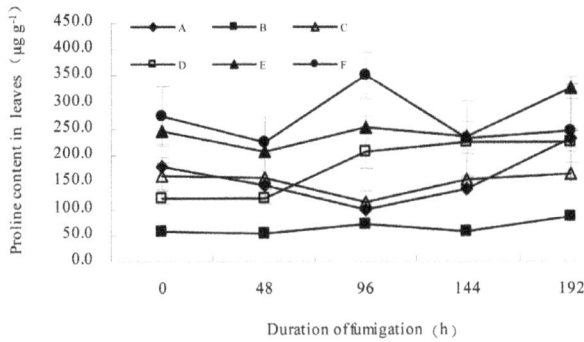

Figure 3.
Changes in proline content of leaves during sulfur dioxide fumigation. A, *Cinnamomum camphora*; B, *Ilex rotunda*; C, *Lysidice rhodostegia*; D, *Ceiba insignis*; E, *Cassia surattensis*; F, *Michelia chapensis*. The error bar on each point indicated standard error.

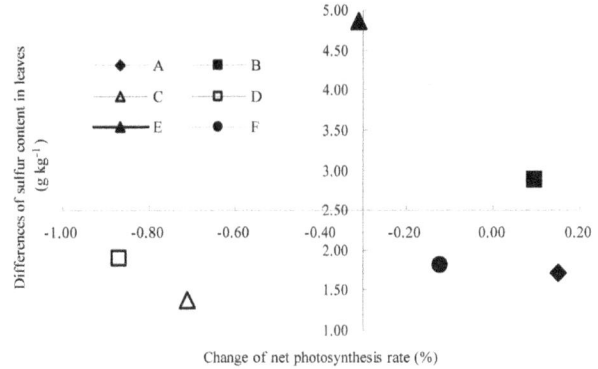

Change of net photosynthesis rate (%)

Figure 4.
Categorization of the six tree species based on their responses to SO_2 pollutant stress. A, *C. camphora*; B, *I. rotunda*; C, *L. rhodostegia*; D, *C. insignis*; E, *C. surattensis*; F, *M. chapensis*.

proline content of *C. insignis* ranged between 120.94 ± 12.49 and 226.16 ± 18.57 $\mu g \cdot g^{-1}$, presenting a significant increase ($P < 0.05$) under SO_2 stress, which corresponded with its remarkably decreased P_n. After fumigation, some leaves were dehydrated, yellowing, coiled and chlorotic, and many dehisced, and the ratio of leaf injury was 100%, which showed it was highly stressed. The proline content in the leaves of *I. rotunda* was relatively low, ranging from 51.44 ± 4.91 to 84.88 ± 7.97 $\mu g \cdot g^{-1}$. Its G_s increased after fumigation from 0.07 ± 0.012 to 0.10 ± 0.014 $mol \cdot m^{-2} \cdot s^{-1}$. Some leaves had rusty piebald patches or dehisced under SO_2. The ratio of leaf injury was 30%. The proline content of *L. rhodostegia* was not initially high nor did it increase much during treatment.

These six species could be divided into four types based on their responses to SO_2 fumigation, as shown in **Figure 4**. The horizontal axis is the change of P_n (R, %). The smaller the R value, the lesser the impact on P_n and the better the plant can adapt to the SO_2 environment. The vertical axis shows the absolute values of the differences in leaf sulfur content ($g \cdot kg^{-1}$) between 0 and 192 hours of fumigation; the larger this value, the greater the tree's capacity to absorb SO_2.

Discussion

The results could be used for environmental design, because they indicated that these six species have different tolerances and could be planted for different purposes, as summarized in **Figure 4**. *Ilex rotunda*, in the first quadrant, was the most appropriate of the six species to be planted near industry complexes and along roads, because it could absorb SO_2 while still growing robustly. To clean the air, *C. surattensis*, in the second quadrant, was a good choice for its strong SO_2 absorption ability. In contrast, if the goal was simply greenery near factories to improve aesthetics, the species in the fourth quadrant, *C. camphora* and *M. chapensis*, could be selected due to their healthy growing condition under high concentration of sulfur dioxide. The net photosynthesis of the plants in the third quadrant, *L. rhodostegia* and *C. insignis*, decreased a relatively large amount after fumigation, showing their poor growth condition under SO_2 stress. They absorbed almost no SO_2, so did not improve the environment. Thus, these two species were not highly tolerant to SO_2.

Many studies have investigated which pollutant-tolerant plants can be grown near areas with severe air pollution (Go-

vindaraju, 2011; Lee, 2004; Sharma, 2008). However, since they chose the sampling sites and took the sample leaves to test different parameters, the variables, such as the concentration of pollutants and the existence of particles, could not be really controlled. The actual reason for a given species' biochemical characteristics could not be well explained, because they might have resulted from any number of factors. Also, climate factors, such as temperature, latitude, or humidity, might have affected the plants, so the results might not apply to other places or seasons. In this experiment, the environmental variables were all controlled, so that the impact of SO_2 on plants was isolated and repeatable.

Jim (2007) demonstrates that *C. camphora* tolerates SO_2, NO_x and particulates well. Nevertheless, according to our results, although *C. camphora* could survive in a severely polluted area, it could not extract atmospheric SO_2 to improve air quality. However, *C. camphora* did not change substantially in either P_n or sulfur content in the leaves, nor in the values of other measurements such as chlorophyll relative content, relative water content, and electrolytic leakage. Because this experiment focused only on the effects of SO_2, the tolerance of *C. camphora* to other air pollutants is still unknown. A previous study (Lyu, 2003) showed that *C. camphora* can absorb a small amount of SO_2, consistent with the results of this experiment. In addition, this species has a strong ability to extract hydrogen fluoride and is suitable for mildly polluted areas.

Wen et al. (2003) found that *L. rhodostegia* was highly sensitive to air pollutants. The authors concluded that the P_n of this species was reduced by pollutant stress, and the amount of decrease was much greater than the decrease in T_r. Thus, the plant suffered both weaker growth due to less photosynthesis and excessive water loss. That observation accorded with our data. Although Wen et al. investigated multiple air pollutants and we only studied the effects of SO_2, both studies concluded that *L. rhodostegia* can neither resist nor adapt to air pollution.

Surprisingly, the data for *I. rotunda* differed between our study and that of Wen et al. (2003). The latter showed that both P_n and G_s decreased nearly 40% under air pollution, while T_r declined by about 50%. In contrast, we found out that G_s increased and the values of the other measurements did not change significantly. These differences may result from the different environments of the two experiments; the other air pollutants in the study of Wen et al. might offset the effects of SO_2 and change the response of *I. rotunda*, leading to different results.

Because *I. rotunda* was the most suitable species for SO_2 polluted areas according to our results, it should receive more attention to fully elucidate the mechanisms by which it adapts to SO_2 stress. We believe that other species with similar behavioral responses can be found and planted to mitigate air pollution. In addition, more parameters of *C. camphora* can be tested in order to further understand the adaptation of it under sulfur dioxide. This species remains poorly understood because most of the data did not change significantly or show obvious trends in this experiment. Therefore, future experiments must be conducted to investigate the effects on plants of other air pollutants, individually and in combination, to permit the selection of the optimal species for severely polluted regions. To fully understand how plants change under pollution stress, more parameters should be evaluated, including stomatal density, chlorophyll fluorescence, and carbon fixation efficiency. To identify more air pollution tolerant plants, other common subtropical species, which were also mentioned in Wen et al. (2003), like *Ficus microcarpa*, *Camellia japonica* L., and *Tutcheria spectabilis* (Benth.) Dunn, can also be tested.

Conclusion

In this study, the performance of six common landscape tree species under SO_2 fumigation was studied to help select trees for greenbelts near industrial complexes in subtropical area, especially where SO_2 is the main emission. *Ilex rotunda*, which remained green and extracted a great amount of SO_2, is recommended as a key species for greenbelts. *Cassia surattensis* can be used to improve air quality in polluted areas. Both *C. camphora* and *M. chapensis* are also recommended for planting in severely polluted areas because of their high aesthetic values. From an economic and management perspective, *L. rhodostegia* and *C. insignis* are more suitable for cleaner, less-polluted environments. Integrating different tree species into a landscape can both contribute to greenery near factories and maintain biodiversity. As more studies are conducted on the appropriate species to grow in heavily industrial areas, the problem of air pollution can be effectively controlled.

Acknowledgements

This study was supported by funding from the Guangdong Forestry Science and Technology Innovation Project (2010 KJCX012-01) and the Guangzhou Science and Technology Project (2012Y2-00011), China.

REFERENCES

Chen, J. X., & Wang, X. F. (2006). Ch. 29: The effects of stress on the free proline content in plants. In *Phytophysiological experiment manual* (2nd ed., pp. 66-67). Guangzhou: South China University Technology Press.

Chung, Y. C., Chung, P. L., & Liao, S. W. (2010). Carbon fixation efficiency of plants influenced by sulfur dioxide. *Environmental Monitoring and Assessment, 173,* 701-707.

Claussen, W. (2004). Proline as a measure of stress in tomato plants. *Plant Science, 168,* 241-248.

Govindaraju, M., Ganeshkumar, R. S., Muthukumaran, V. R., & Visvanathan, P. (2011). Identification and evaluation of air-pollution-tolerant plants around lignite-based thermal power station for greenbelt development. *Environmental Science and Pollution Research.*

Guangdong Provincial Environmental Protection Monitoring Centre & Environmental Protection Department, HKSAR (2006). *Pearl river delta regional air quality monitoring network: A report of monitoring results in 2006.* Report numbers: PRDAIR-2006-2.

Guangdong Provincial Environmental Protection Monitoring Centre & Environmental Protection Department, HKSAR (2007). *Pearl river delta regional air quality monitoring network: A report of monitoring results in 2007.* Report numbers: PRDAIR-2007-2.

Guangdong Provincial Environmental Protection Monitoring Centre & Environmental Protection Department, HKSAR (2008). *Pearl river delta regional air quality monitoring network: A report of monitoring results in 2008.* Report numbers: PRDAIR-2008-2.

Guangdong Provincial Environmental Protection Monitoring Centre & Environmental Protection Department, HKSAR (2009). *Pearl river delta regional air quality monitoring network: A report of monitoring results in 2009.* Report numbers: PRDAIR-2009-2.

Guangdong Provincial Environmental Protection Monitoring Centre & Environmental Protection Department, HKSAR (2010). *Pearl river delta regional air quality monitoring network: A report of monitoring results in 2010.* Report numbers: PRDAIR-2010-2.

Haworth, M., Kingston, C. E., Gallagher, A., Fitzgerald, A., & McElwain, J. C. (2012). Sulphur dioxide fumigation effects on stomatal density and index of non-resistant plants: Implications for the stomatal palaeo-[CO_2] proxy method. *Review of Palaeobotany and Palynology, 182,* 44-54.

Jim, C. Y., & Chen, W. Y. (2007). Assessing the ecosystem service of air pollutant removal by urban trees in Guangzhou (China). *Journal of Environmental Management, 88,* 665-676.

Lee, C. S., Lee, K. S., Hwangbo, J. K., You, Y. H., & Kim, J. H. (2004). Selection of tolerant plants and their arrangement to restore a forest ecosystem damaged by air pollution. *Water, Air, and Soil Pollution, 156,* 251-273.

Lu, Z., Streets, D. G., Zhang, Q., Wang, S., Carmichael, G. R., Cheng, Y. F., Wei, C., Chin, M., Diehl, T., & Tan, Q. (2010). Sulfur dioxide emissions in China and sulfur trends in East Asia since 2000. *Atmospheric Chemistry and Physics, 10,* 6311-6331.

Lyu, H. Q. & Liu, F. P. (2003). A review of phytoremediation to chemical air pollution and plant species selection for greening. *Subtropical Plant Science, 32,* 73-77.

Sha C. Y., Wang, T. H., & Lu, J. J. (2010). Relative sensitivity of wetland plants to SO_2 pollution. *Wetlands, 30,* 1023-1030.

Sharma, A. P., & Tripathi, B. D. (2008) Biochemical responses in tree foliage exposed to coal-fired power plant emission in seasonally dry tropical environment. *Environmental Monitoring and Assessment, 158,* 197-212.

Swanepoel, J. W., Krüger, G. H. J., Heerden, & van P. D. R. (2007) Effects of sulphur dioxide on photosynthesis in the succulent *Augea capensis* Thunb. *Journal of Arid Environments, 70,* 208-221.

Wen, D. Z., Lu, Y. D., Kuang, Y. W., Hu, X. C., Zhang, D. Q., Xue, K. N., & Kong, G. H. (2003). Ecophysiological responses and sensitivity of 39 woody species exposed to air pollution. *Journal of Tropical and Subtropical Botany, 11,* 341-347.

World Health Organization Regional Office for Europe (2000). Ch. 10: Effects of sulfur dioxide on vegetation: Critical levels. In *Air quality guidelines* (2nd ed.). Copenhagen: WHO Regional Publications.

22

Birds of Dhorpatan Hunting Reserve, Nepal

Saroj Panthi[1*], Sher Singh Thagunna[2]

[1]Department of National Park and Wildlife Conservation, Dhorpatan Hunting Reserve, Baglung, Nepal
[2]Department of National Park and Wildlife Conservation, Api-Nampa Conservation Area, Darchula, Nepal

Nepal is rich in bird diversity and recorded 871 bird species. Our study aimed to update bird diversity of Dhorpatan Hunting Reserve (DHR) of Nepal which is an Important Birds Area (IBA) out of 27 IBAs of Nepal. One hundred forty nine species of birds were recorded in DHR; out of them *Catreus wallichii* was globally threatened bird which has faced different anthropogenic problem. Twelve species of birds were included in CITIES Appendices and 3 species were nationally threatened. Most of the species of order *Passeriformis* were recorded. Only one species of order *Upupiformes* i.e. *Upupa epops* was recorded during the study.

Keywords: Bird; Checklist; Threatened; Dhorpatan Hunting Reserve

Introduction

Birds (class Aves) are feathered, winged, bipedal, endothermic, egg-laying, vertebrate animals (Wikipedia, 2013). Although Nepal covers just 0.1% of the global land mass, nearly 9% of the world's bird species are found here (BCN, 2012). With the latest record of ashy minivet (*Pericrocotus divaricatus*) and long-billed vulture (*Gypus indicus*) Nepal's bird diversity has reached 871 species (BCN, 2012; BCN and DNPWC, 2012). A total of 29 species recorded in Nepal were identified as globally threatened by BirdLife International in 1999 (Grimmett et al., 2000). Zoo-geographically, Nepal falls between two great regions: the Palaeartic in the north and Oriental to the south so Nepal has one of the world's richest avian fauna (Shrestha, 1981).

A total of 27 Important Birds Areas (IBAs) had been identified in Nepal. Thirteen IBAs are within protected areas (Thakuri & Thapa, 2012). Dhorpatan Hunting Reserve (DHR) is one of the IBAs (BCN and DNPWC, 2011). DHR was a habitat of 137 species of bird out of them cheer pheasant (*Catreus wallichii*) is listed in endangered category of IUCN red data list (Thapa, 2007). Here we updated the checklist of bird species of DHR. DHR was good habitat of birds (Thapa, 2007) but checklist update was lacking. Our aim was to update the checklist of bird species in DHR and to identify their order, family, conservation status and threats.

Material and Methods

Study Area

The Dhorpatan Hunting Reserve (DHR) is the only one hunting reserve in Nepal located in Dhaulagiri Himal range of the eastern part of the country (Aryal & Kreigenhofer, 2009; Panthi et al., 2012). The reserve was established in 1983 and gazetted in 1987. The reserve covers 1325 km² and ranges from 2000 to 7246 m (altitude). It falls within the Rukum, Baglung and Myadgi districts of Nepal (**Figure 1**). The reserve supports 14 ecosystems types represented in the mid hill to higher Himalayan ecosystem (Shrestha et al., 2002; Lilleso et al., 2005; Bhuju et al., 2007).

The reserve provide prime habitat for fauna such as barking deer (*Munticus muntjak*), Himalayan thar (*Hemitragus jemlachicus*), rhesus macaque (*Macaca radiata*) wolf (*Canis lupus*), red panda (*Ailurus fulgens fulgens*), wild boar (*Sus scorfa*), blue sheep (*Pseudois nayaur*), Himalayan black bear (*Ursus thibetans*), common leopard (*Panthera pardus*), goral (*Naemorhedus goral*), serow (*Capricornis thar*) and variety of avifauna including cheer pheasant (*Catrus wallichii*) and danphe (*Lopophorus impegians*) (Aryal & Kreigenhofer, 2009; Aryal et al., 2010a; Aryal et al., 2010b; Panthi et al., 2012).

Methods: Checklist of Birds in DHR

Intensive survey was conducted throughout the reserve to update the checklist of birds in study area. We used The books "Birds of Nepal" prepared by Richard Grimmett, Carol Inskipp and Tim Inskipp (2000) and "Birds of Nepal: An Official Checklist" (2012) prepared by Bird Conservation of Nepal (BCN) and Department of National Parks and Wildlife Conservation (DNPWC) to identify the birds during the study. We visited all seven hunting blocks of DHR viz. Barse, Dogadi, Fagune, Ghustung, Seng, Sundaha and Surtibang blocks and observed the birds in two season viz. winter and summer throughout the year 2012. We observed the birds carefully by the help of binoculars took photographs with digital camera and compared that photo with the photos of book to identify the birds. We also interviewed local people and DHR staffs to indentify birds and their threats during the study.

Results and Discussion

We recorded one hundred forty nine species of birds in DHR (**Table 1**). Out of them, *Catreus wallichii* was globally threat-

*Corresponding author.

Table 1.
Bird of Dhorpatan Hunting Reserve (DHR), identified during the survey periods in 2012.

S.N.	Common name	Scientific name	Order	Family	Conservation status	CITIES Appendix
1	Mallard	Anas platyrhynchos	Anseriformes	Anatidae		
2	Ruddy Shelduck	Tadorna ferruginea	Anseriformes	Anatidae		
3	Black Eagle	Ictinaetus malayensis	Ciconiformis	Accipitridae		II
4	Cinereous Vulture	Aegypius monachus	Ciconiformis	Accipitridae		II
5	Himalayan Griffon	Gyps himalayensis	Ciconiformis	Accipitridae		II
6	Pallas's Fish Eagle	Haliaeetus leucoryphus	Ciconiformis	Accipitridae		II
7	Ibisbill	Ididorhyncha struthersii	Ciconiformis	Charadriidae		
8	Common Krestrel	Falco tinnunculus	Ciconiformis	Falconidae		II
9	Little Stint	Calidris minuta	Ciconiformis	Scolopacidae		
10	Curlew Sandpiper	Calidris ferruginea	Ciconiformis	Scolopacidae		
11	Hill Pigeon	Columba rupestris	Columbiformes	Columbidae		
12	Oriental Turtle Dove	Streptopelia orientalis	Columbiformes	Columbidae		
13	Rock Pigeon	Columba livia	Columbiformes	Columbidae		
14	Snow Pigeon	Columba palumbus	Columbiformes	Columbidae		
15	Speckled Wood Pigeon	Columba hodgsonii	Columbiformes	Columbidae		
16	Spotted Dove	Streptopelia chinensis	Columbiformes	Columbidae		
17	Blood Pheasant	Ithaginis cruentus	Galliformis	Phasianidae		II
18	Cheer Pheasant	Catreus wallichii	Galliformis	Phasianidae	GT, NP	I
19	Hill Partridge	Arborophila torqueola	Galliformis	Phasianidae		
20	Himalayan Monal	Lophophorus impejanus	Galliformis	Phasianidae	NP	I
21	Kalij Pheasant	Lophura leucomelanos	Galliformis	Phasianidae		
22	Koklas Pheasant	Pucrasia macrolopha	Galliformis	Phasianidae		
23	Satyr Tragopan	Tragopan satyra	Galliformis	Phasianidae	NP	III
24	Snow Patridge	Lerwa lerwa	Galliformis	Phasianidae		
25	White-Throated Tit	Aegithalos niveogularis	Passeriformes	Aegithalidae		
26	Oriental Skylark	Alauda gulgula	Passeriformes	Alaudidae		
27	Winter Wren	Troglodytes troglodytes	Passeriformes	Certhiidae		
28	Brown Dipper	Cinclus pallasii	Passeriformes	Cinclidae		
29	Hill Prinia	Prinia atrogularis	Passeriformes	Cisticolidae		
30	Striated Prinia	Prinia criniger	Passeriformes	Cisticolidae		
31	Zitting Cisticola	Cisticola juncidis	Passeriformes	Cisticolidae		
32	Ashy Drongo	Dicrurus leucophaeus	Passeriformes	Corvidae		
33	Common Raven	Corvus corax	Passeriformes	Corvidae		
34	Common Iora	Aegithina tiphia	Passeriformes	Corvidae		

Continued

35	Eurasian Golden Oriole	*Oriolus oriolus*	*Passeriformes*	*Corvidae*
36	House Crow	*Corvus splendes*	*Passeriformes*	*Corvidae*
37	Large-Billed Crow	*Corvus macrorhynchos*	*Passeriformes*	*Corvidae*
38	Rufous Treepie	*Dendrocitta vagabunda*	*Passeriformes*	*Corvidae*
39	Short-Billed Minivet	*Pericrocotus flammeus*	*Passeriformes*	*Corvidae*
40	Spotted Nutcracker	*Nucifraga caryocatactes*	*Passeriformes*	*Corvidae*
41	Yellow-Billed Blue Magpie	*Urocissa flavirostris*	*Passeriformes*	*Corvidae*
42	Yellow-Billed Chough	*Pyrrhocorax graculus*	*Passeriformes*	*Corvidae*
43	Yellow-Billed Fantail	*Rhipidura hypoxantha*	*Passeriformes*	*Corvidae*
44	White-Throated Fantail	*Rhipidura albicollis*	*Passeriformes*	*Corvidae*
45	Beautiful Rosefinch	*Carpodacus pulcherrimus*	*Passeriformes*	*Fringillidae*
46	Blanford's Rosefinch	*Carpodacus rubescens*	*Passeriformes*	*Fringillidae*
47	Collard Grosbeak	*Mycerobas affinis*	*Passeriformes*	*Fringillidae*
48	Common Rosefinch	*Carpodacus erythrinus*	*Passeriformes*	*Fringillidae*
49	Creasted Bunting	*Melophus lathami*	*Passeriformes*	*Fringillidae*
50	Dark-Rumpled Rosefinch	*Carpodacus edwardsii*	*Passeriformes*	*Fringillidae*
51	Fire-Fronted Serin	*Serinus pusillus*	*Passeriformes*	*Fringillidae*
52	Little Bunting	*Emberiza pusilla*	*Passeriformes*	*Fringillidae*
53	Pine Bunting	*Emberiza leucocephalos*	*Passeriformes*	*Fringillidae*
54	Pink-Browed Rosefinch	*Carpodacus rodochrous*	*Passeriformes*	*Fringillidae*
55	Red-Fronted Rosefinch	*Carpodacus puniceus*	*Passeriformes*	*Fringillidae*
56	Rock Bunting	*Emberiza cia*	*Passeriformes*	*Fringillidae*
57	Scarlet Finch	*Haematospiza sipahi*	*Passeriformes*	*Fringillidae*
58	Streaked Rosefinch	*Carpodacus rubicilloides*	*Passeriformes*	*Fringillidae*
59	Twite	*Carduelis flavirostris*	*Passeriformes*	*Fringillidae*
60	Vinaceous Rosefinch	*Carpodacus vinaceus*	*Passeriformes*	*Fringillidae*
61	White-Winged Grosbeak	*Mycerobas carnipes*	*Passeriformes*	*Fringillidae*
62	Yellow-Breasted Greenfinch	*Carduelis spinoides*	*Passeriformes*	*Fringillidae*
63	Barn Swallow	*Hirundo rustica*	*Passeriformes*	*Hirundinidae*
64	Black-Backed Forktail	*Enicurus immaculatus*	*Passeriformes*	*Muscicapidae*
65	Black Redstrat	*Phoenicurus ochruros*	*Passeriformes*	*Muscicapidae*
66	Blue-Capped Redstrat	*Phoenicurus coeruleocephalus*	*Passeriformes*	*Muscicapidae*
67	Blue-Fronted Redstrat	*Phoenicurus frontalis*	*Passeriformes*	*Muscicapidae*
68	Common Stonechat	*Saxicola torquata*	*Passeriformes*	*Muscicapidae*
69	Dark-Sided Flycatcher	*Muscicapa sibirica*	*Passeriformes*	*Muscicapidae*
70	Dark-Thorated Thrush	*Turdus ruficollis*	*Passeriformes*	*Muscicapidae*
71	Dusky Thrush	*Turdus naumanni*	*Passeriformes*	*Muscicapidae*

Continued

72	Eurasian Blackbird	*Turdus merula*	*Passeriformes*	*Muscicapidae*
73	Golden Bush Robin	*Tarsiger chrysenus*	*Passeriformes*	*Muscicapidae*
74	Grandala	*Grandala coelicolor*	*Passeriformes*	*Muscicapidae*
75	Grey Bushchat	*Saxicola ferra*	*Passeriformes*	*Muscicapidae*
76	Grey-Winged Blackbird	*Turdus boulboul*	*Passeriformes*	*Muscicapidae*
77	Hodgson's Redstrat	*Phoenicurus hodgosni*	*Passeriformes*	*Muscicapidae*
78	Indian Blue Robin	*Luscinia brunnea*	*Passeriformes*	*Muscicapidae*
79	Kessler's Thrush	*Turdus kessleri*	*Passeriformes*	*Muscicapidae*
80	Mistle Thrush	*Turdus viscivorus*	*Passeriformes*	*Muscicapidae*
81	Orange-Flanked Bush Robin	*Tarsiger cyanurus*	*Passeriformes*	*Muscicapidae*
82	Orental Magpie Robin	*Copsychus saularis*	*Passeriformes*	*Muscicapidae*
83	Pied Bushchat	*Saxicola caprata*	*Passeriformes*	*Muscicapidae*
84	Plumbeous Water Redstart	*Rhyacornis fuliginosus*	*Passeriformes*	*Muscicapidae*
85	Purple Cochoa	*Cochoa purpurea*	*Passeriformes*	*Muscicapidae*
86	Red-Thorated Flycatcher	*Ficedula parva*	*Passeriformes*	*Muscicapidae*
87	Rufous-Brested Bush Robin	*Tarsiger hyperythrus*	*Passeriformes*	*Muscicapidae*
88	Rufous-Gorgeted Flycatcher	*Ficedula strophiata*	*Passeriformes*	*Muscicapidae*
89	Rusty-Tailed Flycatcher	*Muscicapa ruficauda*	*Passeriformes*	*Muscicapidae*
90	White-Browed Bush Robin	*Tarsiger indicus*	*Passeriformes*	*Muscicapidae*
91	White-Bellied Redstrat	*Hodgsonius phaenicuroides*	*Passeriformes*	*Muscicapidae*
92	White-Capped Water Redstart	*Chaimarronis leucocephalus*	*Passeriformes*	*Muscicapidae*
93	White-Collard Blackbird	*Turdus albocinctus*	*Passeriformes*	*Muscicapidae*
94	White-Gorgeted Flycatcher	*Ficedula monileger*	*Passeriformes*	*Muscicapidae*
95	White-Tailed Robin	*Myiomela leucura*	*Passeriformes*	*Muscicapidae*
96	White-Throated Redstrat	*Phoenicurus schisticeps*	*Passeriformes*	*Muscicapidae*
97	Black-Lored Tit	*Parus xanthogenys*	*Passeriformes*	*Paridae*
98	Coal Tit	*Parus ater*	*Passeriformes*	*Paridae*
99	Fire-Capped Tit	*Cephalopyrus flammiceps*	*Passeriformes*	*Paridae*
100	Green-Backed Tit	*Parus monticolus*	*Passeriformes*	*Paridae*
101	Rufous-Napped Tit	*Parus rufonuchalis*	*Passeriformes*	*Paridae*
102	Rufous-Vented Tit	*Parus rubidiventris*	*Passeriformes*	*Paridae*
103	Alpine Accentor	*Prunella collaris*	*Passeriformes*	*Passeridae*
104	Altai Accentor	*Prunella himalayana*	*Passeriformes*	*Passeridae*
105	Brown Accentor	*Prunella fulvescens*	*Passeriformes*	*Passeridae*
106	Citrine Wagtail	*Motacilla citreola*	*Passeriformes*	*Passeridae*
107	Grey Wagtail	*Motacilla cinerea*	*Passeriformes*	*Passeridae*
108	House Sparrow	*Passer domesticus*	*Passeriformes*	*Passeridae*

Continued

109	Olive-Backed Pipit	*Anthus hodgsoni*	*Passeriformes*	*Passeridae*
110	Paddyfield Pipit	*Anthus rufulus*	*Passeriformes*	*Passeridae*
111	Red-Thorated Pipit	*Anthus cervinus*	*Passeriformes*	*Passeridae*
112	Robin Accentor	*Prunella rubeculoides*	*Passeriformes*	*Passeridae*
113	Rosy Pipit	*Anthus rosestus*	*Passeriformes*	*Passeridae*
114	Tree Pipit	*Anthus trivialis*	*Passeriformes*	*Passeridae*
115	Upland Pipit	*Anthus sylvanus*	*Passeriformes*	*Passeridae*
116	Yellow Wagtail	*Motacilla flava*	*Passeriformes*	*Passeridae*
117	Long-Tailed Shrike	*Lanius schach*	*Passeriformes*	*Pittidae*
118	Black Bulbul	*Hypsipetes leucocephalus*	*Passeriformes*	*Pycnonotidae*
119	Himalayan Bulbul	*Pycnonotus leucogenys*	*Passeriformes*	*Pycnonotidae*
120	Striated Bulbul	*Pycnonotus melanicterus*	*Passeriformes*	*Pycnonotidae*
121	Goldcrest	*Regulus regulus*	*Passeriformes*	*Regulidae*
122	Black-Faced Laughinthrush	*Garrulax affinis*	*Passeriformes*	*Sylviidae*
123	Black-Throated Parrotbill	*Paradoxornis niplensis*	*Passeriformes*	*Sylviidae*
124	Brownish-Flanked Bush Warbler	*Cettia fortipes*	*Passeriformes*	*Sylviidae*
125	Chesnut-Crowned Bush Warbler	*Cettia major*	*Passeriformes*	*Sylviidae*
126	Great Parrotbill	*Conostoma oemodium*	*Passeriformes*	*Sylviidae*
127	Greenish Warbler	*Phylloscopus trochiloides*	*Passeriformes*	*Sylviidae*
128	Grey-Hooded Warbler	*Seicercus xanthoschistos*	*Passeriformes*	*Sylviidae*
129	Hoary-Thorated Barwing	*Actinodura nipalensis*	*Passeriformes*	*Sylviidae*
130	Hume's Warbler	*Phylloscopus humei*	*Passeriformes*	*Sylviidae*
131	Jungle Babbler	*Turdoides striatus*	*Passeriformes*	*Sylviidae*
132	Large-Billed Leaf Warbler	*Phylloscopus magnirostris*	*Passeriformes*	*Sylviidae*
133	Rufus-Vented Yuhina	*Yuhina occipitalis*	*Passeriformes*	*Sylviidae*
134	Slender-Billed Scimitar Babbler	*Xiphirhynchus Superciliaris*	*Passeriformes*	*Sylviidae*
135	Straited Laughinthrush	*Garrulax striatus*	*Passeriformes*	*Sylviidae*
136	Streaked Laughinthrush	*Garrulax lineatus*	*Passeriformes*	*Sylviidae*
137	Variegated Laughinthrush	*Garrualx variegates*	*Passeriformes*	*Sylviidae*
138	Whistler's Warbler	*Seicercus whistleri*	*Passeriformes*	*Sylviidae*
139	White-Browed Tit Warbler	*Leptopoecile sophiae*	*Passeriformes*	*Sylviidae*
140	White-Creasted Laughingthrush	*Garrulax monileger*	*Passeriformes*	*Sylviidae*
141	White-Throated Laughingthrush	*Garrulax albogularis*	*Passeriformes*	*Sylviidae*
142	Yellow-Bellied Warbler	*Abroscopus supercilliaris*	*Passeriformes*	*Sylviidae*
143	Crimson-Brested Woodpecker	*Dendrocopos cathpharius*	*Piciformes*	*Indicatoridae*
144	Greater Flameback	*Chrysocolaptes lucidus*	*Piciformes*	*Indicatoridae*
145	Rufous-Bellied Woodpecker	*Dendrocopos hyperythrus*	*Piciformes*	*Indicatoridae*

Continued

146	Little Owl	*Athene noctua*	*Strigiformes*	*Strigidae*	II
147	Spotted Owlet	*Athene brama*	*Strigiformes*	*Strigidae*	II
148	Tawny Owl	*Strix aluco*	*Strigiformes*	*Strigidae*	II
149	Common Hoopoe	*Upupa epops*	*Upupiformes*	*Upupidae*	

Note: GT: Globally Threatened, NP: Nationally Protected.

Figure 1.
Dhorpatan hunting reserve, Nepal.

ened bird which has faced different anthropogenic problem.

C. wallichii Lophophorus impejanus Tragopan satyra are nationally protected birds found in the reserve. *C. wallichii* and *L. impejanus* are included in CITIES appendix I (BCN and DNPWC 2012).

We recorded 9 species which are included in CITIES appendix II and one species which is included in CITIES appendix III. Most of the species of order *Passeriformis* were recorded. Only one species of order *Upupiformes* i.e. *Upupa epops* was recorded during the study.

This study showed that snares were main threats of birds. Local people used snares to capture large sized birds such as *Arborophila torqueola, Catreus wallichii, Ithaginis cruentus Lophophorus impejanus, Lophura leucomelanos, Pucrasia macrolopha* and *Tragopan satyra* for fulfillment of meat.

We recorded snares in two places viz. Karichaur in Fagune block and Rukhacahur in Barse block. Small children killed small sized forest birds due to lack of awareness. The present study provided basic information on birds of DHR. However further species wise information is needed. Distribution and detail threats identification of globally threatened and nationally protected species are further recommended study in DHR.

Acknowledgements

The study was possible after the financial and logistic support from the DHR office and Department of National Parks and Wildlife Conservation (DNPWC). We thanked to all staffs of DHR and DNPWC who helped for field survey. We also thank to Dr. Achyut Aryal (Institute of Natural and Mathematical Sciences, Massey University, New Zealand) for his guidance to complete this study.

REFERENCES

Aryal, A., & Kreigenhofer, B. (2009). Summer diet composition of the common leopard *Panthera pardus* (Carnivora: Felidae) in Nepal. *Journal of Threatened Taxa, 1,* 562-566.

Aryal, A., Gastaur, S., Menzel, S., Chhetri, T. B., & Hopkins, J. (2010a). Estimation of blue sheep population parameters in the Dhorpatan Hunting Reserve, Nepal. *International Journal of Biodiversity and Conservation, 2,* 51-56.

Aryal, A., Sathyakumar, S., & Kreigenhofer, B. (2010b). Opportunistic animal's diet depends on prey availability: Spring dietary composition of the red fox (*Vulpes vulpes*) in the Dhorpatan Hunting Reserve, Nepal. *Journal of Ecology and the Natural Environment, 2,* 59-63.

Bhuju, U. R., Shakya, P. R., Basnet, T. B., & Shrestha, S. (2007). Nepal biodiversity resource book: Protected areas, ramsar sites, and world heritage sites. ICIMOD and MOEST, Government of Nepal in cooperation with UNEP, Kathmandu.

Bird Conservation Nepal (2012). Annual Report 2011-12: Bird conservation Nepal, Lazimpat Kathmandu Nepal.

Bird Conservation of Nepal and Department of National Parks and Wildlife Conservation (2012). Birds of Nepal: An Official Checklist, Kathmandu, Nepal.

Bird Conservation of Nepal and Department of National Parks and Wildlife conservation (2011). The state of Nepal's birds 2010: Bird conservation of Nepal and Department of National Parks and Wildlife Conservation, Kathmandu Nepal.

Grimmett, R., Inskipp, C., & Inskipp, T. (2000). Birds of Nepal: Christopher Helm, an imprint of A& C Black Publishers Ltd., 36 Soho Square, London WID 3QY.

Lillesø, J. P. B., Shrestha, T. B., Dhakal, L. P., Nayaju, R. P., & Shrestha, R. (2005). The map of potential vegetation of Nepal—A forestry/agroecological/biodiversity classification system: Forest & Landscape. Development and Environment Series 2-2005 and CFC-TIS Document Series No.110. Denmark.

Panthi, S., Aryal, A., Lord, J., Adhikari B., & Raubenheimer, D. (2012). Summer diet and habitat ecology of red panda (*Ailurus fulgens fulgens*) in Dhopatan Hunting Reserve, Nepal. *Zoological Studies, 51,* 701-709.

Shrestha, T. B., Lillesø, J. P. B., Dhakal, L. P., & Shrestha, R. (2002). Forest and vegetation types of Nepal: MFSC, HMG/Nepal, Natural Resource Management Sector Assistance Programme (NARMSAP) and Tree Improvement and Silviculture Component (TISC), Kathmandu.

Shrestha, T. K., (1981). *Wildlife of Nepal.* Kathmandu: Curriculum Development Centre Tribhuvan University.

Thakuri, J. J., & Thapa, I. (2012). Assessment of Reshunga forest for an important bird area status. *Danphe, 21,* 1-4.

Thapa, B. B. (2007). Dhorpatan sikar aararksha ko parichaya. Baglung, Nepal: Department of National Park and Wildlife Conservation, DHR Office.

Wegge, P. (1976). Himalayan shikar reserves; surveys and management proposals. Field Document No. 5. FAO/NEP/72/002 Project, Kathmandu, 96 p.

Wikipedia (2013). Birds.
http://en.wikipedia.org/wiki/List_of_birds_of_Nepal

23

Modeling the Distribution of Marketable Timber Products of Private Teak (*Tectona grandis* L.f.) Plantations

Noël H. Fonton[1*], Gilbert Atindogbé[1], Arcadius Y. Akossou[2], Brice T. Missanon[1],
Belarmain Fadohan[1], Philippe Lejeune[3]

[1]Laboratory of Study and Research in Applied Statistics and Biometry,
University of Abomey-Calavi, Abomey-Calavi, Benin
[2]Faculty of Agronomy, University of Parakou, Parakou, Benin
[3]Unit of Forest and Nature Management, Gembloux Agro-Bio Tech, University of Liege, Gembloux, Belgium

Management of marketable products of private plantations will not be sustainable without class girth being identifiable readily. Modeling marketable products is a key to obtain good fitness between observed and theoretical girth distribution. We determine the best parameter recovery method with the Weibull function for two sylvicultural regimes (coppice and high forest). Data on stand variables were collected from 1101 sample plots. The three Weibull function parameters were estimated with three parameters recovery methods: the maximum likelihood method, the method of moments and the method of percentiles. Stepwise regression and the simultaneously re-estimated parameter using the Seemingly Unrelated Regression Estimation were applied to model each parameter. The results indicated that the three methods successfully predicted girth size distributions within the sample stands. The method of moments was the best one with lowest values of Reynolds error index and Kolmogorov-Smirnov statistic however the sylvicultural regimes. The Weibull parameter distribution model developed for each of the two sylvicultural regimes was quite reliable.

Keywords: Weibull; Parameter Recovery Method; Reynolds Index; Sylvicultural Regime; Poles; Logs

Introduction

The multipurpose management of small woodlots by smallholder forestry has been gaining more importance (Harrison et al., 2002). The growing of the demand for forest products (Scheer, 2004) explained the importance of their management mainly for the smallholder farmer to generate substantial income (Aoudji et al., 2012). Teak (*Tectona grandis* L.f.) is the most important reforestation and commercial plant species in coastal West Africa due to its fast growing potential (Niskane, 1998), good-quality timber (Louppe, 2008). Reforestation with this specie has increased the above ground biomass and carbon stock at 10-year-old about 45% higher than a nearby degraded secondary forest (Odiwe et al., 2012). In Benin, the success of state-owned plantations has encouraged farmers to invest in teak sylviculture, establishing plantations on small plots ranging from 0.05 ha to 28.10 ha (Atindogbé, 2012). Various problems constrain both traders and smallholder farmers (Aoudji et al., 2012). These include the lack of market information, high transaction costs, difficulties for traders to get timber supplies (Anyonge and Roshetko, 2003; Nawir et al., 2007), and the low return to smallholder farmers (Maldonado & Louppe, 1999; Nawir et al., 2007). According to the above problems, efforts were needed to have information on the different classes of the merchants products on stand before harvesting. Forest owners and managers have no reliable tools to provide them with a comprehensive scheme of resources available and monitoring, harvesting and sales operations. Therefore, one challenge is to determine the minimum level of information required to characterize harvests (Lafond et al., 2012).

Stand tables for total or marketable volume are based on the distribution of tree diameters using traditionally probability density functions (PDFs) (Parresol et al., 2010). Many functions have been suggested for establishing tree diameters size class distribution (e.g., normal, exponential, beta, Johnson's S_B, Gamma, Weibull, logit-logistic). However, the Weibull function appears the most often used (Little, 1983; Rennolls et al., 1985; Rondeux et al., 1992; Lindsay et al., 1996; Liu et al., 2004; Newton & Amponsah, 2005; Lei, 2008) owing to its flexibility (Hafley & Schreuder, 1977; Kilkki et al., 1989) and the best description of diameter structure. This function can also model many types of failure rate behaviors when appropriate parameters are included.

Many techniques for estimating Weibull function parameters have been developed: the graphical methods and the analytical methods. The accuracy of the estimate depends on the size of the sample and the method used. Graphical methods tend to provide crude estimates, while analytical methods provide better estimates that include confidence limits (Murthy et al., 2004) and are reported to be more accurate (Razali et al., 2009). The common analytical methods are the method of moments (*MOM*),

*Corresponding author.

the maximum likelihood method (*MLM*), the method of percentiles (*MOP*) and the method of least squares (*MLS*). However, the most suitable method depends on the stands characteristics (Liu et al., 2004; Lei, 2008).

The aim of this study was to determine the best estimator method for Weibull function parameters for two different sylvicultural regimes: the coppice and the high forest. The Modeled parameters were then used to predict the distribution of marketable products of the private teak plantations as a useful management tool.

Methods

Study Site and Data

This study was carried out in the Guinea-Congo zone of Benin (West Africa) located between 6°17' and 6°58'N, and 1°56' and 2°31'E. The region has a bimodal rainfall regime, with a mean annual precipitation of 1100 mm and a daily mean temperature of 29.9°C over the period 1971-2009 (www.Worldclim.org, 2005). Clayey-sand and vertisol are the dominant soil types. The original native vegetation, a semi-evergreen dry forest, was strongly influenced by human activities and is now reduced to a few relict forests and forest reserves.

Data were collected using a snowball sampling method which yielded 1101 private teak plantations: 844 coppices and 257 high forests. The size of each plantation (area) was measured. Then five (for plantations <0.5 ha) or ten (for plantations ≥0.5 ha) replicates strips of five trees were randomly sampled. On each strip, the planting space between trees (*e*) and between lines (*l*), the survival rate (*t*), and the girth at breast height (cbh) for all trees over 10 cm (lowest girth size of the marketable products) were measured. Timber merchants use height classes of marketable products based on girth classes: small poles (10 - 19 cm), medium poles (20 - 39 cm), large poles (40 - 49 cm), small posts (50 - 64 cm), large posts (65 - 79 cm), small logs (80 - 109 cm) and large logs (≥110 cm).

Statistical Parameter Modeling

The complete three-parameter Weibull probability density function of trees girth x is given by (Bailey & Dell, 1973)

$$f(x;\theta) = \frac{\gamma}{\beta}\left(\frac{x-\alpha}{\beta}\right)^{\gamma-1}\exp\left[-\left(\frac{x-\alpha}{\beta}\right)^{\gamma}\right] \quad (1)$$

for $x \geq \alpha$, $\alpha \geq 0$, $\beta > 0$, $\gamma > 0$

where $\theta = (\alpha,\beta,\gamma)'$, α is a location parameter, β is a scale parameter, and γ is a shape parameter. The recovery methods based on maximum likelihood, the maximum likelihood method (*MLM*), on moments, the method of moment (*MOM*) and on percentiles, the method of percentiles (*MOP*) were compared.

In relation to the unknown parameters α, β, γ and *n* the number of trees, the logarithm of the likelihood function, $\log(L(\theta))$ of Equation (1) is given by:

$$\log L(\theta) = \sum_{i=1}^{n}\log\left[\frac{\gamma}{\beta}\left(\frac{x_i-a}{\beta}\right)^{\gamma-1}\exp\left[-\left(\frac{x_i-a}{\beta}\right)^{\gamma}\right]\right] \quad (2)$$

For estimating these parameters with the *MLM*, the Equation (2) was maximized with a three-equation system as follows:

$$\begin{cases} \hat{\beta} = \left[(n-1)(x_i - \hat{\alpha})^{\hat{\gamma}}\right]^{\hat{\gamma}-1} \\ \hat{\gamma} = \left[\left(\sum_{i=1}^{n}(x_i-\hat{\alpha})^{\gamma}\log(x_i-\hat{\alpha})\right)\left(\sum_{i=1}^{n}(x_i-\hat{\alpha})^{\hat{\gamma}}\right)^{-1}\right. \\ \qquad \left. -n^{-1}\sum_{i=1}^{n}\log(x_i-\hat{\alpha})^{-1}\right] \\ (\hat{\gamma}-1)\sum_{i=1}^{n}(x_i-\hat{\alpha}) - \hat{\gamma}\hat{\beta}^{-\hat{\gamma}}\sum_{i=1}^{n}(x_i-\hat{\alpha})^{\hat{\gamma}} = 0 \end{cases} \quad (3)$$

where *n* is the number of trees in the plantation and x_i the girth of tree *i*. The SAS software (SAS 9.2) was used to solve iteratively the equation system (3).

The moment order k (μ_k) of the Weibull function is given by:

$$\mu_k = \beta^k\sum_{j=0}^{k}(-1)^j\binom{k}{j}\Gamma\left(\frac{k-1}{\gamma}+1\right)\left[\Gamma\left(\frac{1}{\gamma}+1\right)\right]^j \quad (4)$$

with Γ the gamma function written for a real value s as: $\Gamma(s) = \int_0^{\infty}x^{s-1}e^{-x}dx$, $(s>0)$. The parameters α, β and γ were estimated by *MOM* with two processes. The moments of order 1 (μ), order 2 (σ^2) and order 3 (μ_3) (Razali et al., 2009) of the equation 4 were computed as follows:

$$\begin{cases} \mu = \hat{\alpha}\hat{\beta} + \Gamma\left(1+\frac{1}{\hat{\gamma}}\right) \\ \sigma^2 = \hat{\beta}^2\Gamma\left(1+\frac{2}{\hat{\gamma}}\right) - \left[\Gamma\left(1+\frac{1}{\hat{\gamma}}\right)\right]^2 \\ \mu_3 = \hat{\beta}^3\left\{\Gamma\left(1+\frac{3}{\hat{\gamma}}\right) - 3\Gamma\left(1+\frac{1}{\hat{\gamma}}\right)\Gamma\left(1+\frac{2}{\hat{\gamma}}\right) + 2\left[\Gamma\left(1+\frac{1}{\hat{\gamma}}\right)\right]^3\right\} \end{cases} \quad (5)$$

The system (5) was solved using the *R* package rootSolve (R2.14.1).

The parameter recovery method based on percentiles (*MOP*) requires computation of the 0th (minimum girth), 25th, 50th, and 95th percentiles of the distribution of the girth as x_0, x_{25}, x_{50}, and x_{95}, respectively. The three parameters were estimated by solving the following three equations simultaneously (6) (Borders et al., 1987):

$$\begin{cases} \hat{\alpha} = \dfrac{n^{1/3}x_0 - x_{50}}{n^{1/3} - 1} \\ \hat{\gamma} = \dfrac{\ln\left[\dfrac{\ln(1-0.95)}{\ln(1-0.25)}\right]}{\ln(x_{95}-\hat{\alpha}) - \ln(x_{25}-\hat{\alpha})} \\ \hat{\beta} = -\dfrac{\hat{\alpha}\Gamma\left(1+\dfrac{1}{\hat{\gamma}}\right)}{\Gamma\left(1+\dfrac{2}{\hat{\gamma}}\right)} \\ \qquad + \sqrt{\dfrac{\alpha^2}{\Gamma^2\left(1+\dfrac{2}{\hat{\gamma}}\right)}\left[\Gamma^2\left(1+\dfrac{1}{\hat{\gamma}}\right) - \Gamma\left(1+\dfrac{2}{\hat{\gamma}}\right)\right] + \dfrac{\overline{x}_q^2}{\Gamma\left(1+\dfrac{2}{\hat{\gamma}}\right)}} \end{cases} \quad (6)$$

where *n* is the number of trees in the plantation, Γ is the gamma function, $\overline{x}_q$ is the quadratic mean girth of the plantation, and ln is the natural logarithm.

Comparison Criteria

Two statistics were used to assess the goodness of fit of the three methods: the Kolmogorov–Smirnov statistic (KS) and the prediction index error of Reynolds (e) (Reynolds et al., 1988). The optimal recovery parameter method is the one with low value for the two criteria. The prediction index error of Reynolds (Equations (7) and (8)) was computed (Pauwels, 2003) as:

$$e(\%) = \frac{\sum_{j=1}^{m} \left| N_j - \hat{N}_j \right|}{N} \times 100 \qquad (7)$$

where N_j is the observed and $\hat{N}_j$ is the estimated frequency of trees in girth size class j, N is the total number of trees, and m is the number of classes. $\hat{N}_j$ is estimated as follows:

$$\hat{N}_j = N \int_{l_j}^{u_j} f(x)\, dx \qquad (8)$$

where u_j and l_j are the upper and lower limits of class j.

Modeling Weibull Parameters

Multiple regression method was used to establish the relationship between the estimated parameters α, β, and γ (dependent variables) and dendrometric characteristics of stands. The predictor variables were density ($N \cdot ha^{-1}$), surface, and basal area of stand ($G \cdot m^2 \cdot ha^{-1}$); the girth of the tree of mean basal area (x_g cm); the mean, maximum, and minimum of girth; and the 25th (P_{25}), 50th (P_{50}), and 75th (P_{75}) percentiles of girth distribution. Models were established and tested for each set of dendrometric characteristics. A stepwise regression was then used to select the best subset of two variables. The estimated parameters were simultaneously re-estimated using the seemingly unrelated regression estimation (SURE). This can account for correlation errors between equations and is asymptotically efficient in the absence of specific errors (Liu et al., 2004).

Results

Data Summary

Overall, values for girth, density and basal area of trees av-

eraged 21.3 cm, 5170 stems/ha and 18.4 $m^2 \cdot ha^{-1}$, respectively (**Table 1**). While the mean girth was larger in high forests than in coppices, the reverse trend was obtained for the density and basal area of trees.

Optimal Method

Descriptive statistics of the estimated parameters for coppices are presented (**Table 2**). For *MLM*, the parameters $\hat{\alpha}$, $\hat{\beta}$ and $\hat{\gamma}$ averaged 8.14, 12.78, and 2.34, respectively. For *MOM*, mean values were 8.97, 11.79, and 2.20, respectively. For *MOP*, the three parameters averaged 9.20, 12.17, and 2.41, respectively. Statistics on the estimated parameters for high forests were presented (**Table 2**). For *MLM*, mean values for the parameters $\hat{\alpha}$, $\hat{\beta}$ and $\hat{\gamma}$ varied 9.89, 15.97, and 3.07, respectively. For *MOM*, these values were 11.28, 13.94, 2.52 respectively. For *MOP*, the three parameters means were 10.14, 18.50, 3.72, respectively.

For coppices, the percentages of plantations that fitted the Weibull distribution were 0.95% for *MLM*, 0.97% for *MOM*, and 0.91% for *MOP*. For high forests, these percentages were 0.94% for *MLM*, 0.99% for *MOM*, and 0.96% for *MOP*. The *MOM* method showed the lowest values for the error index of Reynolds ($e\%$) and the Kolmogorov-Smirnov statistic (KS) for both coppices and high forests (**Table 3**).

Table 1.
Density, N (/ha), quadratic mean of girth, x_g (cm) and basal area (G $m^2 \cdot ha^{-1}$) of the study plantations.

Sylvicultural regimes	Variables	Mean	Min.	Max.	SE
Coppices (n = 844)	x_g (cm)	20.2	10.7	51.6	0.2
	$N \cdot ha^{-1}$	5952	289	22,458	104
	$G \cdot m^2/ha^{-1}$	20.1	1.0	133.6	0.50
High forests (n = 257)	x_g (cm)	24.4	11.3	58.4	0.43
	$N \cdot ha^{-1}$	2798.9	632.5	8590.1	69.1
	$G \cdot m^2 \cdot ha^{-1}$	13.2	1.8	58.1	0.45

Min. and Max. are minimal and maximal values of the dendrometric parameters; SE is the standard error of the mean.

Table 2.
Estimated parameters of Weibull for the two sylvicultural regimes; α, β, and γ are the Weibull position, scale, and shape parameters, respectively. *MLM* is the maximum likelihood method; *MOM* the method of moments and *MOP* the method of percentiles.

Method	$\hat{\alpha}$ Mean	Min.	Max.	SE	$\hat{\beta}$ Mean	Min.	Max.	SE	$\hat{\gamma}$ Mean	Min.	Max.	SE
a) Coppices regime												
MLM	8.14	0.00	17.05	0.08	12.78	0.01	55.58	0.23	2.34	0.06	8.43	0.03
MOM	8.97	0.00	19.00	0.10	11.79	1.18	46.70	0.22	2.20	0.56	3.50	0.02
MOP	9.20	0.00	20.00	0.12	12.17	0.66	129.38	0.35	2.41	1.00	32.00	0.08
b) High forests regime												
MLM	9.89	0.00	19.88	0.22	15.47	1.91	65.11	0.53	3.07	0.92	6.30	0.06
MOM	11.28	0.00	19.96	0.25	13.94	2.29	67.08	0.49	2.52	0.68	3.52	0.03
MOP	10.14	0.00	19.62	0.32	18.50	1.32	180.15	1.31	3.72	1.00	39.98	0.27

MLM is the maximum likelihood method, *MOM* the method of moments and *MOP* the method of percentiles; Min and max are the minimum and the maximum respectively; SE is the standard error.

Table 3.
Comparison of the efficiency of the three parameters estimation methods: values of the error index of Reynolds (e%) and the Kolmogorov-Smirnov statistic (KS) for the two sylvicultural regimes.

	Coppices				High forests			
	e%		KS		e%		KS	
Methods	Mean	SE	Mean	SE	Mean	SE	Moy	SE
MLM	13.03	0.40	0.29	0.01	12.60	0.77	0.27	0.02
MOM	11.35	0.31	0.26	0.01	10.12	0.56	0.21	0.01
MOP	19.26	1.15	0.43	0.03	13.12	1.28	0.28	0.03

Consequently, *MOM* was chosen as the most appropriate method for modeling the distribution parameters of the Weibull function for private teak plantations. The models revealed significant differences between coppices and high forests for all three parameters estimated using *MOM* (**Table 4**).

Parameters' Model Development

The results of the SURE analysis for coppices are presented in **Table 5**. Stepwise regression revealed that the best subset of two stand characteristics is P_{25} and P_{75} for $\hat{\gamma}$, P_{75} and G^2 for $\hat{\beta}$, and LP_{50} and P_{75} for $\hat{\gamma}$ with $\left(LP_{50} = \ln\left(P_{50}/x_g\right)\right)$. The final regression equations are:

$$\hat{\alpha} = 6.736 + 0.405P_{25} - 0.167P_{75}$$
$$\hat{\beta} = 5.5895 + 0.243P_{75} + 0.001G^2$$
$$\hat{\gamma} = 1.989 + 0.024P_{75} + 8.116\ln\left(P_{50}/x_g\right)$$

The associated P_{value} are less than 0.001. The model to predict marketable products from coppices is:

$$F(x) = 1 - \exp\left[-\left(\frac{x - 6.736 - 0.405P_{25} + 0.167P_{75}}{5.895 + 0.243P_{75} + 0.001G^2}\right)^v\right]$$

where $v = 1.989 + 0.024P_{75} + 8.116LP_{50}$, $F(x)$ is the Weibull function, x is the girth of the tree, G is the basal area, P_{25}, P_{50}, and P_{75} are the 25th, 50th, and 75th are the *th* percentiles, and LP_{50} the weighted percentile. Some dependent variables such as P_{25} and P_{75} were computed with the best adjustment as follows:

$$P_{25} = -9.79 + 5.69\sqrt{x_g} \quad \text{and} \quad P_{75} = -26.6 + 11.1\sqrt{x_g}$$

with $P_{value} = 0.000$. The results of the SURE analysis for high forest are presented in **Table 6** and the best models equations with P_{value} less than 0.023 are:

$$\hat{\alpha} = 5.249 + 0.476x_{min} - 0.111G$$
$$\hat{\beta} = 10.567 + 0.004G^2 - 23.519\ln\left(P_{50}/x_g\right)$$
$$\hat{\gamma} = 1.867 + 8.283\ln\left(P_{50}/x_g\right) + 0.014N^{0.5}$$

The equation to predict the marketable products for high forests is then:

$$F(x) = 1 - \exp\left[\left(\frac{x - 5.249 - 0.476x_{min} + 0.111G}{10.567 + 0.004G^2 - 23.519LP_{50}}\right)^w\right]$$

where $w = 1.867 + 8.283LP_{50} + 0.014\sqrt{N}$, $F(x)$ is the Weibull

Table 4.
Comparison of the sylvicultural regimes according to estimated parameters of Weibull function by the best estimator method *MOM*.

	Coppices		High forests			
Parameters	Mean	SE	Mean	SE	t	P
$\hat{\alpha}$	8.97	0.10	11.28	0.25	8.58	0.000
$\hat{\beta}$	11.79	0.22	13.94	0.49	4.00	0.000
$\hat{\gamma}$	2.20	0.02	2.52	0.03	8.88	0.000

$\hat{\alpha}$, $\hat{\beta}$, and $\hat{\gamma}$ are the Weibull position, scale, and shape parameters, respectively. *MOM* is the method of moments; *t* is the statistic of student, and *P* the associated probability value.

distribution function, x is the girth of the tree, x_{min}, is the minimum girth, G is the basal area, N is the density, P_{50} is the 50th percentile or median, and LP_{50} the weighted 50th percentile.

Discussion

Efficiency of the Parameter Recovery Methods

Regardless of the parameter recovery method used, the observed and theoretical distributions of the plantations were much closed according to the Kolmogorov-Smirnov test. Moreover, for both sylvicultural practices, coppices and high forests, the average value of the shape parameter γ of the Weibull distribution was lower than 3.6, suggesting that the distribution of trees is left-skewed. Using the optimal method (*MOM*), the parameter α, whose value is associated with the minimum girth, was 8.97 for coppices and 11.28 for high forest, both were close to the minimum girth of this study (10 cm).

The parameter β, which gives an idea of the central value of samples, had a maximum value of 46.70 for coppices and 67.08 for high forests, while the values measured, with the completed inventory of 18 plantations, were 51.6 cm for coppices and 58.4 cm for high forests. These results are similar to previous findings, which illustrated that α is a good predictor of the minimum diameter (Frazier, 1981; Knoebel et al., 1986; Leduc et al., 2001). They also support those of Lei (2008), who demonstrated that the two-parameter Weibull distribution and *MOM* provided the best estimation of the diameter distribution of Chinese pine (*Pinus tabulaeformis*). These results are also consistent with those found by Liu et al. (2004) in their study of the diameter distribution of unthinned plantations of black spruce (*Picea mariana*) in central Canada, although in their study *MOP* was the preferred method. Zhang et al. (2003) previously dem-

Table 5.
Regression coefficients of the predictors and statistics resulting from the SURE (Seemingly Unrelated Regression Estimation) analysis for coppices for the response variables i.e. $\hat{\alpha}$, $\hat{\beta}$, and $\hat{\gamma}$ respectively the Weibull position, scale and shape parameters.

	$\hat{\alpha}$			$\hat{\beta}$			$\hat{\gamma}$		
	Constant	P_{25}	P_{75}	Constant	P_{75}	G^2	Constant	P_{75}	LP_{50}
Coef.	6.74	0.405	−0.167	5.895	0.243	0.001	1.989	0.024	8.116
P	<0.00	<0.001	<0.001	<0.001	<0.001	<0.001	<0.001	<0.001	<0.001
t	16.46	8.69	−6.100	7.720	7.130	4.650	25.410	6.810	10.060
SE	0.41	0.05	0.027	0.763	0.034	0.000	0.078	0.004	0.807

Coef. is the regression coefficients, G is the basal area, P_{25} and P_{75} are the 25th and 75th percentiles, and $LP_{50} = \ln(P_{50}/x_g)$ where x_g is the girth of the tree of mean basal area and P_{50} is the 50th percentile, t is the statistic of Student and P the associated probability value.

Table 6.
Regression coefficients values and their significant appreciation statistics resulting from the SURE (Seemingly Unrelated Regression Estimation) analysis for high forests with the response variables $\hat{\alpha}$, $\hat{\beta}$, and $\hat{\gamma}$ respectively the Weibull position, scale and shape parameters.

	$\hat{\alpha}$			$\hat{\beta}$			$\hat{\gamma}$		
	Constant	x_{min}	G	Constant	G^2	LP_{50}	Constant	LP_{50}	$N^{1/2}$
Coef	5.249	0.476	−0.111	10.567	0.004	−23.519	1.867	8.283	0.014
P	<0.001	<0.001	< 0.001	<0.001	0.003	0.023	<0.001	0.003	<0.001
t	6.040	7.440	−4.060	15.000	3.640	−2.290	9.800	3.700	4.320
SE	0.870	0.064	0.027	6.705	0.001	23.519	0.191	2.241	0.003

Coef. is the regression coefficients, x_{min} is the minimum girth, G is the basal area, N is the density and $LP_{50} = \ln(P_{50}/x_g)$ where x_g is the girth of tree of mean basal area and P_{50} is the 50th percentile, t is the statistic of Student and P the associated probability value, SE is the standard error.

onstrated the effectiveness of the Weibull distribution for describing the diameter distribution of natural stands of red spruce (*Picea rubens*) and balsam fir (*Abies balsamea*) in the north-east of North America. Meanwhile, Bailey & Dell (1973) have shown that *MOM* is more efficient than *MOP* for estimating parameters of the Weibull distribution, but it requires very complex calculations. Nanang (1998) in a study on the diameter distribution of *Azadirachta indica* plantations in Ghana asserted the same thing. It was also argued that *MOM* assures compatibility between the characteristics of the observed population used in parameter recovery and those obtained through simulation (Mateus & Tomé, 2011). The differences observed between coppices and high forests for all the three parameters estimated by *MOM* confirm the need to build separate models for different sylvicultural regimes.

Predicting the Weibull Parameters

Parameters of the Weibull distribution were functions of most of the dendrometric characteristics. In all cases the regression coefficients were statistical significant with P value less than 0.001 for coppices and 0.023 for high forests. These findings are in agreement with Liu et al. (2004), who modeled the three parameters of a Weilbull distribution using four characteristics of black spruce stands (age, basal area, average height, and site index). In most cases, the probability values associated to the regressions were less than 0.0001. Since the distribution model of marketable products from coppices differed from high

forests, a global model combining data from these two sylvicultural regimes would not be suitable. This is supported by the observed differences between the estimated parameters. The parameter of the distribution shape, γ, is more influenced by LP_{50} for both sylvicultural practices. The location parameter α depends on the positional parameters P_{25} and P_{75} in coppices, and on the minimum girth and basal area in high forests. β is more influenced by the square of the basal area for both sylvicultural regimes. These results differ from those of Torres-Rojo et al. (2000), who found that the shape of the distribution is strongly influenced by the diameter, mean basal area, density, and dominant height of the trees; and that β is influenced by diameter and mean basal area of trees. Moreover, several studies have previously found that the minimum girth most often influences the value of α (Frazier, 1981; Knoebel et al., 1986; Lejeune, 1994; Leduc et al., 2001).

Conclusion

In this study, models have been developed to assess marketable teak resource produced by private teak plantations. The main advantage of modeling parameter with stand characteristics is the compatibility between the characteristics of the observed populations and those obtained through theoretical distributions. Results indicated that the three methods compared were generally suitable for modeling the distribution of marketable products. However, the relative performance of each method depends on its ability to predict the observed girth size class frequencies. The method based on moment (*MOM*) appears to be the most appropriate.

Distribution models for marketable products were developed for coppices and for high forests using stand variables and *MOM*.

Acknowledgements

This study was sponsored by the Belgiun "*Commission Universitaire pour le Développement (CUD)*"—through le Projet Interuniversitaire Ciblé: Contribution au développement d'une filière du teck au départ des forêts privées du Sud-Bénin (Département de l'Atlantique)".

REFERENCES

Anyonge, C. H., & Roshetko, J. M. (2003). Farm-level timber production: Orienting farmers towards the market. *Unasylva, 54,* 48-56.

Aoudji, A. K. N., Adégbidi, A., Agbo, V., Atindogbé, G., Toyi, M. S. S., Yêvidé, A. S. I., Ganglo, J. C., & Lebailly, P. (2012). Functioning of farm-grown timber value chains: Lessons from the smallholder-produced teak (*Tectona grandis* L.f.) poles value chain in Southern Benin. *Forest Policy and Economics, 15,* 98-107.

Atindogbé, G. (2012). *Evaluation et caractérisation de la ressource en teck (Tectona grandis L.f.) dans les plantations privées du Sud-Bénin.* Thèse de doctorat, Bénin: Université d'Abomey-Calavi.

Bailey, R. L., & Dell, T. R. (1973). Quantifying diameter distributions with Weibull function. *Forest Sciences, 19,* 97-104.

Borders, B. E., Souter, R. A., Bailey, R. L., & Ware, K. D. (1987). Percentile-based distributions characterize forest stand tables. *Forest Sciences, 33,* 570-576.

Frazier, J. R. (1981). *Compatible whole-stand and diameter distribution models for loblolly pine stands.* Ph.D. Thesis, Blacksburg: Virginia Polytechnic Institute.

Hafley, W. L., & Schreuder, H. T. (1977). Statistical distributions for fitting diameter and height data in even-aged stands. *Canadian Journal of Forest Research, 7,* 481-487.

Harrison, S. R., Herbohn, J. L., & Niskanen, A. J. (2002). Non-industrial, smallholder, small-scale and family forestry: What's in a name? *Small-Scale Forest Economics. Management and Policy, 1,* 1-11.

Kilkki, P., & Paivinen, R. (1986). Weibull function in the estimation of the basal area dbh-distribution. *Silva Fennica, 20,* 149-156.

Kilkki, P., Maltamo, M., Mykkanen, R., & Paivinen, R. (1989). Use of the Weibull function in estimating the basal area dbhdistribution. *Silva Fennica, 23,* 311-318.

Knoebel, B. R., Burkhart, H. E., & Beck, D. E. (1986). A growth and yield model for thinned stands of yellow-poplar. *Forest Sciences Monograph, 27,* 64.

Lafond, V., Cordonnier, T., De Coligny, F., & Courbaud, B. (2012). Reconstructing harvesting diameter distribution from aggregate data. *Annals of Forest Science, 69,* 235-243.

Leduc, D. J., Matney, T. G., Belli, K. L., & Baldwin, V. C. (2001). *Predicting diameter distributions of longleaf pine plantations: A comparison between artificial neural networks and other accepted methodologies.* USDA For Serv Res Pap SRS-25.

Lei, Y. (2008). Evaluation of three methods for estimating the Weibull distribution parameters of Chinese pine (*Pinus tabulaeformis*). *Forest Sciences, 54,* 566-571.

Lejeune, P. (1994). Construction d'un modèle de répartition des arbres par classes de grosseur pour des plantations d'épicéa commun (*Picea abies* L Karst) en Ardenne belge. *Annals of Forest Science, 51,* 53-65.

Lindsay, S. R., Wood, G. R., & Woollons, R. C. (1996). Stand table modeling through the Weibull distribution and usage of skewness information. *Forest Ecology and Management, 81,* 19-23.

Little, S. N. (1983). Weibull diameter distributions for mixed stands of western conifers. *Canadian Journal of Forest Research, 13,* 85-88.

Liu, C., Zhang, S. Y., Lei, Y., Newton, P. F., & Zhang, L. (2004). Evaluation of three methods predicting diameter distributions of black spruce (*Picea mariana*) plantations in central. *Canadian Journal of Forest Research, 34,* 2424-2432.

Louppe, D. (2008). *Tectona grandis* (L. f). In D. Louppe, A. A. Oteng-Amoako, & M. Brink (Eds.), *Ressources végétales de l'Afrique Tropicale. Bois d'oeuvre 1.* [Traduction de: *Plant Resources of Tropical Africa. Timbers 1.* 2008]. Wageningen, Pays-Bas: Fondation PROTA; Leiden, Pays-Bas: Backhuys Publishers; Wageningen, Pays-Bas: CTA.

Maldonado, G., & Louppe, D. (1999). Les plantations villageoises de teck en Côte d'Ivoire. *Bois et Forêts des Tropiques, 262,* 19-30.

Mateus, A., & Tomé, M. (2011). Modelling the diameter distribution of eucalyptus plantations with Johnson's SB probability density function: Parameters recovery from a compatible system of equations to predict stand variables. *Annals of Forest Science, 68,* 325-335.

Murthy, P. D. N., Xie, M., & Jiang, R. (2004). *Weibull models.* Wiley series in probability and statistics. Hoboken.

Nanang, D. M. (1998). Suitability of the normal, log-normal and Weibull distributions for fitting diameter distributions of neem plantations in Northern Ghana. *Forest Ecology and Management, 103,* 1-7.

Nawir, A. A., Kassa, H., Sandewall, M., Dore, D., Campbell, B., Ohlsson, B., & Bekele, M. (2007). Stimulating smallholder tree planting—Lessons from Africa and Asia. *Unasylva, 58,* 53-59.

Newton, P. F., & Amponsah, I. G. (2005). Evaluation of Weibull-based parameter prediction equation systems for black spruce and jack pine stand types within the context of developing structural stand density management diagrams. *Canadian Journal of Forest Research, 35,* 2996-3010.

Niskanen, A. (1998). Financial and economic profitability of reforestation in Thaïland. *Forest Ecology and Management, 104,* 57-68.

Odiwe, A. F., Adewumi, R. A., Alami, A. A., & Ogunsanwo, O. (2012). Carbon stock in topsoil, standing floor litter and above ground biomass in *Tectona grandis* plantation 10-years after establishment in Ile-Ife, Southwestern Nigeria. *International Journal of Biological and Chemical Sciences, 6,* 3006-3016.

Parresol, B. R., Fonseca, T. F., & Marques, C. P. (2010). *Numerical details and SAS programs for parameter recovery of the SB distribution.* Forest Service, Southern Research Station, General Technical Report SRS-122, United States Department of Agriculture, 31 p.

Pauwels, D. (2003). *Conception d'un systeme d'aide à la decision pour le choix d'un Scenario sylvicole: Application aux peuplements de mélèze en Région wallonne.* Thèse de Doctorat, Gembloux: Faculté Universitaire des Sciences Agronomiques.

Razali, A. M., Salih, A. A., & Mahdi, A. A. (2009). Best estimate for the parameters of the three parameter Weibull distribution. *Proceedings of the 5th Asian Mathematical Conference,* Malaysia, 2009.

Rennolls, K., Geary, D. N., & Rollinson, T. J. D. (1985). Characterizing diameter distributions by the use of the Weibull distribution. *Forestry, 58,* 57-66.

Reynolds, M. R., Burk, T. E., & Huang, W. C. (1988). Goodness of fit tests and model selection procedures for diameter distribution models. *Forest Sciences, 34,* 373-399.

Rondeux, J., Laurent, C., & Thibaut, A. (1992). Construction d'une table de production pour le douglas (*Pseudotsuga mensiesii* Mirb. Franco) en Belgique. *Bulletin des Recherches Agronomiques de Gembloux, 27,* 327-347.

Scherr, S. J. (2004). Building opportunities for small-farm agroforestry to supply domestic wood markets in developing countries. *Agroforestry Systems, 61-62,* 357-370.

Torres-Rojo, J. M., Magaña, O. S., & Acosta, M. (2000). Metodología para mejorar la predicción de parámetros de distribuciones diamétricas. *Agrociencia, 34,* 627-637.

WorldClim (2005). *WorldClim Global Climate Data* (GIS Data). http://www.worldclim.org/

Zhang, L., Packard, K. C., & Liu, C. (2003). A comparison of estimation methods for fitting Weibull and Johnson's SB distributions to mixed spruce-fir stands in northeastern North America. *Canadian Journal of Forest Research, 33,* 1340-1347.

24

An Evaluation Model for Improving Biodiversity in Artificial Coniferous Forests Invaded by Broadleaf Trees

Yozo Yamada, Sayumi Kosaka

Graduate School of Bio-Agricultural Sciences, Nagoya University, Nagoya, Japan

Increasing attention is being paid to the various functions of forests, especially the conservation of biodiversity. In Japan, 67% of national land is covered by forest, 41% of which is artificial forest (i.e., plantations). Therefore, efforts to conserve forest biodiversity should also target artificial forests. In this study, we investigated the increase in biodiversity resulting from broadleaf tree invasion of artificial coniferous forests. We examined diversity indices and combinations of indices to identify which ones can aid forest managers in evaluating forest diversity. We also studied classification according to the richness of diversity, which corresponded to the growth stages of *Chamaecyparisobtusa* and *Cryptomeria japonica* plantation forests. Moreover, we developed a model that will contribute to sustainable forest management and biodiversity over an entire area. The model, based on a specific rotation scenario in a geographic information system, is easy to use and presents spatial and temporal changes at sites visually.

Keywords: Artificial Forest; Invading Broadleaf Trees; Broadleaf Tree Diversity; Sustainable Forestry Management; Basin Scale

Introduction

The importance of forest biodiversity has been discussed since the United Nations Conference on Environment and Development (UNCED) in 1992. The UNCED adopted the Declaration on Forest Principles in Agenda 21, which is a series of principles for sustainable forest use (Lund et al., 2004). In 1994, the Montréal Process Working Group agreed on seven criteria of sustainable forest management and 54 indicators of the criteria. Criterion 1 stipulates the conservation of biological diversity and states that conserving the diversity of organisms and their habitats supports forest ecosystems and their ability to function, reproduce, and remain productive (Montreal Process, 2009). Many studies have confirmed the importance of forest diversity, although most have focused on natural forests (Burslem, 2004; Ehrlich, 1996; Kondoh, 2002; Noss, 1990).

Artificial forests account for approximately 6.5% of the global forested area. The percentage of artificial forest varies by country. In Japan, 68.5% of the land area is covered by forest, 41.3% of which is artificial forest (FAO, 2012). By age, artificial forests in Japan form a nearly perfect normal distribution, with a top age of 50 years. Thus, vast areas of forest require immediate thinning (Forestry Agency, 2012). However, many of the artificial forests that require thinning have been deteriorating, mainly because of reductions in forestry surplus. In countries such as Japan that have vast areas of artificial forest, biodiversity loss has become a problem, even in the artificial forests since last few decades. Many artificial forests are designed as monoculture plantations with the aim of growing high-quality timber. Thus, these forests are generally dark with sparse understory vegetation. Brockerhoff et al. (2008) noted that artificial forests usually have less habitat diversity and complexity. Yamaura (2007) also suggested that artificial forests have been treated as a homogeneous non-habitat (matrix) and ignored in biodiversity conservation.

However, the environment in an artificial forest can be altered to become more favorable to biodiversity. Hartley (2002) indicated that earlier thinning schedules or longer rotations can strongly affect biodiversity, as can reserve trees that are left after plantation harvest and remain through a second rotation. Busing and Garman (2002) argued that proportional thinning retains understory stems, thereby expediting the recruitment of shade-tolerant trees. El-Keblawy (2005) suggested the importance of reducing forest crowns to promote species growth. Yamaura (2007) proposed that the negative effects of artificial forest could be mitigated by increasing the complexity of the structure and composition of a plantation through extended rotation, strong thinning, wider tree spacing, and the retention of broadleaf trees and coarse woody debris in clear-cuts. Brockerhoff et al. (2008) also stressed that to sustain native biodiversity within an artificial forest, managers should consider using a greater diversity of planted species, extending rotation lengths in some stands, and adopting a variety of harvesting approaches.

The overall aim of the present study was to promote the diversity of invading broadleaf trees in artificial coniferous forests. We examined effective diversity indices and combinations of these indices that can be used by forest managers to evaluate forest diversity. In addition, we studied classification according to the richness of diversity, which was found to correspond to growth stages in *Chamaecyparisobtusa* and *Cryptomeria japonica* artificial forests. Moreover, to help forest managers maintain sustainable forests with rich biodiversity over the entire area, we developed an easy-to-use model that shows spatial and temporal changes at forest sites visually. The model, implemented in a geographic information system (GIS), is based on a specific rotation scenario.

Methods

Survey Location

Our survey was conducted at Danto National Forest (35°6'N, 137°28'E), located in Shitara-cho, Aichi Prefecture, Japan. The forest covers 5303 ha, 93% of which is artificial forest dominated by hinoki (C. obtusa). It ranges from 400 to 1150 m above sea level, with an average annual temperature of 11.7°C and annual precipitation of 2036 mm.

The Chubu Regional Forest Office of Forestry Agency has performed systematic thinning and promoted understory vegetation in their forests, with the farsighted (100-year) goal of creating diversified, vital, and sound forests for the coexistence of log production and various public benefit functions (Chubu Regional Forest Office, 2011).

Danto National Forest is divided into about 1300 sub-compartments, and the Forest Office has created specific management plans and concrete management schedules for each sub-compartment. As part of this management, the sub-compartments are arranged to achieve a proper balance of vegetation types and different stages of succession to aid in the conservation of biodiversity in the entire forest.

Vegetation Survey

We chose six hinoki (C. obtusa) sub-compartments (aged 13, 36, 44, 56, 76, and 118 years) and eight sugi (C. japonica) sub-compartments (aged 26, 45, 55, 60, 72, 79, 98, and 116 years). We also selected four additional plots: two hinoki (aged 34 and 55 years) and two sugi (aged 34 and 55 years) sub-compartments in the Inabu Experimental Forest of Nagoya University, located near the Danto National Forest. These plots served as controls, representing forests in which management has been insufficient.

As in our previous study (Kosaka & Yamada, 2013), we established 10 m × 10 m plots within areas representative of each sub-compartment. We examined all broadleaf and planted trees that were taller than 1 m. For the broadleaf trees, we determined the number of species and population size. We also measured the tree height and diameter at a height of 50 cm from the ground surface. For planted trees, we measured the height and diameter and counted the number of planted trees in 20 m × 20 m areas to calculate the stand density.

Analysis at the Sub-Compartment Scale

For the analysis of broadleaf tree diversity within sub-compartments, we used the number of species, population size, proportion of basal area, and two species diversity indices: the Shannon-Wiener index and inverse Simpson index. We clarified the characteristics and diversity of invading broadleaf trees during each growth stage in artificial forests to discuss practical ways of retaining broadleaf trees to encourage biodiversity

Analysis at the Basin Scale

Using a GIS and the forest age data from the forest register,

we determined the distribution of broadleaf tree diversity across the entire Danto forest by sub-compartment. Broadleaf tree diversity was classified into three levels based on forest age, which was estimated from our analysis at the sub-compartment scale.

We set up a rotation scenario in which all forests more than 80 years old are entirely cut and the plots are then replanted with the same tree species. That is, clear cutting and reforestation are repeated every 80 years in each sub-compartment. We simulated the change in broadleaf tree diversity in Danto forest from the present to 40 years later and discussed the spatiotemporal evaluation of broadleaf tree diversity shown by the model.

Results and Discussion

Analysis at the Sub-Compartment Scale

The invading broadleaf trees were divided into the following three categories by height:

Lower layer (Shrub): tree height is more than 1 m and less than 4 m.

Middle layer (Sub-tree): tree height is more than 4 m and less than 8 m.

Upper layer (Tree): tree height is more than 8 m.

We analyzed every diversity index and its layer composition by forest age. All diversity indices indicated a similar trend of change in forest age. For hinoki forests, **Figure 1** shows the relation between forest age and number of species, while **Figure 2** shows the relation between forest age and proportion of basal area. Both indices increased significantly after an age of 76 years. After that, in the mature stage, the number of species decreased slightly at 118 years old, but the proportion of basal area continued to increase. The young stage represented by the 13-year-old forest also had abundant species and was jungle-like in its appearance, but its proportion of basal area was not very large. The middle stage, from 36 to 56 years old, was the stage of height growth, when the crowns closed entirely. Thus, the lack of sunlight in this stage led to a decrease in invading broadleaf trees.

For the sugi forests, **Figure 3** shows the relation between forest age and number of species, while **Figure 4** shows the relation between forest age and proportion of basal area. The sugi forests exhibited a slightly different tendency from hinoki forests. As shown in the figures, both indices increased remark-

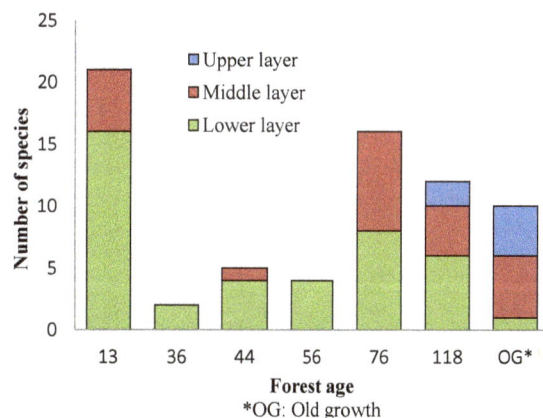

Figure 1.

Forest age and number of species in hinoki (C. obtusa) forests.

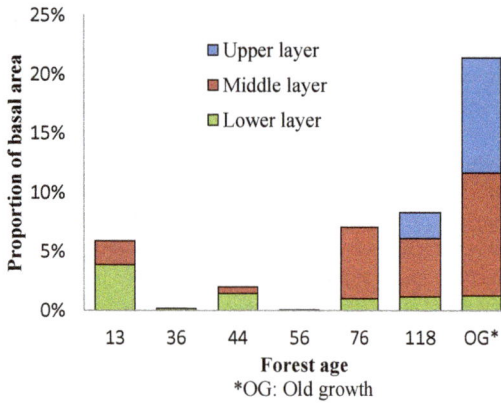

Figure 2.
Forest age and proportion of basal area in hinoki (*C. obtusa*) forests.

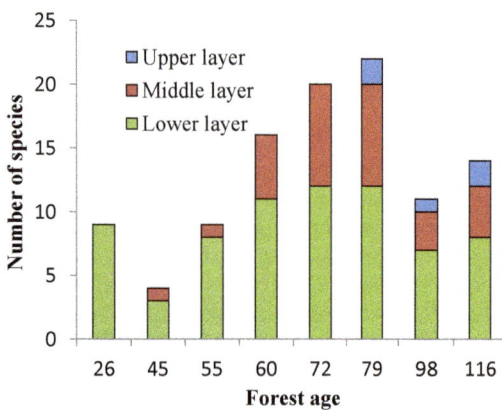

Figure 3.
Forest age and number of species in sugi (*C. japonica*) forests.

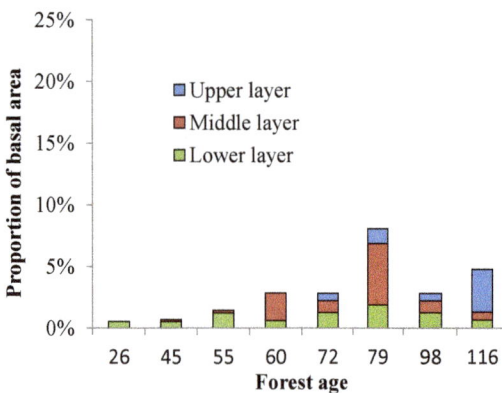

Figure 4.
Forest age and proportion of basal area in sugi (*C. japonica*) forests.

ably from 60 years of age and kept increasing until 79 years. The proportion of basal area also reached its maximum at 79 years. The young stage, represented by the 26-year-old forest, had fewer species compared to the young-stage hinoki forest, whereas the middle stage (from 45 to 55 years) retained a relatively large number of invading broadleaf species.

The layer composition developed toward the upper layer as both hinoki and sugi forests reached the mature stage. In the

Danto forest, invading broadleaf trees remained, because they were not cut as long as they impeded thinning or threatened safety.

From these results, we can define three stages on the basis of the diversity of invading broadleaf trees: young, middle, and mature. The young stage includes forests less than 25 years old that have relatively rich diversity of invading broadleaf trees, unless those trees were cut by intense weeding and cleaning operations.

The middle stage includes forests more than 26 and less than 60 years old. These forests have poor diversity of invading broadleaf trees because of a lack of sunlight. Although this stage does not have good conditions for understory vegetation, it is indispensable for the production of high-quality logs. These low-light conditions will not be improved until comercial thinning begins. Commercial thinning usually starts at age 40, but its effect on improving the light conditions does not appear until a few decades later.

The mature stage includes forests more than 61 years old and has rich diversity and a developed layer of invading broadleaf trees. Commercial thinning is performed repeatedly at 15- to 20-year intervals; thus, favorable light conditions for invading broadleaf trees are kept and improved unless broadleaf trees are cut during thinning.

In comparison, the four control plots in the Inabu Experimental Forest of Nagoya University were typical examples of poorly tended forests. The understory was very dark, with insufficient light for the invasion of understory vegetation. Consequently, we found no invading broadleaf trees in these plots.

To promote understory vegetation, management measures such as thinning must be implemented at appropriate times based on a specific density control plan. **Figure 5** shows the relation between forest age and stand density in Danto forest, Inabu Experimental Forest, and Hayami forest, which we examined in our previous study (Kosaka & Yamada, 2013). In forests less than 40 years old, before launching commercial thinning, we found no remarkable differences in stand density among the plots, including the two plots at Inabu. However, the stand densities differed after 50 years. As shown in **Figure 5**, Hayami forest has maintained the lowest density and is considered to have ideal density control. Danto forest has also maintained a low density, showing good performance and careful tending. In Inabu, 55-year-old plots have higher densities than

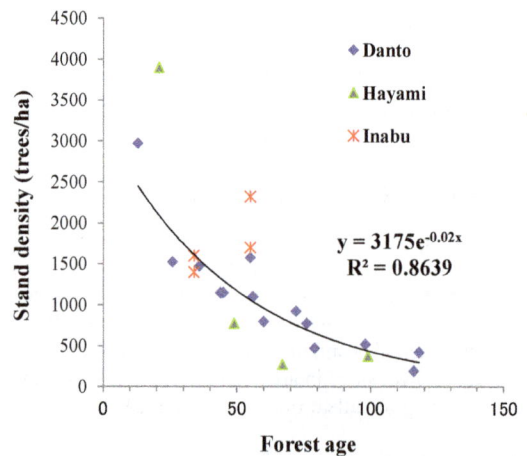

$$y = 3175e^{-0.02x}$$
$$R^2 = 0.8639$$

Figure 5.
Forest age and stand density.

plots of that age in the other two forests and are clearly too late for thinning.

We should also consider the traditional practice of removing all understory vegetation to provide safe and convenient working conditions for thinning. In this case, invading broadleaf trees will be cut as often as thinning takes place and will not be able to grow to create a layered forest structure. Both Hayami forest and Danto forest have their own guidelines, which say that invading broadleaf trees should remain as long as they do not impede thinning or threaten safety.

Selection of Diversity Indices for Broadleaf Trees

The five indices recorded in this research showed a similar trend of change in forest age, but we found that each index had certain advantages and disadvantages. For example, although the number of species is easy to survey, this measure cannot be used to estimate the degree of uniformity. Moreover, the number of species may increase in proportion to plot size. Population size can be used to estimate the degree of uniformity, but if one species dominates a plot, uniformity may be overestimated. Otherwise, this measure may tend to show a trend similar to that of the number of species. The proportion of basal area can indicate the species dominance in each layer, but the difference between dominance and diversity requires careful consideration. The Shannon-Wiener index is popular and easy to understand because it includes both the number of species and the population size.

The inverse Simpson index has the same characteristics and can indicate species uniformity. However both indices are influenced largely by the observed parameter.

Each index has own characteristics, making it difficult to choose a single index as the best for estimating the diversity of invading broadleaf trees. By using a few indices in combina-

tion, we can evaluate the different perspectives each provides and they can compensate for deficits in each other.

To determine the most effective combination of indices, we created a precedence matrix for each index and calculated the rank difference of each sub-compartment between pairs of indices. **Table 1** shows the average value. Smaller values indicate that both indices of a combination estimate diversity similarly, so we could use either one of the two. Meanwhile, a larger value means that the indices estimate diversity from different perspectives, so it is effective to use both indices. The inverse Shannon index is different from the other indices, with the exception of the Shannon-Wiener index.

Moreover, we cannot determine the best indices based on statistical results alone; other features should also be considered. **Table 2** shows the advantages, disadvantages, and applications of each index. On the basis of our results, to evaluate broadleaf tree diversity at the sub-compartment level, we recommend using the proportion of basal area to evaluate the species dominance in each layer and the inverse Simpson index to evaluate diversity from a different perspective than the proportion of basal area.

However, the total number of species did not largely differ among forest ages. On the other hand, when considering layer structure, the number of trees located lower in the canopy decreased with age, whereas the number of trees at intermediate levels increased with age. After age 67, an upper layer began to form, indicating that layer composition becomes more complex as the forest matures. In particular, layer composition within forests at age 99 approaches that of natural forests.

The number of species is an effective index for forest management, as species richness is a good indicator of current forest conditions. However, the evenness of species must also be considered. Furthermore, the number of species tends to in-

Table 1.
Average rank difference between pairs of indices.

	Number of species	Population size	Proportion of basal area	Shannon-Wiener index	Inverse Simpson index
Number of species	-	1.068	2.006	1.961	4.187
Population size	-	-	2.442	2.637	4.679
Proportion of basal area	-	-	-	2.004	3.510
Shannon-Wiener index	-	-	-	-	2.270
Inverse Simpson index	-	-	-	-	-

Table 2.
Merits, defects, and uses of each index.

Indexes	Merits	Defects	Applications
Number of species	High versatility	Ambiguous uniformity, Increasing in association with expansion of sample size	Provides a tentative evaluation of diversity
Population size	Degree of uniformity	Apt to show a similar trend to number of species	Can be substituted by number of species
Proportion of basal area	Dominancy, insusceptible to population size	Dominancy ≠ diverseness	For evaluating dominance
Shannon-Wiener index	Popular and easy to follow index	Large influence from parameter	For evaluating both the number of species and population size
Inverse Simpson index	Evaluation including uniformity of species	Large influence from parameter	For equivalent evaluations of several invading species

crease as plot size increases.

Analysis at the Basin Scale

To evaluate diversity of invading broadleaf trees at the basin level, it is important to consider fragmentation and networks of sub-compartments having rich diversity within the whole subject area or forest. Fragmentation is important for avoiding overly large expanses of even-aged plantations, as large monoculture areas may not have good effects in terms of biodiversity and ecosystem functioning. Meanwhile, the network should be maintained for all living organisms to move around and live in.

Every management plan must start with an understanding of the present condition of the entire subject forest. We propose the following steps to estimate invading broadleaf diversity in a whole forest. Here, "site" refers to the minimum scale of management, which in the Danto forest equals the sub-compartment scale.

Step 1 (GIS mapping): Specify the location of each site using a GIS. Next, add site-specific data obtained from the forest register, including the main planted species, its percentage of dominance, and forest age.

Step 2 (Categorization by growth stage): Classify all sites into three stages (determined previously): young stage (less than 25 years), middle stage (more than 26 and less than 60 years), and mature stage (more than 61 years).

Step 3 (Visual display): In the GIS, color-code each site according to its classification: young stage as green, middle stage as light green, and mature stage as dark green. That is, a darker color means richer broadleaf tree diversity.

Figure 6 shows the present condition of broadleaf tree diversity in Danto forest. Most of the area is light in color, indicating large homogeneous groups, with dark areas scattered throughout the forest. This situation is considered insufficient to maintain the diversity of invading broadleaf trees. To evaluate the potential for improving the diversity within Danto forest, we simulated changes in conditions up to 40 years in the future.

Step 4 (Rotation scenario): All sites more than 80 years old are entirely cut and then replanted with the same tree species the following year.

Step 5 (Simulation): The change in broadleaf tree diversity from the present to 40 years later is simulated and the spatio-temporal changes shown by the diversity model are discussed.

Figures 7 and **8** show the simulation results after 20 and 40 years, respectively. After 20 years, the overall size of the dark area increased and the different dark areas began to coalesce, while the light areas became more fragmented. After 40 years, the dark areas increased in size further and were observed

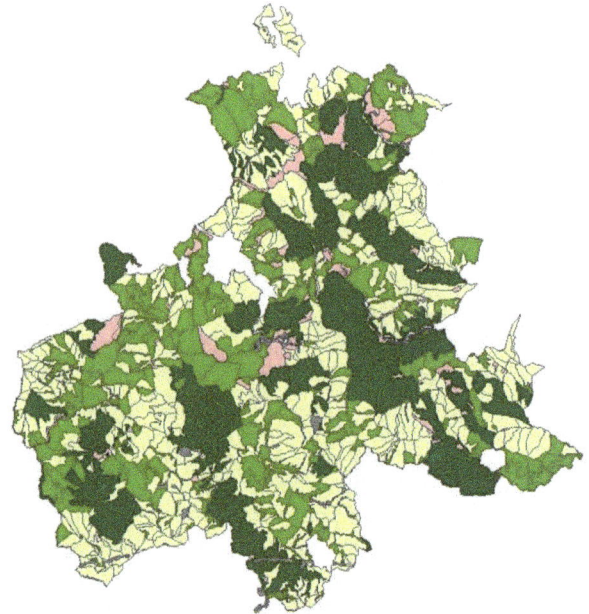

Figure 7.
Diversity of invading broadleaf trees in Danto forest (after 20 years).

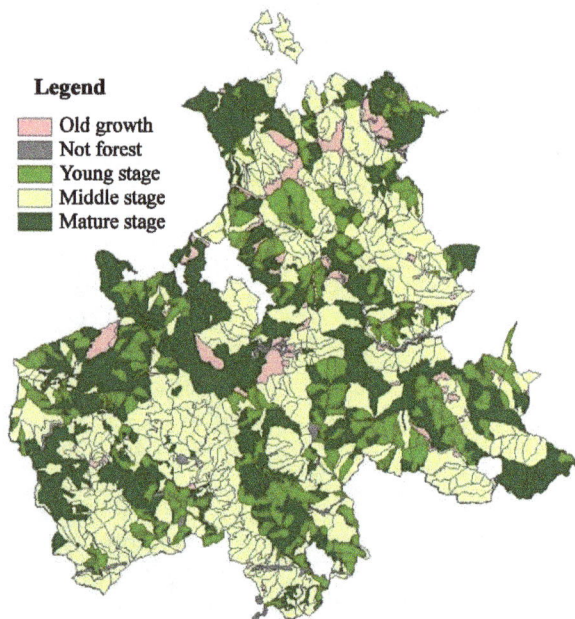

Figure 6.
Diversity of invading broadleaf trees in Danto forest (present).

Figure 8.
Diversity of invading broadleaf trees in Danto forest (after 40 years).

throughout the entire forest, while the fragmentation of the light areas continued. The network of dark areas at 40 years appeared to be strengthened and expanded compared to the present and after 20 years.

Figure 9 shows the change in proportion of the sub-compartments by stage. The proportion of mature stage sub-compartments decreased to 22% after 10 years under the 80-year rotation scenario and then recovered gradually to 42% after 40 years. Meanwhile, the proportion of young stage sub-compartments increased gradually to 35% after 30 years and then decreased. Middle stage sub-compartments occupied 44% of the area at present, increased to 48% after 10 years, and thereafter showed a downward trend. The overall percentages of high-diversity stages (i.e., the sum of the young and mature stages) were 56% at present, 51% after 10 years, 57% after 20 years, 72% after 30 years, and 65% after 40 years.

In this study, we created a simple scenario in which all sites more than 80 years old were cut completely; thus, the first decade was on the overcutting side, and, after 80 years, the sites returned to the starting point. Therefore, this scenario is strongly affected by the present site age and location, and we cannot change the arrangement of sites to improve the spatiotemporal diversity of invading broadleaf trees. Clearly, if we set a different rotation scenario in Step 4, the results may change.

Conclusion

Our sub-compartment-scale analysis suggested that we can classify each site into one of three stages according to the richness of diversity of the invading broadleaf trees: young stage, less than 25 years with rich diversity but small height; middle age, more than 26 years and less than 60 years with poor diversity; and mature stage, more than 61 years with rich diversity and a developed layer. For a Scots pine (Pinussylvestris) stand in northern Scotland, Mason (2004) recognized four phases of stand development: 1) stand initiation, from 0 to 20 years; 2) stem exclusion, from 20 to 80 years; 3) understory re-initiation, from 80 to 150 years; and (4) old growth, from 150 to 350 years. Busing and Garman (2002) concluded that most wood quantity, wood quality, and ecological objectives can be met with long rotations of approximately 260 years. The ranges of our proposed classification are shorter than those in previous studies. Invading broadleaf trees may still be in the process of growing even in our mature stage. In Japan, there are artificial forests

older than 200 years, but the conventional long-rotation system in Japan is more than 70 and less than 100 years, considering economic efficiency and management conditions. Busing and Garman (2002) noted that certain objectives could be met with shorter rotations of 80 - 150 years, when treatments of thinning and canopy tree retention were applied. Hartley (2002) argued that during establishment, forest managers should consider innovations in snag and reserve tree management, in which mature native trees and/or understory vegetation were left unharvested or allowed to regenerate. The same idea applies to our study area: invading broadleaf trees should be retained as long as they do not impede thinning or risk safety.

In this study, we examined five diversity indices for broadleaf trees and studied the features of each index. We then examined combinations of the diversity indices. The proportion of basal area and inverse Simpson index were identified as the index pair for evaluating the diversity of invading broadleaf trees at a single site. In future, we plan to test additional diversity indices to identify and confirm the most suitable indices for evaluating broadleaf tree diversity in artificial forests.

From the results of our basin-scale analysis, we proposed a model for evaluating the diversity richness of invading broadleaf trees in whole forests. The simulation showed a spatiotemporal change in richness. Regarding the importance of a basin-scale perspective, Kupfer et al. (2006) noted that the study of forest fragmentation effects was shifting away from a patch-based perspective focused on factors to a landscape-mosaic perspective that recognizes the importance of gradients in habitat conditions. Fischer et al. (2006) suggested that a landscape should include structurally characteristic patches of native vegetation, corridors, and stepping stones between them, a structurally complex matrix, and buffers around sensitive areas. Although our GIS-based model can show a basin-scale perspective, we could not quantitatively evaluate the fragmentation of sites as a structurally complex matrix and network of rich diversity sites as corridors in this study. In future, we plan to improve the rotation scenario of our proposed model by incorporating a road-construction plan and felling plan that specifies when and which sites to fell and plant.

REFERENCES

Brockerhoff, E., Jactel, H., Parrotta, J., Quine, C., & Sayer J. (2008). Plantation forests and biodiversity: Oxymoron or opportunity? *Biodiversity Conservation, 17,* 925-951.

Burslem, D. F. R. P. (2004). Plant diversity in forests. *Encyclopedia of Forest Sciences, 1,* 40-44.

Busing, R. T., & Garman, S. L. (2002). Promoting old-growth characteristics and long-term wood production in Douglas-fir forests. *Forest Ecology and Management, 160,* 161-175.

Chubu Regional Forest Office (2011). Forest for people; National Forest.
http://www.rinya.maff.go.jp/chubu/introduction/gaiyou_kyoku/pamphlet/pdf/20110223kyoku-panhu.pdf

Ehrlich, P. R. (1996). Conservation in temperate forests: What do we need to know and do? *Forest Ecology and Management, 85,* 9-19.

El-Keblawy, A. (2005). Artificial forests as conservation sites for the native flora of the UAE. *Forest Ecology and Management, 213,* 288-296.

FAO (2012). Global forest resources assessment 2010.
http://www.fao.org/forestry/fra/fra2010/en/

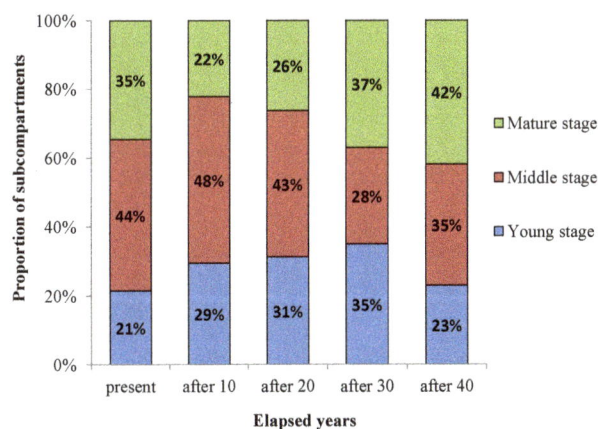

Figure 9.
Change in the proportion of sub-compartments by stage.

Fischer, J., Lindenmayer, D. B., & Manning, A. D. (2006). Biodiversity, ecosystem function, and resilience: Ten guiding principles for commodity production landscape. *Frontiers in Ecology and the Environment, 4,* 80-86.

Forestry Agency (2012). Annual report on forest and forestry in Japan, fiscal year 2011.

Hartley, M. J. (2002). Rationale and methods for conserving biodiversity in plantation forests. *Forest Ecology and Management, 155,* 81-95.

Kondoh, M. (2002). What determines the biodiversity pattern? Unifying "intermediate disturbance hypothesis" and "procuctively hypothesis". *Biological Science, 53,* 195-203.

Kosaka, S., & Yamada, Y. (2013). Evaluation of broadleaf tree diversity at the basin scale: In case of artificial Chamaecyparisobtusa forests. *Open Journal of Forestry, 3,* 62-65.

Kupfer, J. A., Malanson, G. P., & Franklin, S. B. (2006). Not seeing the ocean for the islands: The mediating influence of matrix-based processes on forest fragmentation effects. *Global Ecology and Biogeography, 15,* 8-20.

Lund, H., Dallmeier, F., & Alonso, A. (2004). Biodiversity in forests. *Encyclopedia of Forest Sciences, 1,* 33-40.

Mason, W. (2004). Multiple-use silviculture in temperate plantation forestry. *Encyclopedia of Forest Sciences, 2,* 859-865.

Montreal Process (2009). Criteria and indicators for the conservation and sustainable management of temperate and boreal forests. 4th Edition, 29 p.

Noss, R. F. (1990). Indicators for monitoring biodiversity: A hierarchical approach. *Conservation Biology, 4,* 355-364.

Yamaura, Y. (2007). Mitigating effects of broadleaved forest fragmentation on birds: Proposal of plantation matrix management. *Journal of Japanese Forest Society, 89,* 416-430.

25

Recent Trends of Fire Occurrence in Sumatra (Analysis Using MODIS Hotspot Data): A Comparison with Fire Occurrence in Kalimantan

Nina Yulianti[1], Hiroshi Hayasaka[1], Alpon Sepriando[2]
[1]Graduate School of Engineering, Hokkaido University, Sapporo, Japan
[2]Meteorological, Climatology and Geophysical Agency of Central Kalimantan, Palangkaraya, Indonesia

MODIS (Moderate Resolution Imaging Spectroradiometer) hotspot and precipitation data for the most recent 11-year period (2002 to 2012) were analyzed to elucidate recent trends in the seasonal and spatial fire occurrence in Sumatra and the relationship with precipitation. Using a latitude line of S 0.5°, Sumatra was divided into two regions, N. (north) and S. (south) Sumatra. Different trends in seasonal fire occurrence were discussed and further defined by considering two different precipitation patterns. Analysis of hotspot (fire) data was carried out using $0.5° \times 0.5°$ grid cells to evaluate recent trends of spatial fire occurrence. Analysis results of hotspot and precipitation data were also tallied every 10-day to find the relationship between seasonal fire occurrence and the dry season. Standard deviation (SD) and variance (V) were then used to evaluate fire occurrences in Sumatra and Kalimantan objectively. The relatively mild fire occurrence tendency in Sumatra compared to Kalimantan could be the result of different stages of forest development or the high deforestation rate in Sumatra compared with Kalimantan. This paper also shows that the two different seasonal fire activities in N. and S. Sumatra were closely related to the two different dry season types: a winter and summer dry season type (W_D & S_D) in N. Sumatra, and a summer dry season type (S_D) in S. Sumatra. Extreme fire occurrences in the Dumai region in 2005 and Palembang region in 2006 could be partially explained by a severe drought occurrence enhanced by two different kinds of El Niño events.

Keywords: Drought; Dry Season; El Niño; MODIS Hotspot; Peatland Fire

Introduction

During 1997 and 1998, areas of forest measured approximately 45,000 km^2 were destroyed by extreme fires in Kalimantan and Sumatra, releasing more than 2.6 gigatonnes of CO_2 to the atmosphere (Tacconi, 2003), and causing a transboundary haze event in South East Asia. An estimate of the size of the area burned in East Kalimantan alone, between January and April 1998, was 52,000 km^2 (Siegert et al. 2001). In addition to this vast amount of damage, the fires caused serious health and economic problems within the region (Dennis, 1999). Since then, Sumatra and Kalimantan have experienced frequent fire occurrences, but as discussed in our preliminary paper, (Yulianti et al., 2012), the number has diminished. It is notable that the frequent fire occurrence in Sumatra coincides with the rapid rate of deforestation occurring over the past 10-years (2000-2012), which has been at a rate of about 23.7%/(10-yr) for all forests and 5.2%/(yr) for peat swamp forest (Miettinen et al., 2011a).

Since 1990, deforestation in Sumatra has contributed to a loss of 7.45 million hectares in an area of primary forest measured 75,000 km^2 (Margono et al., 2012); a faster rate than in Kalimantan. In other words, the present area of forest in Sumatra (120,000 km^2) is smaller than that of Kalimantan (250,000 km^2), as officially reported by the Ministry of Forestry of Indonesia (MoF, 2012). It is also evident that most fires are related to agriculture activity, and this is consistent with Sumatra being home to the largest industrial plantations in Indonesia (15,280 km^2 Joosten et al., 2012), as well as the most prolific cultivation of palm oil (10,467 km^2, Miettinen et al., 2012).

During development of the Tidal-coast Development Project, which began in the late 1960s in Sumatra (Furukawa, 2004), for instance, the swamp drainage channels caused a noticeable reduction in the dry season water table (Fearnside, 1997) and provided a low ground water level environment, which is very suitable for the occurrence of peat fires. This also occurred in the MRP area in Kalimantan (Putra & Hayasaka, 2011). Slashing and burning activities associated with the rapidly increasing number of plantation crops and industrial forests, have destroyed the peat swamp forests in both Kalimantan and Sumatra. This has caused serious environmental problems, and long-burning fires occur from January to October every year (Yulianti et al., 2012). For instance, vegetation fires in the southern peatland of Sumatra, often occur in relation to burning activities within the plantations during the dry season (Stolle et al., 2003).

Once a fire begins, its behavior is highly dependent on the prevailing environmental factors, such as the weather conditions and the desiccated state of the material to be burned. Rainfall conditions are an extremely important factor in the tropical zone, and rainfall patterns can be affected by active fire, particularly on peatland areas (Putra & Hayasaka, 2011). The characteristics of fires in the Sumatra peatland areas are also influenced by the rate of decomposition of the peat, as described in Saharjo (2005). Under dry conditions, peat that is more highly decomposed (mature) is very flammable and results in a longer period of smoldering fire, as discussed by Usup et al. (2004). Thus, the emissions of carbon solely from peat fires in Southeast Asia, (mostly from the Sumatran and Kalimantan peatlands) are very similar to emissions from the whole of Sub-Saharan Africa (Harris et al., 2012). However, vegetation fires related to forest clearing activities tend to be more widespread in Sumatra, even outside peatland or undeveloped areas which are far from sea coast. Recent frequent fires related to deforestation in the Riau areas of North Sumatra Province have also become a global environmental issue, due to the amount of CO_2 emitted (3.6 Giga tons CO_2; Uryu et al., 2008).

The use of satellite-based monitoring enables a more comprehensive understanding and evaluation of fire activities inside forest areas, and also in large peatland areas. Since 2002, the MODIS sensor on board the Terra and Aqua satellites has covered the whole of Indonesia. Several earlier studies using MODIS hotspot data have focused on Asia (Vadrevu & Justice, 2011), Southeast Asia (Miettinen et al., 2011b), and Borneo (Langner & Siegert, 2009). Our previous research analyzed the recent fire trends during the MODIS era for Indonesia (Yulianti et al., 2012), and Kalimantan (Borneo part of Indonesia, Yulianti & Hayasaka, 2013). Peat fire activity in the MRP area was discussed in Putra & Hayasaka (2011). However, there have only been a few reports related to fires in Sumatra.

In this study, we carefully analyzed the seasonal and spatial occurrence of fire using the last 11-year (2002-2012) of MODIS hotspots (fire) and the last 12-year (2001-2012) of precipitation data. We attempted to determine trends in the seasonal and spatial occurrence of fire under severe drought conditions in north and south Sumatra separately. Analysis of seasonal fire occurrence was carried out using hotspot and precipitation data at 10-day intervals from January to December. Fire (hotspot) distribution maps were drawn to identify active fire areas in Sumatra for the analysis of spatial fire occurrence. The results of similar research that was conducted on Kalimantan fires by Yulianti and Hayasaka (2013), were compared to the above-mentioned analyses, to observe typical trends in the occurrence of peatland fires in Sumatra and to enable an effective, future peatland fire forecast.

Methodology

Study Area and Peatland

The study area covers Sumatra Island from N 6° to S 6° and from E 95° to 108° (**Figure 1**). Sumatra covers a total land area of about 474,000 km² (approximately 89.1% of the size of Kalimantan which covers 532,000 km²). The total peatland area of Sumatra is about 72,000 km², which is approximately 15% of the land area of Sumatra; (Wetlands, 2003), (120% larger than that of Kalimantan at 58,000 km²). (The distribution of peatland is shown in a dark color (brown) in **Figure 1**). The main area of peatland in Sumatra has developed in a large area of lowland on

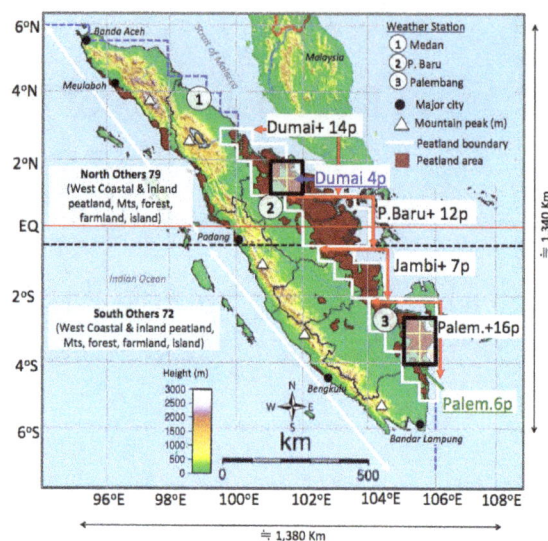

Figure 1.
Map of peatland, weather stations, and the eight regions used for analysis in Sumatra.

the east coast, covering a distance of roughly 1200 km from the north of Dumai in Riau (~N 2.6°) to the north of Lampung (~S 5°), and covering a distance of nearly 200 km (maximum distance) inland from the coast. Other several small areas of peatland are independently located on the west coast and inland, (west of Riau, Jambi and Palembang). North Sumatra has the deepest area of peatland, at a depth of around 9 m (Jauhiainen et al., 2011); in S. Sumatra the depth is shallower, at about 4 m (Wetlands, 2003). Nowadays, major areas of peatland are maintained by large plantation companies by the provision of drainage systems. However, many of these use unsuitable systems for the preservation of peatlands, and cause a very low ground water level).

In this paper, in order to clarify spatial and seasonal fire occurrences in Sumatra, the authors have given suitable names to regions. The borders of these regions are different from those used in conventional political and geographical maps. We achieved this by dividing Sumatra into two regions, N. (north) and S. (south) Sumatra, by using a latitude line of S 0.5° and with consideration of the following: two different precipitation patterns (Aldrian & Susanto, 2003), an historical climatic map showing precipitation patterns in Sumatra (Oldeman et al., 1979), and a previous study by Chang et al. (2005), (as shown in **Figure 1**).

Analysis using a 0.5° grid cell provided approximately 105 and 95 cells for N. and S. Sumatra respectively. These cells in N. and S. Sumatra were then grouped into regions, to clarify the spatial and seasonal fire occurrence, particularly for areas of peatland (shown in **Figure 1**). The borders of these regions were constructed on the connecting lines from the rectangular analysis cells (0.5° grid cell). We then allocated suitable regional names, such as "Dumai + 14" (14 cells, a subset of North Sumatra and Riau Province), "Pekan Baru + 12" (12 cells for a subset of Riau Province) in N. Sumatra, "Jambi + 7" (7 cells, a subset of south Riau and Jambi Province) and "Palembang + 16" (16 cells, a subset of South Sumatra and north Lampung Province) in S. Sumatra. Borders of these four regions are shown by the lines in **Figure 1**. We put "+" after each region to show the expanded region of each local area name (such as "Dumai+"), so as to avoid confusion with the conventional administrative name. The number provided at the end of the names is used to describe the number of cells that the region occupies. The other two regions, and the remaining cells from the above four regions, were simply named N. Others (79 cells), and S. Others (72 cells). Finally, two special areas, Dumai 4 in N. Sumatra and Palembang 6 in S. Sumatra, were introduced to highlight the extreme fire activities in these areas in 2005 and 2006.

MODIS Hotspot Data and Grid Analysis

Daily MODIS hotspot data (Collection 5.1 active fire product) from the most recent 11-year period (2002 to 2012) were automatically extracted from the FIRMS website (Fire Information for Resources Management System, http://earthdata.nasa.gov/ data/near-real-time-data/data/firms) and were then analyzed in this paper. Approximately 250,000 hotspots were recovered from the 11-year data for Sumatra, (which includes Insular Malaysia).

In this paper, 0.5° grids were used to clarify the spatial trend of fire occurrence. Hotspots were tallied depending on their longitude and latitude. For a hotspot density, units of "hotspots/cell," and "hotspots/km²," were used. The approximate area of these grid cells at the equator is 3070 km², and this was used for the conversion to hotspot density, "hotspots/km²".

Weather and Precipitation Data

To determine the relationship between fire activity on peatland and weather patterns, we attempted to obtain weather data from many places in Sumatra, but were only able to purchase daily precipitation data from six stations. In this paper, we analyzed precipitation data for the most recent 12-year period (2001 to 2012) measured at three airports (see **Figure 1**) in Medan (Polania, N 3.66°, E 98.60°, North Sumatra), Pekan Baru (Sultan Syarif Kasim II, N 0.47°, E 101.45°, Riau), and Palembang (Sultan Mahmud Badaruddin II, S 2.90°, E 104.70° South Sumatra). All data belongs to the Indonesia Meteorological, Climatology and Geophysical Agency (BMKG).

Precipitation data measured at three weather stations in Sumatra were tallied every 10-day to define the dry season period and to explain the different trends in fire occurrence in N. and S. Sumatra. Data processing from every 10-days was applied instead of using other data smoothing periods of a 10-day (week) and 30-day (month), because the smoothing of data over a period of 10-day is less difficult than over other time periods. In addition, the 10-day period will enable us to label seasons by using expressions such as, "early June," instead of "DN = 152 to 161".

Results and Discussions

Dry Season Type, Period, and Lowest Daily Mean Precipitation

Precipitation patterns for three weather stations (Medan, Pekan Baru (known as P. Baru), and Palembang) were calculated using the recent 12-year interval from 2001 to 2012 (see **Figures 2(a)-(c)**). In total, 35 mean daily precipitation values for every 10-day interval, were calculated for each station, and are shown by the solid lines in the graphs in **Figure 2**. Each daily mean precipitation value was placed on its representative

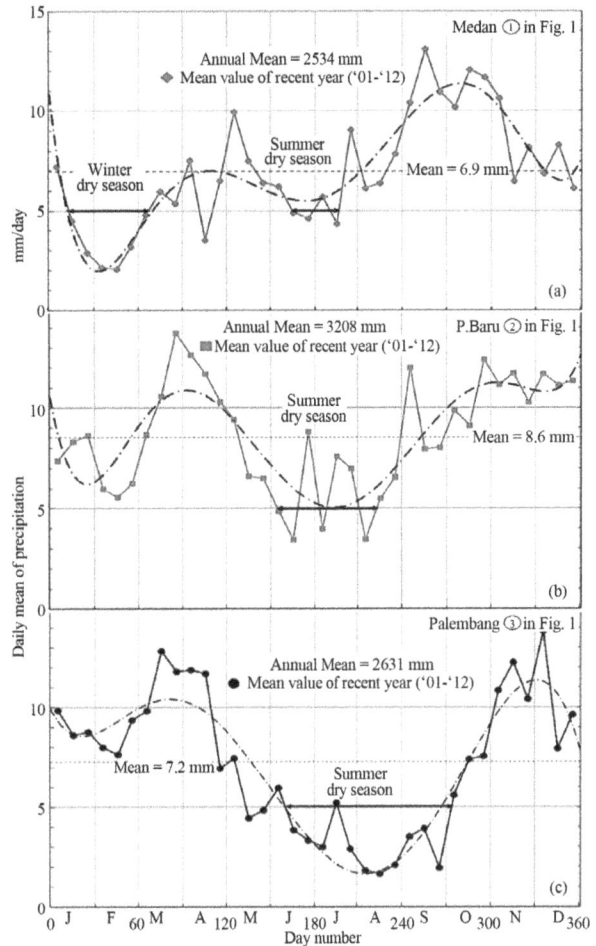

Figure 2.
Dry season period in: (a) Medan; (b) Pekan Baru; and (c) Palembang derived from 12-year (2001-2012) data.

point from DN = 5 to DN = 355 within the 10-day interval. In addition, simple smoothed curves (dotted lines) have been added to the graph, representing the daily mean precipitation. With the help of these two lines, we were able to ascertain the dry season period and the low precipitation values during each dry season, for each station.

The definition of a dry season period for each area in this paper, follows previous analysis of Kalimantan by Yulianti & Hayasaka, (2013), and the daily mean precipitation of 5 mm/day was used as a threshold value for the dry season. By using the line for daily mean precipitation in our graph in **Figure 2**, we can see that threshold value is apparent in Medan (N 3.66°) in the Northern Hemisphere, revealing an approximately two month winter dry season in January and February ($W_{D(2)}$) and a summer dry season ($S_{D(2)}$) in June and July. In Palembang (S 2.90°) in the Southern Hemisphere, there was a relatively long dry season, about 4-month $S_{D(4)}$ from June to September, and in P. Baru (N 0.47°) near the Equator, (located between Medan and Palembang, see **Figure 1**), there was an intermediate dry season, or a one-month quasi-winter dry season with rain (daily mean precipitation in February higher than 5.5 mm/day), and a dry season of about 2-months $S_{D(2)}$ in June and August.

The lowest values of daily mean precipitation during the

above-mentioned dry seasons were an indication of fire activities near each weather station. In Medan, the lowest values of daily mean precipitation in $W_{D(2)}$ and $S_{D(2)}$ found in **Figure 2(a)** were about 2.1 mm/day in February and 4.3 mm/day in July respectively. In Palembang, the lowest value in **Figure 2(c)** was 1.7 mm/day in $S_{D(4)}$ in August. In P. Baru, the lowest value was 3.4 mm/day, from only one period in each of June and August, as shown in **Figure 2(b)**. Moreover, the annual mean precipitation of P. Baru (about 3200 mm) was higher than the amounts of about 2500 and 2600 mm from the other two weather stations. Reasons for this were that we could estimate that the heavy precipitation in P. Baru would mainly be due to the effect of the mountain behind the area (see **Figure 1**).

Fire Prone Areas and Peatland

Fire prone areas (>100 hotspots/(yr. cell) in the most recent 11-year period (2002 to 2012) are highlighted using different colors in **Figure 3**. Most of these areas are located on the eastern side of Sumatra, but some are in the northern part. There were 12 cells showing a very high hotspot density (>400 hotspots/(yr. cell) = 0.129 hotspots/(yr. km^2)). These cells were named H1, H2, …, H12, in descending order of hotspot density. The 7 highest hotspot density cells (H1~4, H8, H11 and H12), located in the Dumai region or Dumai + 14p (Dumai4p: H1, H3, H4, and H12), cover most of the northeast coast peatland in N. Sumatra (Riau Province). According to the peatland map (Wetlands, 2003), the Dumai region contains a peat layer that is relatively deeper (~8 mm) than other places in S. Sumatra and Kalimantan (except the MRP area). It is of note that Dumai + 14p belongs to a different climate zone than most of S. Sumatra and Kalimantan (see **Figures 2(a)** and **3**).

The seventh highest cell, H7, is located on the eastern side in Pekan Baru (capital of Riau Province) near the Equator and on the coast. The other highest cells, H5, H6, H9, and H10, are located in the region of S. Sumatra. H5 and H6 are located on the eastern side at Palembang (capital of the South Sumatran Province), and two cells in Palembang 6p. H9 in the south in Pekan Baru, and H10 in southeast Jambi are two cells included in Jambi + 7p (see **Figure 3**).

From the distribution of these highest hotspot cells, it is evident that the most recent fires in Sumatra have occurred mainly on the coastal peatland. Many of the fires on peatland can be explained by the history of development as in the MRP area of Kalimantan (Furukawa, 2004). Areas in the MRP with dense hotspots were related to high human activity such as deforestation, slash and burn clearing, and the maintenance of plantations (Yulianti et al., 2012).

Recent Fire Occurrence Trends in Sumatra

Annual and Average Seasonal Fire Occurrence in N. Sumatra

The stacked bar graph in **Figure 4** shows the number of fires that have occurred in three regions in N. Sumatra (from bottom to top: Dumai + 14, P. Baru + 12p, and North Others 79). The annual mean numbers of fires in these three regions and those in S. Sumatra, (and the annual fire in Kalimantan), are shown in the two bars on the far right in **Figure 4**. The unit of the Y-axis is the number of hotspots in each region.

The annual mean number of hotspots in N. Sumatra over the past 10-years was about 8600, and 60% of these fires occurred in peatland areas (Dumai + 14p and P. Baru + 12p). A comparison of the two bars in **Figure 4**, shows that the number of fires in 2005, (about 27,000 hotspots/yr.), is larger than the average number in Kalimantan (23,000 hotspots/yr.). About 80% of the extreme fires in 2005 occurred mainly in peatland in the Dumai and P. Baru regions; areas responsible for approximately 16,000 and 5400 fires (60.3% and 20.4%) respectively. Fire occurrences in both peatland areas were 4.2 times larger than in the North Others 79, which is predominantly non-peatland.

The occurrence of fires in most years (6 out of the 10 years) in N. Sumatra was in the range of +1σ to −1σ. The variance of fire occurrence (V) of about (3900)^2 was smaller than that of about (16,000)^2 in Kalimantan. This relatively stable fire occurrence in N. Sumatra may suggest that the occurrence of so-called routine fires on plantations and developed land are not so strongly related to the weather conditions, and the yearly fire

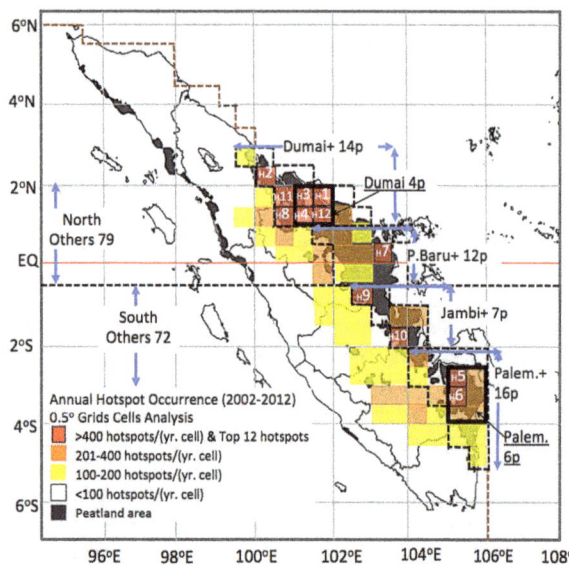

Figure 3.
Map of the 12-highest hotspot cells, fire prone cells, and peatland in Sumatra.

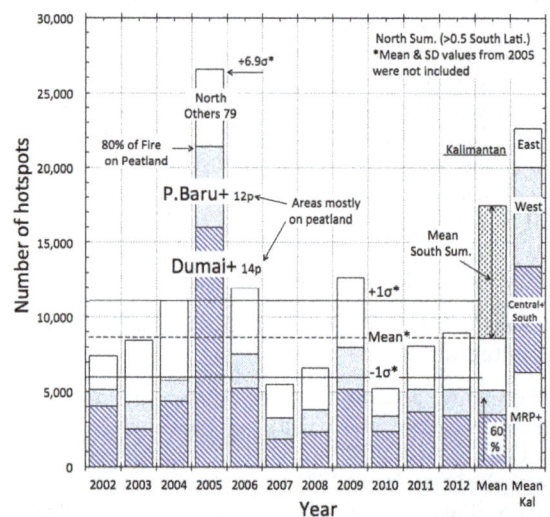

Figure 4.
Recent trends in annual fire occurrence in N. Sumatra.

Recent Trends of Fire Occurrence in Sumatra (Analysis Using MODIS Hotspot Data): A Comparison with Fire Occurrence in Kalimantan

159

occurrences could be related to agriculture activities such as agricultural residue burning. Small-sized fire occurrence could, therefore, imply that large land clearing or deforestation activities are no longer taking place in N. Sumatra.

The extreme fire occurrences in Dumai + 14p in 2005 was an exceptional case, and the main cause of these severe fires was due to the expansion of fires lit for land clearing and agricultural activities. On inspection of weather data measured at Medan, (located about 400 km northeast of Dumai), these fires may also have occurred during severe drought conditions. In Medan, the drought lasted for about two months, from mid-January to mid-March in 2005, and overlapped with the dry season in N. Sumatra.

In this paper, to clarify the average seasonal fire occurrence tendency in N. Sumatra, the average number of hotspots every 10-days (1/3-month) were calculated. In doing so, one active fire area named "Dumai 4p" was extracted from a part of Dumai + 14p, to illustrate the seasonal fire occurrence clearly.

From **Figure 5**, it can be seen that the seasonal fire occurrence trend in N. Sumatra displays two major fire seasons: one fire peak in February and one in March, and another from May to August. Such seasonal fire peaks can be partially explained by the two precipitation patterns and the dry season periods for Medan (N 3.5°) and P. Baru (N 0.5°) in **Figure 2**. We have observed that, firstly, the winter fire peak in February (in **Figure 5**) is mainly due to fires occurring in Dumai 4p (N 1.8°) and P. Baru + 12p. These fires correspond to the winter dry seasons WD found in **Figures 2(a)** and **(b)**. The highest fire peak for Dumai 4p (mid-February) corresponds to the lowest daily mean precipitation (about 2 mm/day) in **Figure 2(a)**. Secondly, the early summer peak in June in **Figure 5** is mainly due to fires occurring in North Others 79 and Dumai + 14p. These fires can be mainly explained by the summer dry season in P. Baru in **Figure 2(b)**, because the low latitude 32 cells (<N 2.5) in North Others 79 are responsible for about 80% of the fires, and these cells have a similar rainfall pattern to P. Baru. Thus, most fires in N. Sumatra tend to occur under a rainfall condition of about 3.5 mm/day in mid-June, as seen in **Figure 2(b)**. Thirdly, the highest fire peak in the whole of N. Sumatra occurs in early August. This peak in August is mainly due to the increased number of fires in P. Baru under the lowest daily rainfall of about 3.5 mm/day. Such a tendency may, therefore, imply the occurrence of peat fires in P. Baru+.

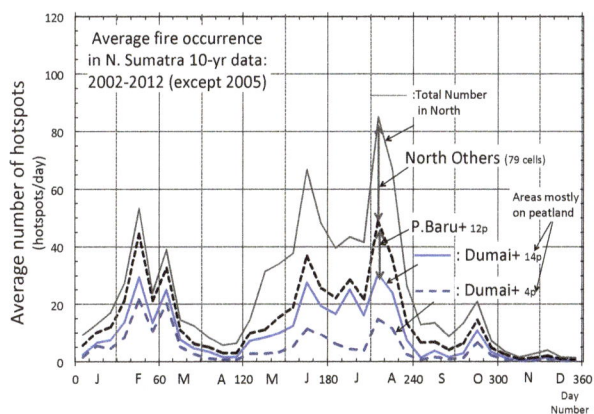

Annual and Average Seasonal Fire Occurrence in S. Sumatra

The annual mean number of fires in S. Sumatra is shown in **Figure 6** (using a similar graph format to **Figure 4**). The stacked bar graph in **Figure 6** shows the number of fires in three regions in S. Sumatra, from bottom to top: Palembang + (16p), Jambi + (7p), and South Others 72. The annual mean number of fires in the three regions (excluding 2006), and the annual number of fires in Kalimantan, are shown by the two bars on the far right in **Figure 6**.

The annual mean number of hotspots in S. Sumatra during the last 10-years was about 8900, and, approximately 47% of the fires occurred in peatland areas (Palembang+ and Jambi+). Fire data from 2006 were excluded from the statistics calculation, due to severe fire occurrence (+5.3σ, +2.6σ if the data from 2006 are included) under the drought enhanced by El Niño in 2006. A comparison of the two bars in **Figure 6**, shows that the number of fires in 2006, (about 29,000 hotspots/yr.), was larger than that of the average number in Kalimantan (23,000 hotspots/yr.). About 57% of these extreme fires in 2006 occurred mainly in peatland areas in the Palembang+ and Jambi+ regions, which are responsible for around 13,300 and 3300 fires (45.5% and 11.4%), respectively. The occurrence of fires in both these peatland areas was 1.3 times larger than that in South Others (mostly non-peatland).

The fire occurrence in the majority of years in S. Sumatra, (6 out of the 10 years), was within the range of +1σ to −1σ and the variance of fire occurrence (V) of an area of around $(7200)^2$ was smaller than that of an area of about $(16,000)^2$ in Kalimantan, but larger than that of an area of about $(5900)^2$ in N. Sumatra. The relatively stable fire occurrence in S. Sumatra may also suggest that continual and small-size fire occurrences are not strongly related to weather conditions, as in N. Sumatra. It is evident that the occurrence of small fires in heavy rain conditions during 2010 could have made the variance of fire occurrence in S. Sumatra larger than that in N. Sumatra.

The extreme fire occurrence in Palembang+ in 2006 was an exceptional case and the main cause of these severe fires was due to an expansion of fires for land clearing and agricultural

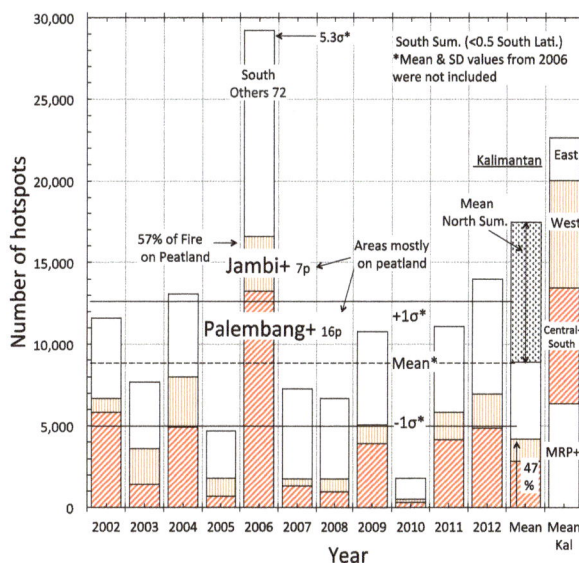

Figure 5.
Recent trends of average seasonal fire occurrence in N. Sumatra.

Figure 6.
Recent trends in annual fire occurrence in S. Sumatra.

activities, and which occurred under severe drought conditions lasting three months from early August to late October in 2006, and which overlapped with the dry season in S. Sumatra.

Figure 7 is similar to that of **Figure 5** for N. Sumatra, and was used to clarify the average seasonal fire occurrence tendency in S. Sumatra. By using this graph, one fire active area named "Palembang 6p," was defined and extracted from one part of Palembang + 16p, to clearly illustrate the seasonal fire occurrence.

From **Figure 7**, the trends of seasonal fire occurrence in S. Sumatra show one fire season from May to late October, with a few peaks in September. This one-fire-season pattern in S. Sumatra almost corresponds to the one-summer-dry-season pattern in **Figure 2(c)**. The difference in the fire peak in September of about one month in **Figure 7** and the low of precipitation in August in **Figure 2(c)** can be estimated using the author's previous research related to Kalimantan, where we found there was a time lag of approximately one month between precipitation and a rising ground water level (Putra & Hayasaka, 2011). This result was obtained from measurement results in peatland in Block C, the MRP, in Central Kalimantan (Takahashi et al., 2007).

From **Figure 7** the highest peak can be seen in late September for all three regions in S. Sumatra. Palembang 6p had the same values as the MRP area in Central Kalimantan, (Yulianti & Hayasaka, 2013). This could suggest that peat fires in S. Sumatra and south Kalimantan occur under a similar weather condition created by the southern monsoon during the summer months (between May and October).

Recent Extreme Fire Occurrences in Sumatra

Highest Hotspot Cell in Sumatra

Extreme fire occurrences were observed in both N. and S. Sumatra in 2005 and 2006 respectively. The cells with the top 12 highest hotspot densities (>400 hotspots/yr.) in **Figure 3** were mainly due to the occurrence of these extreme fires. The top four highest hotspot cells from H1 to H4 are located near

Dumai, and two cells of H5 and H6 are near Palembang. A comparison of the seven highest cells from the top 12 highest hotspot density cells in Kalimantan (Yulianti & Hayasaka, 2013) are listed in **Table 1**.

From **Table 1**, the fires in H1 near Dumai could be referred to as "extreme" fires, because the average and the maximum number of hotspots in H1 were 941 and 4494, respectively. The average number in H1 was almost the same as the value of 971 in the H1k cell in Kalimantan. However, the greatest value in H1 (4494) was the highest overall. SD values of 2.77 - 2.94 for the four highest cells in Sumatra were higher than those values (1.45 - 2.4) in Kalimantan (seen in **Table 1**). These differences could imply that the extreme fires in Sumatra with higher SDs were incidental fires. In contrast, the extreme fires in Kalimantan with lower SDs were routine fires lit for the purpose of development. The seasonal occurrence of these extreme fires in Sumatra is discussed in the sections below.

The Extreme Seasonal Fire Occurrence and Drought in N. Sumatra (2005)

In **Figure 8**, seasonal fire occurrence and drought in N. Su-

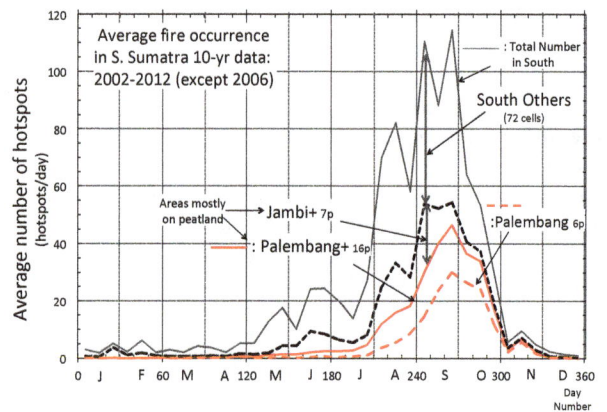

Figure 7.
Recent trends in average seasonal fire occurrence in S. Sumatra.

Table 1.
Comparison of the highest hotspot cells in Sumatra and Kalimantan.

	Rank	Ave.[*]	Max.	SD (σ) for Max.	Year for Max.	Region
Sumatra	H1	941	4494	2.94	2005	Dumai 4p (east)
	H2	686	2385	2.77	2005	Dumai (northwest)
	H3	624	2231	2.85	2005	Dumai4p
	H4	483	772	1.35	2006	Dumai4p (south)
	H5	477	2966	2.95	2006	Palembang (east)
Kalimantan	H1k	971	2417	1.76	2006	MRP (C south)
	H2k	804	2289	1.80	2009	MRP (C north)
	H3k	678	1846	1.65	2002	MRP (C middle)
	H4k	629	629	1.45	2006	MRP (B&D)
	H5k	561	1443	1.51	2002	MRP (A)
	H7k	506	2298	2.40	2006	Sampit

[*]Ave.: 11-year (2002-2012) for Sumatra, 10-year (2002-2011) for Kalimantan.

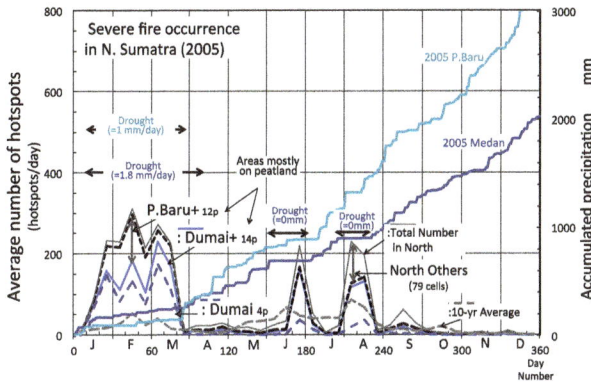

Figure 8.
Fire occurrence and accumulated precipitation in 2005.

matra are found in the various curves of hotspots and precipitation. In the figure, N. Sumatra, fire occurrence in three regions, Dumai + 14p, P. Baru + 12p, and north others (79), are arranged from bottom to top using a variety of lines. The number of hotspots in Dumai 4p is shown independently by a dotted line. The difference between Dumai + 14p and Dumai 4p shows only the fire occurrence in the 10 grid cells in Dumai + 14p, and not that in Dumai 4p or "*Dumai + 10p*". Average seasonal fire occurrence over 10-years is also shown with a dotted line to highlight the severe fire occurrence in 2005. Two accumulated curves for precipitation using two solid lines of different colors, and using daily data for P. Baru and Medan, are also included so as to evaluate the severity of the drought period.

From **Figure 8**, it is evident that the extreme fires of 2005 in N. Sumatra occurred in different regions and in different seasons. The most distinctive feature is that the 2005 extreme fires occurred during the dry seasons. The first extreme fires in Dumai 4p (a part of Dumai + 14p) and P. Baru + 12p occurred during a drought in the winter dry season W_D or between mid-January and mid-March. The value of more than 100 hotspots/day was considerably higher than that of 50 hotspots/day (the 10-year average value for the whole of N. Sumatra). The following extreme fires occurred in late June in Dumai + 14p (mostly in Dumai north in **Table 1**). The third extreme fire peak was mostly due to fires in Dumai north and North Others (79), and occurred during early and mid-August.

The extreme fires could therefore be explained by the drought conditions (using the accumulated precipitation curves in **Figure 8**) or by the flat part of the lines for both Medan (~2 mm/day) and Pekan Baru (~1 mm/day). The drought conditions in 2005 are likely to have been caused by the active winter boreal monsoon occurring under El Niño Modoki or quasi-El Niño conditions (Ashok et al., 2007). This abrupt, wide area drought affected the north of Southeast Asia, and caused a rising by about +0.5 of the average EMI (El Niño Modoki Index) values between late 2004 and early 2005, as recorded by JAMSTEC (2013).

Fires in the Dumai region, (except in Dumai4), showed another two peaks in late June and in early and mid August. These fires can be explained by the short, a devastating summer drought (S_D) in Medan occurring in June and August, giving the lowest precipitation rate over the past 12-years. Correspondingly, fires in North Others (mostly areas of non-peatland) became very active in late June and early August (see **Figure 8**) under drought during S_D.

The Extreme Seasonal Fire Occurrence and Drought in S. Sumatra (2006)

In **Figure 9**, the seasonal fire occurrence and drought in S. Sumatra are found using the various curves for hotspots and precipitation. In S. Sumatra, fire occurrence in three regions, Palembang + 16p, Jambi + 7p, and south others (72), are displayed from bottom to top using a variety of lines (in **Figure 9**). The number of hotspots in Palembang 6p is independently shown by a dotted line. The difference between Palembang + 16p and Palembang 6p shows only the fire occurrence within the 10 grid cells in Palembang + 16p, and not that in Palembang 6p or "*Palembang + 10p*". In **Figure 9**, the average seasonal fire occurrence over 10-years is also shown with a dotted line to illustrate the severe fire occurrence in 2006. An accumulated curve for precipitation using the daily data for Palembang was also added to evaluate the severity of the drought period (using solid lines).

From **Figure 9**, we can see that the 2006 extreme fires in S. Sumatra occurred in different regions and in different seasons. Of particular note, however, is that the 2006 extreme fires in S. Sumatra also occurred during the dry season. The extreme fires in Palembang 6p (a part of Palembang + 16p) and Jambi + 7p occurred either during a drought in summer dry season S_D or from mid September to mid October. Their numbers (more than 200 hotspots/day) are considerably higher than that of the 100 hotspots/day of the 10-year average values for the whole of S. Sumatra. A catastrophic fire (about 700 hotspots/day), can be ascertained by one sharp peak in early October (twice as large as in Sampit). The worst fires in the Palembang and Jambi region coincided with the fire peak in the MRP area. About 60% of these fires in early October occurred in Palembang 6p (mostly in eastern Palembang in **Table 1**). Fire occurrences on non-peatland (or in South Others) were in early July and August, and we are thus able to refer to these fires as "warning fires" for the areas of peatland in S. Sumatra.

The extreme fires occurred in Palembang, under a devastating long-term drought involving a period of no rain lasting about 3-months. About 3-month prior to this drought, a pre-drought with very low rainfall (3.6 mm/day or the same rate as the annual rate in a dry season) was observed (see **Figure 9**). These conditions are likely to be related to the El Niño event with ONI (Ocean Niño index) values in NDJ of about +1, as discussed in relation to Kalimantan (Yulianti & Hayasaka, 2013).

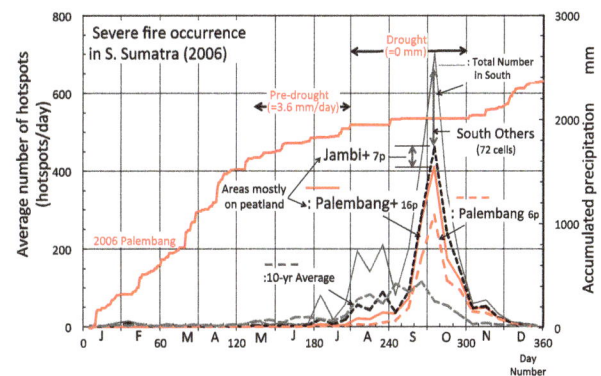

Figure 9.
Fire occurrence and accumulated precipitation in S. Sumatra (2006).

Conclusion

Analysis results using MODIS hotspot data of the most recent 11-year period, clearly show trends of seasonal and spatial fire occurrence in Sumatra. The recent trends in fire occurrence are:

1. Two extreme fires occurred both in N. and S. Sumatra in 2005 and 2006 respectively under enhanced drought or rainless conditions related to two different types of El Niño events.

2. The highest hotspot cell found near east of Dumai in N. Sumatra contained an annual maximum number of hotspots of 4494 in 2005. The 11-year average number is 941/yr., and this average number is slightly smaller than the 971 hotspots observed in one of cells in the MRP region (Block C south) in Kalimantan (where a maximum number of 2417 was observed in 2006). However, the annual maximum number of 4494 within the Dumai cell is the highest recorded.

3. The fifth highest hotspot cell found near eastern Palembang in S. Sumatra had a maximum number of 2966 hotspots in 2006, when the average number was 477/yr. This maximum number was also higher than that of 2417 in the cell in the MRP referred to above.

4. Extreme fire occurrences in N. Sumatra in 2005 and in S. Sumatra in 2006 could be partially explained by an enhanced drought occurrence due to El Niño events. But their relatively high standard deviation (σ) values of 2.7σ in N. Sumatra, and 2.6σ in S. Sumatra (higher than 1.8σ in Kalimantan in 2006 which was the worst recent fire year for Kalimantan), suggest that both extreme fire occurrences could be classified as accidental fires. The origin of extreme fires could be from intentional fires related to practices such as land clearing and plantation development.

5. The recent fire occurrence in Sumatra was not as intense as in Kalimantan (the annual mean number of fires was about 18,000 in the whole of Sumatra and smaller than that of 23,000 in Kalimantan). Nevertheless, the peatland area in Sumatra (about 70,000 km^2) is larger than that of Kalimantan (about 60,000 km^2).

6. With the exception of data from the extreme fire years, most of the annual fire occurrences in the recent 10-years lay between $+1\sigma$ and -1σ, except for 2010 in S. Sumatra.

7. The results of spatial analysis using $0.5° \times 0.5°$ cells for N. Sumatra, clearly show one extreme fire occurrence cell (centered at N 1.75° and E 101.75°) in the southeast of Dumai (one of the cells in Dumai + 4). Within this cell, the 11-year average number of fires was 944 hotspots/year. This result was much higher than the result of the second highest cell, at 686 hotspots/year (centered at N 2.25° and E 100.25°) in the northeast of Dumai (one of the cells in Dumai + 14).

8. In 2005, (an extreme fire year in N. Sumatra), a few clusters of hotspots were found in the southeast of Dumai (centered at around N 1.55° and E 101.7°, 30 km and 115 deg. from Dumai, N 1.7° and E 101.5°). From these hotspot clusters, the burnt area in the southeast of Dumai was roughly estimated. During January to March (DN = 1 - 90) in 2005, the burnt area of these few clusters was estimated to be more than 700 km^2.

9. In late June 2005 (DN = 170 - 180) and in August 2005 (DN = 210 - 230), fires in the northwest of Dumai (centered at around N 2.0° and E 100.4°, 120 km and 290 deg. from Dumai) also became active. The burnt area of these few clusters was estimated to be about 400 km^2.

10. In 2006, (an extreme fire year for S. Sumatra), a few clusters of hotspots were also found in eastern Palembang (centered at around S 2.85° and E 105.43°, 76 km and 78 deg from Palembang, S 3° and E 104.8°). The total burned area of these hotspots clusters was estimated to be about 500 km^2.

11. The above-mentioned recent fire occurrences in Sumatra were partially explained by using various precipitation patterns and dry season periods for both N. and S. Sumatra.

Acknowledgements

This research was partially supported by the JST-JICA Science and Technology Research Partnership for Sustainable Development (SATREPS) project on "Wild Fire and Carbon Management in Peat-Forests in Indonesia."

REFERENCES

Aldrian, E., & Susanto, R. D. (2003). Identification of three dominant rainfall regions within Indonesia and their relationship to sea surface temperature. International *Journal Climatology, 23,* 1435-1452.

Ashok, K., Behera, S. K., Rao, S. A., Weng, H., & Yamagata, T. (2007). El Niño Modoki and its possible teleconnection. *Journal of Geophysical Research, 112,* C11007.

Badan Pusat Statistik (2010). *Trends of the selected socio-economic indicators of Indonesia.* Jakarta: BPS-Statistic Indonesia.

Chang, P., Wang, Z., McBride, J., & Liu, C. H. (2005).Annual cycle of Southeast Asia-Maritime Continent rainfall and the asymmetric monsoon transition. *Journal Climate, 18,* 287-301.

Dennis, R. (1999). *A review of fire projects in Indonesia (1982-1998).* Jakarta, Indonesia: Center for International Forestry Research (CIFOR).

Fearside, P. M. (1997). Transmigration in Indonesia: Lessons from its environmental and social impacts. *Environmental Management, 21,* 553-570.

Furukawa, H. (2004). The ecological destruction of coastal peat wetlands in Insular Southeast Asia. In H. Furukawa, M. Nishibuchi, Y. Kono, & Kaida, Y. (Eds.), *Ecology destruction, health, and development: Advancing Asian paradigms* (pp. 31-72). Nagoya: Kyoto University Press.

Harris, N. L., Brown, S., Hagen, S. C., Saatchi, S. S., Petrova, S., Salas, W., Hansen, M. C., Potapov, P. V., & Lotsch, A. (2012). Baseline map of carbon emissions from deforestation in tropical regions. *Science, 336,* 1573-1576.

Indonesia Ministry of Forestry (2012). *Forest statistics in 2011.* Jakarta: Directorate General of Forestry Plantology.

Japan Agency for Marine-Earth Science and Technology (2013). MODOKI ENSO: A new phenomenon is found in the Tropical Pacific.

http://www.jamstec.go.jp/frcgc/research/d1/iod/modoki_home.html.en

Jauhiainen, J., Hooijer, A., & Page, S. E. (2012). Carbon dioxide emissions from an Acacia plantation on peatland in Sumatra, Indonesia. *Biogeosciences, 9,* 617-630.

Joosten, H., Tapio-Biström, M., & Tol, S. (2012). *Peatlands—Guidance for climate change mitigation through conservation, rehabilitation and sustainable use.* Rome: The Food and Agriculture Organization of the United Nations and Wetlands International.

Langner, A., & Siegert, F. (2009). Spatiotemporal fire occurrence in Borneo over a period of 10 Years. *Global Change Biology, 15,* 48-62.

Margono, B. A., Turubanova, S., Zhuravleva, I., Potapov, P., Tyukav-

Recent Trends of Fire Occurrence in Sumatra (Analysis Using MODIS Hotspot Data): A Comparison with Fire Occurrence
in Kalimantan

163

ina, A., Baccini, A., Goetz, S., & Hansen, M. C. (2012). Mapping
and monitoring deforestation and forest degradation in Sumatra (Indonesia) using Landsat time series data sets from 1990 to 2010. *Environmental Research Letters, 7*, 034010.

Miettinen, J., Hooijer, A., Tollenaar, D., Page, S., Malins, C., Vernimmen, R., Shi, C., & Liew, S. C. (2012). *Historical analysis and projection of oil palm plantation expansion on peatland in Southeast Asia*. Washington DC: International Council on Clean Transportation (ICCT).

Miettinen, J., Shi, C., & Liew, S. C. (2011a). Deforestation rates in insular Southeast Asia between 2000 and 2010. *Global Change Biology, 17*, 2261-2270.

Miettinen, J., Shi, C., & Liew. S. C. (2011b). Influence of peat-land and land cover distribution on fire regimes in insular Southeast Asia. *Regional Environmental Change, 11*, 191-2011.

Oldeman, L. R., Darwis, S. N., & Las, I. (1979). *An agro-climatic map of Sumatra*. Bogor: Central Research Institute for Agriculture.

Putra, E. I., & Hayasaka, H. (2011). The effect of the precipitation pattern of the dry season on peat fire occurrence in the Mega Rice Project area, Central Kalimantan, Indonesia. *Tropics, 19*, 145-156.

Saharjo, B. H. (2006). Fire behavior in Pelalawan peatland, Riau Province. *Biodiversitas, 7*, 90-93.

Siegert, F., Ruecker, G., Hinrichs, A., & Hoffmann, A. A. (2001). Increased damage from fires in logged forests during droughts caused by El Nino. *Nature, 414*, 437-440.

Stolle, F., Chomitz, K. M., Lambin, E. F., & Tomich, T. P. (2003). Land use and vegetation fires in Jambi Province, Sumatra, Indonesia.

Forest Ecology and Management, 179, 277-292.

Tacconi, L. (2003). *Fire in Indonesia: Causes, costs and policy implication*. Bogor: Center for International Forestry Research (CIFOR).

Takahashi, H., Usup, A., Hayasaka, H., & Limin, S. H. (2007). *Overview of hydrological aspects for recent 10 years in the basins of River Sebangau and Kahayan*. Environmental Conservation and Land Use Management of Wetland Ecosystem in Southeast Asia. Annual Report for April 2006-March 2007.

Uryu, Y., Mott, C., Foedad, N., Yulianto, K., Budiman, A., Setiabudi, Takakai, F., Nursamsu, S., Purastuti, E., Fadhli, N., Hutajulu, C. M. B, Jaenicke, J., Hatano, R., Siegert, F., & Stuwe, M. (2008). *Deforestation, forest degradation, biodiversity loss and CO₂ emissions in Riau, Sumatra*, Indonesia. Jakarta: World Wide Fund (WWF) Indonesia.

Usup, A., Hashimoto, Y., Takahashi, H., & Hayasaka, H. (2004). Combustion and thermal characteristics of peat fire in tropical peatland in Central Kalimantan. *Tropics, 14*, 1-19.

Vadrevu, K. P., & Justice, C. O. (2011). Vegetation fires in the Asian region: Satellite observational needs and priorities. *Global Environmental Research, 15*, 65-76.

Wetlands (2003). *Maps of area of peatlands distribution and carbon content in Sumatra*. Bogor: Wetlands International-Indonesia Programme.

Yulianti, N., & Hayasaka, H. (2013). Recent active fire under El Niño conditions in Kalimantan, Indonesia. *American Journal of Plants Science, 4*, 685-696.

Yulianti, N., Hayasaka, H., & Usup, A. (2012). Recent forest and peat fire trends in Indonesia, the latest decade by MODIS hotspot data. *Global Environmental Research, 16*, 105-116.

26

Diversity, Population Structure and Regeneration Status of Woody Species in Dry Woodlands Adjacent to *Molapo* Farms in Northern Botswana

John Neelo[1], Demel Teketay[2*], Wellington Masamba[1], Keotshephile Kashe[1]

[1]Okavango Research Institute, University of Botswana, Maun, Botswana
[2]Department of Crop Science and Production, Botswana College of Agriculture, Gaborone, Botswana

The diversity, population structure and regeneration status of woody species were studied at Xobe and Shorobe Villages in northern Botswana. A total of 130 and 111 quadrats of 20 × 20 m size were laid down at 50 m intervals along parallel line transects at Xobe and Shorobe, respectively. A total of 46 woody species, 27 from Xobe and 41 from Shorobe were recorded. Of the 46 woody species, only 22 were recorded at both sites. Ten genera and six families were found only in Shorobe while one genus and one family were found only in Xobe. The diversity and evenness of woody species were 1.5 and 0.5 in Xobe, respectively, and 2.18 and 0.6 in Shorobe, respectively. The similarities of woody species in terms of richness of species, genera and families at the two sites were about 50%, 54% and 56%, respectively. The mean densities of woody species were 2745.7 ± 1.35 and 4269.7 ± 36 individuals ha^{-1} at Xobe and Shorobe, respectively. Despite differences in absolute numbers, the total mean densities of woody species at both sites did not exhibit significant ($P = 0.35$) differences. At both sites, woody species were dominated by individuals of only a few species, which also exhibited the highest values of important value index. The population structure patterns of the woody species were categorized into five groups. The species in the first group exhibited reverse J-shaped distribution, which indicates stable population structures. The species in the second group showed relatively good recruitment but the regeneration is negatively affected. The species in the other three groups exhibited hampered regeneration as a result of disturbances caused by humans, domestic animals and annual fires. The parameters assessed indicate the need for attention and appropriate management interventions by the relevant national authorities at various levels.

Keywords: Density; Dominance; Floristic Similarity; Frequency; Grazing; Important Value Index; Species Richness

Introduction

Dry forests and woodlands, including those in Botswana, form more than 40% of all tropical forests, and Africa and tropical islands of the world house the largest proportion of dry forest and woodland ecosystems where they account for 70% - 80% of the forested area (Murphy & Lugo, 1986). Tropical dry forests and woodlands have been under exploitation for thousands of years since they have often been preferred for human settlement to wetter forest zones for biological and ecological reasons (Murphy & Lugo, 1986; Janzen, 1988). As a result, they are either vanishing or being degraded rapidly due to accelerated growth of human and livestock populations, which result in the conversion of forested land to agriculture and excessive exploitation of forests for fuelwood, construction material and timber for export. For instance, of all the harvested wood in the tropics, 80% is used for fuel purposes, and the proportion is higher (90%) in the African tropics, where dry forests are predominant (Murphy & Lugo, 1986). A typical example is Kenya, which obtains 74% of its energy requirements from wood (Lamprecht, 1989).

Janzen (1988) has also argued that the threat to the tropical dry forests is multiple as well as complex, and tropical dry forests are the most threatened of the major tropical forest types, emphasising the urgent need for studying, conserving and restoring tropical dry forest ecosystems. The loss or degradation of forests results in land degradation in the form of soil erosion and decline of fertility, decline or loss of biodiversity and water bodies, impoverishment of ecosystems and global warming, which affect the welfare of humans, plants, animals and micro-organisms negatively (Teketay, 2004-2005).

The challenge generated by the reduction and degradation of forest cover can be adequately met only if serious efforts are made, on the one hand, to maintain the remaining forests and on the other to restore deforested and degraded areas (Teketay, 1996a). This requires understanding of the diversity and natural dynamics of woody species, i.e. causes, mechanisms and fac-

*Corresponding author.

tors that drive the process of regeneration of woody species as well as population change and replacement through time (Gomez-Pompa et al., 1991; Teketay, 1996a).

Therefore, studies on population structure and density of major canopy tree species can help to understand the status of regeneration of species, and, thereof, management history and ecology of the forest or woodland (Harper, 1977; Lykke, 1998; Sano, 1997; Teketay, 2005a, 2005b; Mwavu & Witkowski, 2008, 2009a, 2009b; Tesfaye et al., 2010). Plant population structure shows whether or not the population has a stable distribution that allows continuous regeneration to take place (Rao et al., 1990; Teketay, 1997a; Tesfaye et al., 2002; Mwavu & Witkowski, 2008, 2009a, 2009b; Tesfaye et al., 2010). If regeneration was taking place continuously, then, the distribution of species cohorts would show reverse J-shaped curve, which is an indicator of stable regeneration (Harper, 1977; Teketay, 1997a, 2005a, 2005b; Mwavu & Witkowski, 2009b). Hence, analyses of population structures, using frequency distribution of diameter classes of naturally regenerated woody species, can provide an insight into their regeneration status (Sano, 1997; Lykke, 1998; West et al., 2000; Obiri et al., 2002; McLaren et al., 2005; Mwavu & Witkowski, 2008, 2009b), which plays a key role in the promotion of their sustainable management, utilization and conservation.

Several studies in Benin (Sokpon & Biaou, 2002), Burkina Faso (Savadogo et al., 2007; Zida et al., 2007; Pare et al., 2009; Bognounou et al., 2010; Sop et al., 2011), Ethiopia (Teketay, 1997a; 2005a, 2005b; Tesfaye et al., 2010; Teketay, 2011; Fiseha et al., 2013), Ghana (Swaine et al., 1990), Oman (El-Sheikh, 2013), South Africa (Shackleton, 1993; Helm & Witkowski, 2012), Tanzania (Luoga et al., 2004), Uganda (Tabuti, 2007; Mwavu & Witkowski, 2008, 2009a, 2009b; Kalema, 2010), South Africa (Obiri et al., 2002; Venter & Witkowski, 2010; Helm & Witkowski, 2012) and West Africa (Poorter et al., 1996) demonstrated the crucial role of natural regeneration in the sustainable management of forest and woodland resources.

Forest (20%) and woodland (60%) resources cover 80% of the total land area of Botswana (FAO, 2010). Despite the relatively high forest and woodland cover, research undertakings on the various attributes of these resources are not adequate. The research activities carried out so far focused on investigations on the various attributes of woody species in relation to elephants and fire (e.g. Ben-Shahar, 1993, 1996a, 1996b, 1998a, 1998b; Ben-Shahar & Macdonald, 2002; Robinson et al., 2002; Heinl et al., 2004; Rutina, 2004; Makhabu, 2005a, 2005b; Rutina et al., 2005; Makhabu et al., 2006; Heinl et al., 2007; Rampart, 2007; Heinl et al., 2008; Tacheba et al., 2009; Kalwij et al., 2010; Aarrestad et al., 2011; Mmolotsi et al., 2012).

A long-term study has been set up and carried out since 1997 on three pairs of permanent (fenced and unfenced) plots established in the Mokolodi Nature Reserve, southeastern Botswana (Skarpe, 1990a, 1990b, 1992; Käller, 2003; Bengtsson-Sjörs, 2006; Leife, 2010; Aarrestad et al., 2011; Herrera, 2011). These studies have been monitoring to assess the dynamics of woody species over time.

Also, although it has never been implemented, an inventory has been carried out to develop the management plan of Chobe Forests in 1993 (NFS, 1993). Other studies have also been conducted focusing on different aspects of the vegetation resources in northern Botswana (Ringrose et al., 1998; Ringrose et al., 1999; Bonyongo et al., 2000; Sekhwela et al., 2000; Sekhwela, 2003; Barnes, 2001; Moleele et al., 2001; Ringrose & Mathe-

son, 2001; Moleele et al., 2002; Mosugelo et al., 2002; Ringrose et al., 2002; Ringrose, 2003; Neudeck et al., 2012; Babitseng & Teketay, 2013).

However, knowledge about diversity, population structure and natural regeneration of woody species is still scanty or lacking in Botswana. Particularly, no studies have been undertaken on these themes in areas where flood recession (locally known as *molapo*, plural *melapo*) farming (Oosterbaan et al., 1986; Bendsen, 2002; Vanderpost, 2009; Motsumi et al., 2012) is practiced by farmers in Ngamiland District, northern Botswana (**Figure 1**). *Molapo* farming is mainly practiced in the Okavango Delta and involves clearing of vegetation from fields in seasonally flooded areas for agriculture as well as lopping branches (**Figure 2(A)**) and cutting trees for fencing farms (**Figure 2(B)**) and kraals (**Figure 3(A)**). The clearing of vegetation may affect the diversity, population structure and natural regeneration of woody species as well as predispose the areas to erosion.

The specific objectives of the study were to: 1) investigate the species richness, diversity and evenness of woody species; 2) assess similarities in composition of woody species between the two study sites; 3) determine the density, frequency and domin-

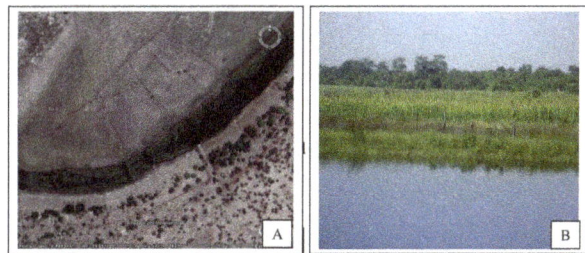

Figure 1.
Molapo farms, partly flooded (A) (VanderPost, 2009) and an example of a *molapo* farm with maize ((B), picture taken by Demel Teketay), in the Okavango Delta, northern Botswana.

Figure 2.
Molapo farms cleared and fenced with lopped branches (A) and stems (B) cut from woody species growing naturally at Xobe (pictures taken by Demel Teketay).

Figure 3.
Woody plants cut and used for constructing kraals (A) and large areas cleared of woody vegetation for dryland rain-fed agriculture (B) at Xobe (pictures taken by Demel Teketay).

ance and important value indices of woody species in the study sites; and (iv) assess the population structure and regeneration status of woody species in the study sites.

Study Sites

The study was carried out in Xobe and Shorobe Villages, located in the Ngamiland District, northern Botswana (**Figure 4(a)**). The two villages have been identified as suitable sites since they have different flooding patterns (**Figure 4(b)**) and local topography and, consequently, variations in *molapo* farming practices (Chimbari et al., 2009). The two villages fall within the Okavango Delta, part of the wetland system, which starts in the highlands of Angola. The *molapo* farming system in these areas is dependent upon the rainfall at the source averaging 1400 mm per year with the Okavango Delta receiving an

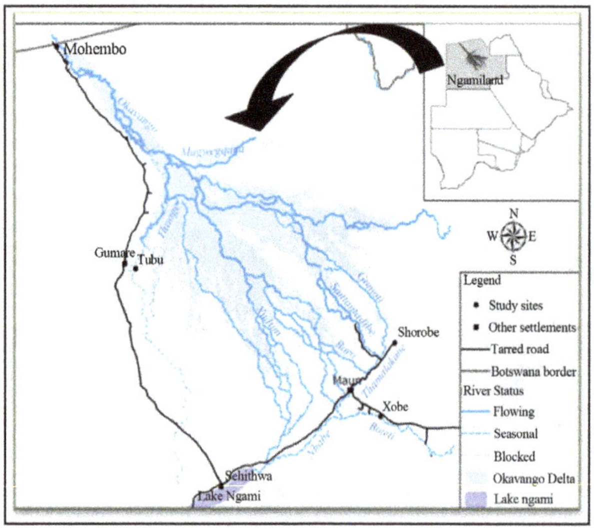

(a)

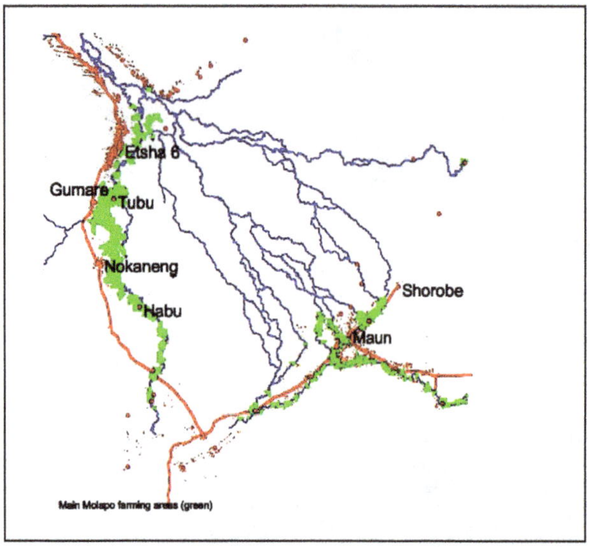

(b)

Figure 4.
Map showing study sites ((a), Chimbari et al., 2009) and the main *molapo* farming areas in the Okavango Delta ((b), VanderPost, 2009), northern Botswana.

average of 450 - 500 mm per year (McCarthy et al., 1998). The vegetation in Shorobe and Xobe is dominated by mopane [*Colophospermum mopane* (J. Kirk ex Benth.) J. Kirk ex Leonard] and mixed species of *Acacia*, respectively (DEA, 2008).

Shorobe Village is located in Ngamiland East Sub-District, about 36 km northeast of Maun. It lies between 19°45'56.71" latitude and 23°40'10.53" longitude and has a population of 1031 (CSO, 2011). To the northwest of the village, there is an extensive network of *molapo* fields fed by the Santantadibe and Gomoti Rivers and by backflow from the Thamalakane River. Soils in *molapo* fields are classified as young alluvial soils. Texture varies from clayey (35% to 60% clay), especially in low-lying areas, through fine loamy (18% to 35% clay) to coarse loamy (<18% clay) (Chimbari et al., 2009). Flooding of most *molapo* fields has not occurred for several years until flood waters returned in 2009. When floods occur, *molapo* fields are cultivated as the flood recedes. The villagers practise both arable and pastoral farming. The main crops planted are maize, sorghum, millets, pumpkins, water melons, melons, sweet reeds, beans and groundnuts. The local people are also engaged in traditional beer brewing and palm wine making, fishing and basket weaving.

Xobe Village, with a population of 418 (CSO, 2011) people, is a cattlepost area on the south bank of the Boteti River, about 13 km east of Maun. The area lies between 20°7'10.26' latitude and 23°27'41.26' longitude. The river normally floods in July, but in dry years (e.g. in the late 1990s), the flood may not reach the settlement. Soils in Xobe settlement *molapo* fields are coarse textured alluvial deposits (Chimbari et al., 2009). Livelihood activities for people in Xobe settlement include rain-fed farming, *molapo* farming, irrigated vegetable production and livestock rearing. *Molapo* farmers cultivate along the river banks as the water flow recedes. Ploughing is done using donkeys, though smaller areas may be cultivated with hand hoes. The main crops planted are maize, pumpkins, sweet sorghum and gourds. Other activities include harvesting of wild plants for sale and fishing.

Data Collection

At both study sites, three *molapo* fields (replicates), at least one kilometer apart, and having adjacent vegetation were randomly selected for the study. To assess plant diversity (species richness and evenness) and density, abundance, frequency, dominance (basal area), population structure, regeneration status and important value indices of the woody species, three one kilometre long (depending on the size of the field) parallel line transects, 50 meters apart, were used in each *molapo* field. Quadrats measuring 20 × 20 m were laid down at every 50 meters interval along the line transects.

A total of 111 quadrats representing a total area of 4.44 ha were sampled to assess the woody vegetation at Shorobe. Similarly, a total of 130 quadrats representing a total area of 5.2 ha were sampled to assess the woody vegetation at Xobe. The number of quadrats in Shorobe were lower than those in Xobe since one of the sites used for the survey had shorter transects because of the surrounding settlements.

In each of the quadrats, the following parameters were recorded: identity of all woody species (WS), number of live individuals of all WS and diameter at breast height (DBH) of all woody species (with DBH >2 cm), except juveniles (seedlings and coppices). In the case of seedlings and coppices, the number of individuals of each species was counted and recorded in each quadrat. A calliper and graduated measuring stick were used to measure diameter, respectively, of the woody plants.

Plant species identification was done first directly in the field using relevant manuals and other reference materials as well as with the help of local people familiar with the flora. Plant nomenclature in this article follows that of Setshogo and Venter (2003) and Setshogo (2005).

Data Analyses

Species richness (*S*) is the total number of different woody species recorded in each of the project sites. It does not take into account the proportion and distribution of each species at the project sites.

The *diversity* of woody species was analysed by using the Shannon Diversity Index (*H*) (also known as the Shannon-Weiner/Weaver Diversity Index in the ecological literature) (Krebs, 1989; Magurran, 2004). The index takes into account the species richness and proportion of each species in all sampled quadrats of each project site. The following formula was used to analyse woody species diversity:

$$H' = \sum_{i=1}^{S} P_i \ln P_i$$

where, *H'* = Shannon index, *S* = species richness, P_i = proportion of *S* made up of the ith species (relative abundance).

Evenness or equitability, a measure of similarity of the abundances of the different woody species in the sampled project sites, was analysed by using *Shannon's Evenness* or *Equitability Index* (E) (Krebs, 1989; Magurran, 2004). Equitability assumes a value between 0 and 1 with 1 being complete evenness. The following formula was used to calculate evenness.

$$J' = H'/\ln S$$

where, *J'* = evenness and *S* = species richness.

The similarity in woody species composition of the two sites was computed by using Jaccard's Similarity Coefficient (*S_J*) (Krebs, 1989). The values of *S_J* range between 0 and 1: 0 indicates complete dissimilarity and 1 indicates complete similarity in species composition. The following formula was used to determine similarity of the woody species in the two sites:

$$S_J = a/(a + b + c)$$

where, *a* = number of woody species common in the two sites; *b* = number of woody species recorded only in Xobe; and *c* = woody species recorded only in Shorobe.

The mean *density* (MDE) of woody species was determined by converting the total number of individuals of each woody species encountered in all the quadrats and all transects of the three replicated areas in each of the two sites to equivalent number per hectare as described in Mueller-Dombois and Ellenberg (1974). Student's t-test, at the confidence level of *P* < 0.05 (Zar, 1999), was used to assess if differences existed between the mean values of total densities of woody species at Xobe and Shorobe.

The *frequency* (MF) was calculated as the proportion (%) of the number of quadrats in which each woody species was recorded from the total number of quadrats in each of the sites. The *dominance* of the woody species, with diameter at breast height (DBH) of >2 cm, was determined from the space occu-

pied by a species, usually its basal area (BA). The mean dominance of each species was computed by converting the total basal area of each woody species to equivalent basal area per hectare (Kent & Coker, 1992).

The *important value index* (IVI) indicates the relative ecological importance of a woody species in each of the project sites (Kent & Coker 1992). It is determined from the summation of the relative values of density, frequency and dominance of each woody species. *Relative density* (RMDE) was calculated as the percentage of the density of each species divided by the total stem number of all species ha^{-1}. *Relative frequency* (RMF) of a species was computed as the ratio of the frequency of the species to the sum total of the frequency of all species. *Relative dominance* (RMDO) was calculated as the percentage of the total basal area of a species out of the total basal areas of all species.

Population structure of each woody species in each of the two project sites was assessed through histograms constructed by using the density of individuals of each species (Y-axis) categorized into ten diameters classes (X-axis) (Peter, 1996), i.e. 1 = <2 cm; 2 = 2 - 5 cm; 3 = 5 - 10 cm; 4 = 10 - 15 cm; 5 = 15 - 20 cm; 6 = 20 - 25 cm; 7 = 25 - 30 cm; 8 = 30 - 35; 9 = 35 - 40; 10 = >40 cm. Then, based on the profile depicted in the population structures, the regeneration status of each woody species was determined.

Results

Species Richness, Diversity and Evenness

A total of 46 different species of woody plants representing 16 families and 24 genera were recorded from the two sites. Only two species, one species at each site, were not identified. Of these, 27 (10 families and 14 genera) and 41 (15 Families and 23 genera) woody species were recorded at Xobe and Shorobe, respectively (**Tables 1** and **2**). Fabaceae (Leguminosae) was the most diverse family at both Xobe and Shorobe represented by 12 (about 46%) and 15 (about 38%) woody species, respectively. The second and third most diverse families were Combretaceae [with 4 species (about 15%) at Xobe and 6 species (15%) at Shorobe] and Tiliaceae [with 3 species (about 12%) at Xobe and 4 species (10%) at Shorobe], respectively. Capparaceae, Ebenaceae and Rhamnaceae were represented by two species (5%) each at Shorobe. All other families contained only one species at both sites. Ten genera and six families were found only in Shorobe while one genus and one family were found only in Xobe (**Tables 1** and **2**).

Of the 24 genera encountered, 14 and 23 were recorded from Xobe and Shorobe, respectively. At both sites, *Acacia* was the most diverse genus with 8 (about 57%) and 9 (about 39%) species of the total number of genera at Xobe and Shorobe sites, respectively. *Combretum* and *Grewia* [with 3 species (about 21%) each] were the second diverse genera followed by *Philoneptera* [with 2 species (about 14%)] at Xobe. Similarly, *Combretum* and *Grewia* [with 4 species (about 17%) each] were the second diverse genera followed by *Philoneptera*, *Albizia* and *Terminalia* [with 2 species (about 9%) each] at Shorobe. All the other genera were represented only by one species at both sites (**Tables 1** and **2**).

The diversity and evenness of woody species were 1.5 and 0.5 in Xobe, respectively, and 2.18 and 0.6 in Shorobe, respectively.

Table 1.
List of woody species recorded from Xobe with their densities (individuals ha⁻¹), frequencies (%), dominance (m²·ha⁻¹), relative values (%) of densities, frequencies and dominance as well as Important Value Index (MDE = mean density, MF = mean frequency, MDO = mean dominance, RMDE = relative mean density, RMF = relative mean frequency, RMDO = relative mean dominance and IVI = important value index).

Species	Family	Local names	MDE	MF	MDO*	RMDE	RMF	RMDO	IVI
Acacia mellifera[F]	Fabaceae	Mongana	1194.3	82.2	33.4	45.8	19.2	25.2	90.2
Acacia tortilis[F]	Fabaceae	Mosu	1068.4	69.9	49.7	37.5	16.2	36.1	89.8
Philenoptera nelsii	Fabaceae	Mohatha	106.1	35.9	22.2	3.7	8.2	16.1	27.9
Acacia luederitzii[F]	Fabaceae	Mokgwelekgwele	49.3	28.3	10.6	1.7	6.5	7.6	15.7
Acacia erubescens[F]	Fabaceae	Moloto	45.9	24.6	2.5	1.5	5.6	1.7	8.9
Albizia anthelmintica	Fabaceae	Monoga	37.5	13.8	2.5	1.2	3.1	2.0	6.3
Acacia erioloba[F]	Fabaceae	Mogotho	37.0	27.0	4.6	1.4	6.5	3.5	11.4
Gymnosporia senegalensis	Celastraceae	Mothono	36.3	20.8	0.3	1.2	4.8	0.2	6.3
Terminalia prunioides	Combretaceae	Motsiara	35.9	21.4	2.3	1.2	4.7	1.8	7.7
Gardenia volkensii	Rubiaceae	Morala	26.1	6.1	0.6	0.9	1.4	0.4	2.7
Mimusops zeyheri	Sapotaceae	Mmupudu	23.9	12.1	0.2	0.8	2.7	0.1	3.6
Combretum albopunctatum	Combretaceae	Motsoketsane	20.4	22.8	1.8	0.7	4.9	1.2	6.8
Ximenia americana	Olacaceae	Moretologa	13.0	8.5	0.0	0.5	2.2	0.0	2.7
Croton megalobotrys[F]	Euphorbiaceae	Motsebe	11.9	2.2	0.5	0.4	0.5	0.3	1.2
Dichrostachys cinerea	Fabaceae	Moselesele	10.5	15.2	0.1	0.4	3.5	0.1	4.0
Grewia flava	Tiliaceae	Moretlwa	7.6	16.8	0.0	0.3	3.7	0.0	3.9
Combretum hereroense[F]	Combretaceae	Mokabi	6.9	2.2	0.4	0.2	0.5	0.2	0.9
Grewia retinervis	Tiliaceae	Mokgomphata	5.5	13.0	0.0	0.2	2.9	0.0	3.1
Acacia galpinii	Fabaceae	Mokala	2.6	0.7	0.2	0.1	0.2	0.1	0.4
Boscia albitrunca	Capparaceae	Motopi	1.8	5.9	2.8	0.1	1.4	2.2	3.7
Ziziphus mucronata[F]	Rhamnaceae	Mokgalo	1.4	2.3	0.0	0.0	0.5	0.0	0.6
Acacia nilotica	Fabaceae	Motlhabakgosi	1.2	0.8	0.0	0.1	0.2	0.0	0.3
Acacia fleckii[F]	Fabaceae	Mohahu	0.6	0.8	0.5	0.0	0.2	0.4	0.6
Combretum mossambicense	Combretaceae	Motsweketsane	0.6	0.8	0.0	0.0	0.2	0.0	0.3
Philenoptera violacea	Fabaceae	Mopororo	0.4	1.6	0.4	0.0	0.3	0.3	0.7
Unidentified species	-	-	0.4	0.8	0.5	0.0	0.2	0.4	0.6
Grewia bicolor	Tiliaceae	Mogwana	0.2	0.8	0.0	0.0	0.2	0.0	0.2
Total			**2745.7**			**100**	**100**	**100**	

*The "0" values indicate several decimal places, i.e. close to zero; [F] = species most preferred for fencing (Chimbari et al., unpublished).

Similarities in Composition of Woody Species

Of the 46 woody species, 22 were recorded at both sites while 5 and 19 woody species were recorded only from Xobe and Shorobe sites, respectively (**Tables 1** and **2**). The similarities of woody species in terms of richness of species, genera and families at the two sites were about 50%, 54% and 56%, respectively.

Density, Frequency and Dominance

The total mean densities of woody species were 2745.7 ± 1.35 (SD) (range = 2222 - 6545) and 4269.7 ± 36 (SD) (range 2155 - 3133) individuals ha⁻¹ at Xobe and Shorobe, respectively, and ranged from 0.2 (*G. discolor*) - 1194.3 (*A. mellifera*) at Xobe and 0.2 (*E. divinorum*) - 1675.4 (*C. mopane*) (**Tables 1** and **2**). Despite differences in absolute numbers, the total mean

Table 2.
List of woody species recorded from Shorobe with their densities (individuals ha^{-1}), frequencies (%), dominance (m$^2 \cdot$ha^{-1}), relative values (%) of densities, frequencies and dominance as well as Important Value Index (MDE = mean density, MF = mean frequency, MDO = mean dominance, RMDE = relative mean density, RMF = relative mean frequency, RMDO = relative mean dominance and IVI = important value index).

Species	Family	Local names	MDE	MF	MDO*	RMDE	RMF	RMDO	IVI
Colophospermum mopane[F]	Fabaceae	Mophane	1675.4	39.9	87.6	25.7	6.2	26.5	58.5
Acacia tortilis[F]	Fabaceae	Mosu	509.8	71.4	34.9	14.7	10.3	16.6	41.6
Acacia erioloba[F]	Fabaceae	Mogotho	468.6	81.6	42.5	10.7	12.0	20.6	43.3
Philenoptera violacea	Fabaceae	Mopororo	258.3	48.9	20.4	7.7	7.0	9.9	24.6
Dichrostachys cinerea	Fabaceae	Moselesele	218.4	59.3	4.0	6.8	8.6	2.2	17.6
Hyphaene petersiana[F]	Arecaceae	Mokolwane	175.7	41.8	0.0	5.1	5.9	0.0	11.0
Grewia bicolor	Tiliaceae	Mogwana	175.1	29.5	2.1	7.1	4.5	0.9	12.5
Combretum imberbe[F]	Combretaceae	Motswere	174.7	40.2	22.8	6.5	5.7	10.3	22.5
Croton megalobotrys[F]	Euphorbiaceae	Motsebi	133.5	21.8	15.9	3.5	3.3	6.9	13.7
Grewia flava	Tiliaceae	Moretlwa	88.1	33.6	0.6	2.0	5.2	0.2	7.4
Grewia retinervis	Tiliaceae	Mokgomphatla	71.0	38.8	0.6	2.2	5.6	0.3	8.2
Acacia mellifera[F]	Fabaceae	Mongana	69.4	24.6	0.5	1.9	3.3	0.2	5.4
Gymnosporia senegalensis	Celastraceae	Mothono	39.9	11.4	0.6	1.0	1.5	0.4	3.0
Ziziphus mucronata[F]	Rhamnaceae	Mokgalo	36.0	25.9	2.0	1.1	3.5	1.1	5.7
Combretum hereroense[F]	Combretaceae	Mokabi	33.1	18.9	1.3	1.0	2.7	0.6	4.3
Rhus tenuinervis	Anacardiaceae	Morupaphiri	24.2	18.9	0.2	0.7	2.8	0.1	3.6
Acacia nilotica	Fabaceae	Motlhabakgosi	12.5	7.0	0.5	0.3	0.9	0.3	1.5
Combretum mossambicense	Combretaceae	Motsweketsane	12.2	5.3	0.4	0.3	0.8	0.2	1.2
Boscia albitrunca	Capparaceae	Motlopi	8.8	13.7	1.2	0.2	2.0	0.4	2.5
Albizia anthelmintica	Fabaceae	Monoga	8.5	2.6	0.5	0.2	0.4	0.4	1.0
Diospyros lycioides	Ebenaceae	Letlhajwa	6.4	6.8	0.0	0.1	0.9	0.0	1.1
Berchemia discolour	Rhamnaceae	Motsentsela	4.8	1.0	0.6	0.2	0.2	0.3	0.6
Acacia hebeclada	Fabaceae	Setshi	4.7	0.9	0.2	0.1	0.1	0.1	0.3
Rhigozum brevispinosum	Bignoniaceae	Lebuta	4.5	0.9	0.0	0.1	0.1	0.0	0.2
Commiphora mossabicensis	Burseraceae	Moroka	4.3	9.4	0.1	0.1	1.5	0.0	1.6
Terminalia prunioides	Combretaceae	Motsiara	4.2	3.7	0.6	0.1	0.5	0.3	1.0
Grewia flavescens	Tiliaceae	Motsotsojane	3.2	6.0	0.0	0.1	0.9	0.0	1.0
Ximenia americana	Olacaceae	Moretologa	3.1	5.3	0.0	0.1	0.8	0.0	0.9
Acacia sieberiana[F]	Fabaceae	Moremostlha	2.4	2.6	0.1	0.1	0.3	0.1	0.5
Albizia harveyi	Fabaceae	Molalakgaka	2.3	2.0	2.2	0.1	0.3	0.9	1.3
Mimusops zeyheri	Sapotaceae	Mmupudu	2.1	0.9	0.0	0.0	0.1	0.0	0.2
Unidentified species	-	Motorokofina	1.9	0.9	0.0	0.0	0.1	0.0	0.2
Acacia nigrescens[F]	Fabaceae	Mokoba	1.7	2.6	0.0	0.0	0.3	0.0	0.4
Philenoptera nelsii	Fabaceae	Mohatha	1.7	1.7	0.0	0.0	0.3	0.0	0.3
Capparis tomentosa	Capparaceae	Motawana	1.5	1.0	0.2	0.1	0.2	0.1	0.3
Combretum collinum	Combretaceae	Modubana	1.3	0.9	0.8	0.0	0.1	0.2	0.4
Acacia fleckii[F]	Fabaceae	Mohahu	0.4	0.9	0.0	0.0	0.1	0.0	0.1
Terminalia sericea[F]	Combretaceae	Mogonono	0.4	1.7	0.0	0.0	0.2	0.0	0.3
Garcinia livingstonei[F]	Clusiaceae	Motsaudi	0.3	1.0	0.0	0.0	0.2	0.0	0.2
Acacia erubescens[F]	Fabaceae	Moloto	0.2	0.9	0.0	0.0	0.1	0.0	0.1
Euclea divinorum	Ebenaceae	Motlhakola	0.2	0.9	0.0	0.0	0.1	0.0	0.1
Total			**4269.7**			**100**	**100**	**100**	

*The "0" values indicate several decimal places, i.e. close to zero; [F] = species most preferred for fencing (Chimbari et al., unpublished).

densities of woody species at both sites did not exhibit significant differences (Students T-Test, $P = 0.35$).

The top 10 densest woody species at Xobe were (in descending order of density) *Acacia mellifera, A. tortilis, Philenoptera nelsii, Acacia luederitzii, A. erioloba, A. erubescens, Terminalia prunioides, Combretum albopunctatum, Albizia anthelmintica* and *Gymnosporia senegalensis* (**Table 1**). The least five densest woody species at Xobe were (in descending order of density) *Acacia fleckii, Combretum mossambicense, Philenoptera violacea*, Unidentified sp. and *Grewia bicolor*. The 10 densest woody species at Shorobe were (in descending order of density) *Colophospermum mopane, Acacia tortilis, A. erioloba, Philenoptera violacea, Dichrostachys cinerea, Hyphaene petersiana, Grewia bicolor, Combretum imberbe, Croton megalobotrys* and *Grewia retinervis* (**Table 2**). The least five densest woody species at Shorobe were (in descending order of density) *Acacia fleckii, Terminalia sericea, Garcinia livingstonei, Acacia erubescens* and *Euclea divinorum*.

Acacia erioloba, A. erubescens, A. fleckii, A. galpinii, A. luederitzii, A. mellifera, A. nilotica, A. tortilis, Albizia anthelmintica and *Boscia albitrunca* were the ten most frequent woody species at Xobe (in descending order of frequency) (**Table 1**). The five least frequent woody species were *Philenoptera violacea, Terminalia prunioides*, Unidentified species, *Ximenia americana* and *Ziziphus mucronata*. *Acacia erioloba, Acacia tortilis, Dicrostachys cinerea, Philenoptera violacea, Hypaene petersiana, Combretum imberbe, Colophospermum mopane, Grewia retinervis, G. flava* and *G. bicolor* were the top 10 most frequent species in Shorobe (in descending order of frequency) (**Table 2**). The five least frequent woody species were *Combretum collinum, Acacia fleckii, Garcinia livingstonei, Acacia erubescens* and *Euclea divinorum*.

The top 10 most dominant woody species at Xobe were (in descending order of dominance) *Acacia tortilis, A. mellifera, Philenoptera nelsii, Acacia luederitzii, A. erioloba, Boscia albitrunca, Acacia erubescens, Albizia anthelmintica, Terminalia prunioides* and *Combretum albopunctatum* (**Table 1**). Seven woody species (22% the total number of species recorded at the site) had very insignificant dominance values, i.e. close to zero. These are *Ximenia americana, Grewia flava, Ziziphus mucronata, Acacia nilotica, Combretum mossambicense, Grewia bicolor* and the unidentified species. The top 10 most dominant woody species at Shorobe were (in descending order of dominance) *Colophospermum mopane, Acacia erioloba, Acacia tortilis, Combretum imberbe, Philenoptera violacea, Croton megalobotrys, Dichrostachys cinerea, Grewia bicolor, Ziziphus mucronata* and *Albizia harveyi* (**Table 2**). Sixteen of the woody species (about 39% of the total number of species recorded at the site) had very insignificant dominance values, i.e. close to zero.

Important Value Index

Based on their Importance Value Index, *Acacia mellifera, A. tortilis, Philenoptera nelsii, Acacia luederitzii, A. erioloba, A. erubescens, Terminalia prunioides, Combretum albopunctatum, Albizia anthelmintica* and *Gymnosporia senegalensis* in Xobe (**Table 1**) and *Colophospermum mopane, Acacia erioloba, Acacia tortilis, Philenoptera violacea, Combretum imberbe, Dichrostachys cinerea, Croton megalobotrys, Grewia bicolor, Hyphaene petersiana* and *Grewia retinervis* in Shorobe (**Table 2**) were the top ten species in their descending order of eco-logical importance. Nine and 16 species had IVI values of less than one in Xobe and Shorobe, respectively, indicating that they are the least ecologically important species (**Tables 1** and **2**).

Population Structure and Regeneration Status

Based on their population structures, the woody species recorded at Xobe and Shorobe could be categorized into five diameter class distribution patterns (**Figure 5**).

The first group was composed of species that exhibited higher number of individuals at the lowest diameter class and progressively declining numbers with increasing diameter classes. To this group belonged *Acacia erioloba, A. tortilis, A. mellifera, Colophospermum mopane, Combretum imberbe, Dichrostachys cinerea, Grewia retinervis* and *Gymnosporia senegalensis* at Shorobeand *Combretum albopunctatum, Dichrostachys cinerea, Gymnosporia senegalensis* and *Ximenia americana* at Xobe (**Figure 5(a)**).

The second group was composed of species with similar diameter class distribution pattern as the first group except that individuals are missing at the higher diameter classes. To this group *Commiphora mossabicensis, Diospyros lycioides, Grewia flavescens* and *Rhus tenunervis* at Shorobe and *Acacia luederitzii, A. tortilis, Albizia anthelmintica, Terminalia prunioides* and *Ziziphus mucronata* at Xobe (**Figure 5(b)**) have been categorized.

The third group consisted of species that showed both hampered seedling/coppice recruitment and missing of individuals at the higher diameter classes. To this group belonged *Acacia hebeclada, A. nilotica, A. sieberiana, Albizia anthelmintica, Combretum hereroense, C. mossambicense, Croton megalobotrys, Grewia bicolor, Mimusops zeyheri* and *Ziziphus mucronata* at Shorobe and *Acacia erubescens, A. galpinii, A. mellifera, Combretum hereroense, C. mossambicense, Croton megalobotrys, Gardenia volkensii, Mimusops zeyheri* and *Philenoptera violacea* at Xobe (**Figure 5(c)**).

The fourth group was composed of species with missing individuals in one or more of the diameter classes. To this group *Albizia harveyi, Berchemia discolor, Boscia albitrunca, Capparis tomentosa, Combretum collinum, Grewia flava, Philenoptera violacea* and *Terminalia prunioides* at Shorobe and *Acacia erioloba, A. fleckii, A. nilotica, Boscia albitrunca, Philenoptera nelsii* and the unidentified species at Xobe (**Figure 5(d)**) were categorized.

The fifth group consisted of species with individuals represented only in the first diameter class or juveniles. This group was composed of *Acacia erubescens, A. fleckii, A. nigrescens, Euclea divinorium, Garcinia livingstonei, Hyphaene petersiana, Philenoptera nelsii, Rhigozum brevispinosum*, "Motorokofina" (unidentified sp.), *Terminalia sericea* and *Ximenia Americana* at Shorobe and *Grewia bicolor, G. flava* and *G. retinervis* at Xobe (**Figure 5(e)**).

Discussion

Forest and woodland resources in Botswana are important in providing socio-economic and ecological services, e. g. timber, food, fuelwood, traditional medicine, fodder, other non-timber forest products, source of grazing areas, wildlife habitats, tourism, watershed regulation, soil protection, carbon sequestration and storage, etc., that sustain livelihoods of communities and

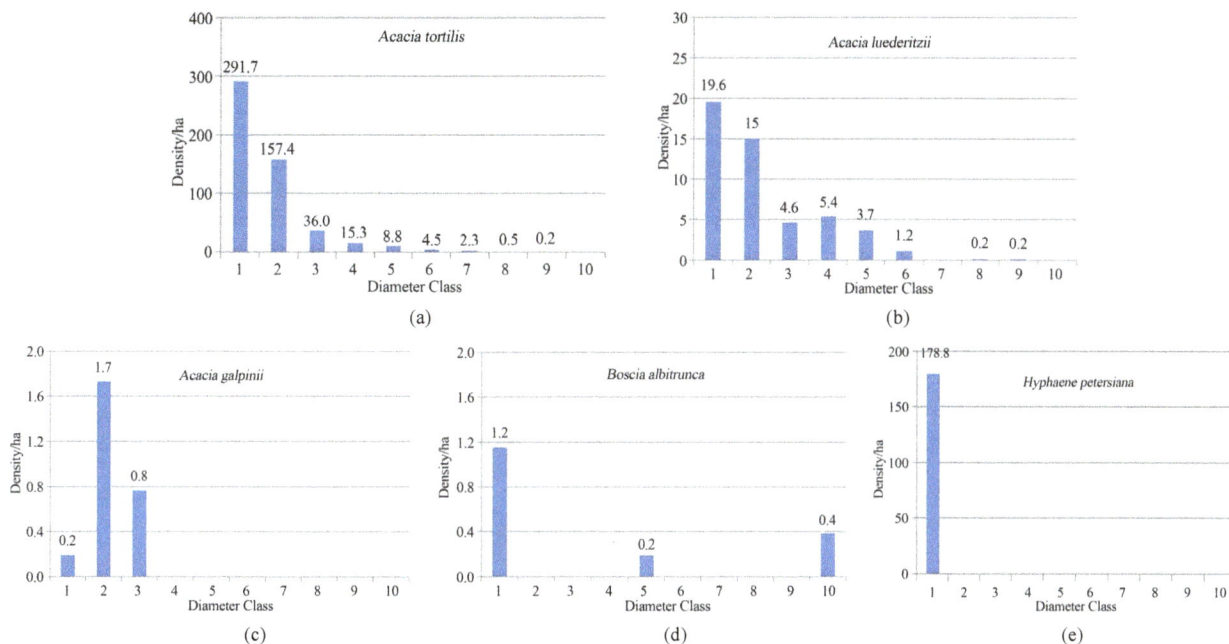

Figure 5.
Population structure of woody species recorded at Shorobe ((a) and (e)) and Xobe ((b), (c) and (d)) [diameter class (DBH): 1 = <2 cm; 2 = 2 - 5 cm; 3 = 5 - 10 cm; 4 = 10 - 15 cm; 5 = 15 - 20 cm; 6 = 20 - 25 cm; 7 = 25 - 30 cm; 8 = 30 - 35; 9 = 35 - 40; 10 = >40 cm.

the national economy. Therefore, their sustainable management, utilization and conservation are crucial.

Measures of species diversity play a central role in ecology and conservation biology (Magurran, 2004) since species diversity is an important parameter of a plant community, one of the major criteria for nature conservation and connected to ecosystem dynamics and environmental quality (Kalema, 2010). A change in species diversity is often used as an indicator of anthropogenic or natural disturbances in an ecosystem (Liu & Brakenhielm, 1996; Kalema, 2010). Therefore, characterization of biodiversity through inventories can be useful in the planning of operations that aim to conserve biodiversity (Belbin, 1995; Faith & Walker, 1996; Kelema, 2010).

Of the two study sites, Shorobe exhibited the highest species richness with 18 of the woody species not recorded from Xobe. Only five of the species recorded at Xobe were not encountered in the studied quadrats at Shorobe. Interestingly, even the densest and most dominant tree species in northern Botswana, namely *C. mopane*, and the common and conspicuous species, such as *H. petersiana* and *Combretum imberbe* at Shorobe were not recorded from the quadrats examined in Xobe, indicating their scarcity at this study site. Conversely, one of the most common species recorded in Xobe, namely *A. leuderitzii*, was not encountered in the quadrats sampled from Shorobe. In addition, the numbers of genera and families were higher in Shorobe than Xobe. With the exception of *A. tortilis* and *A. erioloba*, the highest densities, frequencies, dominances and, hence, IVI values were exhibited by different species in the two sites. In terms of species richness and overall diversity, Shorobe and Xobe had much lower number of woody species compared with reports from studies in the Sudanian savanna in Burkina Faso (Savadogo et al., 2007), dryland forests and woodlands in Ethiopia (Woldemariam et al., 2000; Senbeta & Teketay, 2003; Zegeye et al., 2006, 2011; Worku et al., 2012) as well as woodlands and forests in South Africa (Dovie et al., 2008), Tanzania

(Luoga et al., 2000; Banda et al., 2008) and Uganda (Nangendo et al., 2006; Kalema, 2010). However, the evenness values of woody species in Shorobe and Xobe were comparable with those reported for other dry land forests (Senbeta & Teketay, 2003; Alelign et al., 2007; Zegeye et al., 2006, 2011).

The overall diversity of woody plants was much higher in Shorobe (*H'* = 2.18) than Xobe (*H'* = 1.5), which may be a consequence of the high species richness in Shorobe. It has been noted that the value of *H'* obtained from empirical data usually falls between 1.5 and 3.5, and rarely surpasses 4 (Margalef, 1972; Magurran, 2004). This implies that the diversity of woody species at Xobe falls at the lowest value of the diversity range. Although we have not investigated the causes, the difference in the edaphic factors, especially the big difference in the soil types and moisture availability (Chimbari et al., 2009), may be responsible for the considerable floristic variations (richness of species, genera and families as well as overall diversity) between the two sites. The diversity values of woody species obtained at Shorobe and Xobe are lower than those reported from Miombo Woodlands in Tanzania (Nduwamungu, 1997; Zahabu, 2001) and savanna woodlands in South Africa (Dovie et al., 2008). However, evenness values of the woody species at Shorobe (E = 0.6) and Xobe (E = 0.5) were more or less similar, also with those from other studies (Zegeye et al., 2006, 2011; Worku et al., 2012), suggesting that individuals of the different species recorded exhibited moderately similar abundance at the two sites.

The Jaccard's Similarity Coefficients of about 50% - 56% for the richness of species, genera and families of woody species recorded from the two sites indicate that about half of the total species, genera and families encountered were specific to one or the other site. This reality has to be taken into consideration in any plan aimed at the sustainable management and conservation of these resources.

Shorobe exhibited much higher density of woody species

compared with that of Xobe, which could be associated with the higher number of woody species encountered at Shorobe than Xobe. With the exception of *A. tortilis* and *A. erioloba*, the two study sites differed in their densest woody species. Surprisingly, as stated above, the densest woody species in Shorobe, *C. mopane*, was not recorded from any of the quadrats assessed in Xobe, which may be attributed to its requirement of habitats with heavy textured and poorly drained soils (Ellery & Ellery, 1997).

A. mellifera and *A. tortilis* had much higher densities at Xobe than Shorobe with more than 82% of the total density recorded in Xobe. In fact, more than 87% of the total density in Xobe was represented by the six *Acacia* species, five of which are also among the most dominant and frequently found species with the highest IVI values. This might suggest signs of bush encroachment due to overgrazing and over-exploitation of resources at Xobe (DEA, 2008). *Acacia mellifera* forms impenetrable patches of thickets at Xobe and is known to encroach eroded sites (Ellery & Ellery, 1997) and heavily grazed areas (El-Sheikh, 2013). *Acacia tortilis* is also common and widespread species in Botswana, which occurs on clay or loam soils in a variety of woodlands, generally near floodplains and pans. It tends to encroach heavily grazed sites (Ellery & Ellery, 1997). The domination of *Acacia* species, which are indicative of heavy grazing and encroachment, at Xobe is consistent with the fact that Xobe is used as a cattle post by people living in the nearby Maun Town.

In addition, the relatively high number of species and density of *Acacia* species recorded at Xobe may be attributed to seed dispersal, which is known to be facilitated by ruminants, and the subsequent favourable initial habitat for the developing seedlings within the accompanying droppings of the animals (Schultka & Cornelius, 1997; Teketay, 1996b, 1997b, 2005a; Kalema, 2010). Many *Acacia* species also use the soil seed bank as one route of regeneration after disturbance, especially following fire incidences (Sabiiti & Wein, 1987; Teketay & Granström, 1995, 1997; Teketay, 1998; Witkowski & Garner, 2000; Eriksson *et al.*, 2003; Teketay, 2005a). It has been reported that grazing is a predictable selective form of disturbance with animal behaviour through browse choice playing a significant role in determining which species are impacted (Whelan, 2001; Kalema, 2010). Grazing pressure may also play a significant role in determining plant community structure and composition by facilitating bush encroachment in frequently grazed areas (Witkowski & O'Connor, 1996; Kalema, 2010). Grazing, fire and selective tree harvesting, which are very common in the study sites, are considered major disturbances shaping species diversity and productivity (Savadogo, 2007; Kalema, 2010).

It is interesting to note that *P. violacea* had much higher densities at Shorobe than Xobe while *P. nelsii* exhibited higher density in Xobe than Shorobe. On the one hand, *P. violacea* grows in open woodlands on edges of islands in seasonal swamps, and occasionally on interior regions of islands in permanent swamps (Ellery & Ellery, 1997), habitats common in the surroundings of Shorobe. On the other, *P. nelsii* occurs in deep sand as part of short woodlands (Ellery & Ellery, 1997), which characterize the habitat at Xobe.

The overall density of woody species recorded at Shorobe was higher than those reported from other dry Afromontane forests (Alelign et al., 2007; Zegeye et al., 2006, 2011) and woodlands (Worku et al., 2012). The overall density of woody

species recorded at Xobe was higher than other dry woodlands (Worku et al., 2012), similar to those reported from different forests (Woldemariam et al., 2000; Zegeye et al., 2006, 2011) and lower than other dry Afromontane forests (Alelign et al., 2007; Zegeye et al., 2011). The overall horizontal distribution of the woody species, represented by the frequency of occurrence of the species in the studied quadrats, was relatively low with only 10 (out of 41) and three (out of 27) species having more than 30% frequency values at Shorobe and Xobe, respectively. This implies that the other species have scarce horizontal distribution at both study sites, which requires further investigations that can assist in the future design of appropriate management interventions.

Importance Value Index is an important parameter that reveals the ecological significance of species in a given ecosystem (Lamprecht, 1989; Zegeye et al., 2006; Senbeta & Teketay, 2003; Worku et al., 2012). *Acacia mellifera*, *A. tortilis* and *Philenoptera nelsii* at Xobe and *Colophospermum mopane*, *Acacia erioloba*, *Acacia tortilis* and *Philenoptera violacea* at Shorobe can be considered the most ecologically important woody species with IVI values of more than 20 contributed by their high values of density, frequency and dominance. It is interesting to note that *A. tortilis* is recorded among the most ecologically important woody species at both study sites.

In the absence of long-term demographic data on population trends, the use of diameter class distributions of woody species from a single survey has been shown to be a potential and reliable tool to reveal status of population structures and regeneration of woody species as well as predict responses of the species to disturbance and resultant changes in population structure (Condit et al., 1998; Lykke, 1998; Obiri et al., 2002; Sokpon & Biaou, 2002; Teketay, 2005a, 2005b; Feeley et al., 2007; Tabuti, 2007; Mwavu & Witkowski, 2009a; Tesfaye et al., 2010; Venter & Witkowski, 2010; Sop et al., 2011; Helm & Witkowski, 2012; El-Sheikh, 2013). A population size structure is simultaneously the outcome of past demographic events and an indicator of its demographic future (Wilson & Witkowski, 2003; Kalema, 2010).

Based on the assessment of diameter class distributions, the population structure patterns of the woody species recorded from Shorobe and Xobe were categorized into five groups. The species in the first group exhibited reverse J-shaped distribution, which is widely acknowledged to indicate stable population structure, naturally replacing senesced individuals with seedlings and saplings (Condit et al., 1998; Lykke, 1998; Obiri et al., 2002; Teketay, 1997a; 2005a, 2005b; Tabuti, 2007; Mwavu & Witkowski, 2009a; Tesafye et al., 2010; Sop et al., 2011; Helm & Witkowski, 2012; El-Sheikh, 2013). This appeared to be the case for about 20 and 15% of the woody species recorded at Shorobe and Xobe, respectively. Typical examples are *A. erioloba*, *A. tortilis* and *C. mopane* at Shorobe and *C. albopunctatum* and *G. senegalensis* at Xobe. The species in the second group exhibited relatively good recruitment (of seedlings/coppices) but the regeneration is negatively affected as evidenced from the absence of individuals in progressively higher classes. This may be attributed to either natural- or human-induced hampered regeneration. Pole-sized and mature individuals may have been cut by the local people for various purposes. This group was comprised of about 10 and 19% of the woody species recorded at Shorobe and Xobe, respectively. The species in the other groups exhibited both naturally- and human-induced disturbances leading to their hampered regeneration. This ap-

Diversity, Population Structure and Regeneration Status of Woody Species in Dry Woodlands Adjacent to
Molapo Farms in Northern Botswana

173

peared to be the case for about 70% and 66% of the woody species recorded at Shorobe and Xobe, respectively. In particular, the species in the fifth group, represented by about 27% and 11% of the woody species recorded at Shorobe and Xobe, respectively, exhibited a very serious problem of regeneration with individuals only in the first diameter class.

Some of the causes for the hampered regeneration of the woody species include clearing of the woody vegetation for cultivating crops both by *molapo* (**Figure 2(A)**) and dry land (**Figure 3(B)**) farmers, heavy grazing pressure, cutting of stems and lopping of branches of woody species for fencing of farms (**Figures 2(A)** and **(B)**, **3(B)**), kraals (**Figure 3(A)**) and house compounds and fuel wood. A land use assessment carried out by the University of Botswana on the basis of satellite images found that of the 48,900 ha cleared for cultivation in Ngamiland, 75% consist of dryland fields and 25% of fields in temporarily inundated floodplains (VanderPost, 2009). VanderPost (2009) also indicated that *molapo* farming takes place on small fields separated by strips of "natural" floodplain, and wholesale land-clearing does not usually occur, although removal of some vegetation takes place.

The major impact of the *molapo* farming results from the heavy lopping of branches and cutting of trees of woody species used for fencing the farms. The heavy lopping of branches of woody species will negatively affect the production of fruits/seeds in enough qunatities required for the stable recruitment and regeneration of the species. However, the major impact was observed from the clearing and grazing (**Figure 3(B)**) of considerable areas of woody vegetation for the purpose of establishing and fencing rain-fed dry land farming (**Figure 3(B)**). In addition, the free grazing system in the study sites in particular and Botswana in general leads to the browsing and trampling of seedlings of woody species. At Shorobe, *H. petersiana* was categorized in the fifth group based on its population structure. The trees of this species are cut down at about 30 cm above the ground for tapping the stem sap, which is, then, processed by the local people to produce palm wine (Babitseng & Teketay, 2012). This traditional wine tapping method, which leads to the destruction of the stems of trees, is responsible for the obvious population structure (**Figure 5(e)**) and, hence, hampered regeneration.

The socio-economic survey carried out at the two study sites revealed that 15 woody species (**Tables 1** and **2**) are most preferred for fencing, especially *molapo* farms (Chimbari et al., unpublished). Of these, 10 and five species recorded at Shorobe and Xobe, respectively, belong to the last three groups of population structure patterns. Six of the species recorded at Shorobe belong to the fifth group of the population structure pattern. This suggests that cutting of woody species for fenicng is contributing to the observed hampered regeneration of the species. If it is not properly managed, clearing of woody vegetation, cutting and lopping trees excessively, coupled with annual recurrent fires common in the study sites and elsewhere in Botswana, will affect the population structure of woody species negatively and, hence, reduce or prevent their potential of regeneration and, therefore, perpetuation. This will, in turn, result in the decline or loss of biodiversity ultimately. Whelan (1995) argues that frequent fire outbreak can affect population structure through elimination of certain classes or a delay in the whole recruitment process. Though information on fire tolerances for woody species is not available, fire might have caused the poor representation of individuals of some of the woody species since farmers use fire to clear woody vegetation for grazing and for farming.

Conclusion

The results revealed that Shorobe and Xobe together house 46 different woody species, which provide the local communities with various goods and services. Despite the apparent impacts from humans, domestic animals and recurrent annual fires, some of the woody species exhibited desirable values of density, frequency, dominance, IVI, population structures and regeneration status. Shorobe Village exhibited higher values of species, genus and family richness as well as relatively high diversity and overall density of woody species than Xobe Village. This may be attributed to the difference in the edaphic factors, especially the big difference in the soil types and moisture availability, as well as the relatively higher human- and domestic animal-induced disturbances at Xobe than Shorobe. The similarities of woody species in terms of richness of species, genera and families were medium suggesting that each site has its own characteristic species, genera and families. This is particularly important to consider when planning activities aimed at the responsible management, sustainable utilization and conservation of the woody vegetation at the study sites.

At both sites, the frequency of woody species was relatively low with the exception of a few species, suggesting that individuals of the species are thinly or very thinly spread horizontally. This is also evident from the medium evennes values recorded from the two study sites. Some of the woody species, especially those categorized in population structure Groups 3-5, exhibited negatively affected popultion structures and, as a result, hampered regeneration, which requires special attention and appropriate management intervention by the concerned bodies.

The results revealed relatively low diversity and overall density at Xobe. Also, relatively low evenness indices, which were caused by the sparse to very sparse horizontal distribution of individuals, were recorded at both sites. Many of the woody species exhibited low values of densities, basal areas and IVI values, unstable population structure and hampered regeneration at both study sites. These results indicate the need for attention and appropriate management interventions by the relevant national authorities at various levels, including the Kgosi (Chiefs), Village Development Committees and local communities of the study sites.

The following are a few examples of feasible management interventions to address the problems described above.

1) Reduction of the pressure on regeneration of the woody species from the uncontrolled grazing through matching the carrying capacity of the study sites with appropriate numbers of livestock.

2) Introduction of management plans and appropriate technologies that regulate or promote the type, diameter and height classes, and number of individuals of the available woody species to be harvested for the various needs of the communities. To this effect, we are also undertaking an experiment focusing on the effect of cutting diameter and height on the re-sprouting/coppicing ability of selected woody species at the study sites (Neelo et al., unpublished) with the aim of determining the optimal cutting diameter size(s) and height(s) for maximum re-sprouting ability of the species. Once promising results are achieved, it is hoped that the optimal cutting diameter(s) and

height(s) will be demonstrated and disseminated to the local communities to promote desirable regeneration of the woody species with appropriate and sustainable utilization.

3) Establishment of rotational exclosures (Mengistu et al., 2005; Birhane et al., 2006; Aerts et al. 2009) of the areas covered with woody species at both sites, from both human and animal disturbances, so that the species will get enough time and appropriate environmental conditions to recover from the heavy grazing that affected their regeneration.

4) Research on the life-cycle and propagation (both sexual and asexual) methods of the woody species with hampered regeneration to promote their assisted regeneration, e.g. through enrichment planting.

5) Creation of awareness of the local communities on the status of population structures, regeneration and perpetuation of the woody species in their localities so that they can promote responsible management and utilization as well as conservation of the species.

Acknowledgements

Authors are grateful to the International Development Research Centre (IDRC) for providing financial support for this study through the Botswana Ecohealth Project. We are thankful to *Kgosi* (Chiefs), Village Development Committees and local communities of the study sites, field technicians and Transport Section of the Okavango Research Institute (ORI), University of Botswana, community research assistants and Peter Smith University of Botswana Herbarium (PSUB). We are grateful to Chenamani Ntogwa for allowing us to use his compass and caliper and Amanda Tas for her voluntary help to collect, organise and analyse our data. We would also like to thank ORI and its management for its logistical support.

REFERENCES

Aarrestad, P. A., Masunga, G. S., Hytteborn, H., Pitlagano, M. L., Marokane, W., & Skarpe, C. (2011). Influence of soil, tree cover and large herbivores on field layer vegetation along a savanna landscape gradient in northern Botswana. *Journal of Arid Environments, 75,* 290-297.

Aerts, R., Nyssen, J., & Haile, M. (2009). On the difference between "exclosures" and "enclosures" in ecology and the environment. *Journal of Arid Environments, 73,* 762-763.

Alelign, A., Teketay, D., Yemshaw, Y., & Edwards, S. (2007). Diversity and status of regeneration of woody plants on the peninsula of Zegie, northwestern Ethiopia. *Tropical Ecology, 48,* 37-49.

Babitseng, T. M., & Teketay, D. (2013). Impact of wine tapping on the population structure and regeneration of *Hyphaene petersiana* Klotzsch ex Mart. in northern Botswana. *Ethnobotany Research & Applications, 11,* 9-27.

Banda, T., Mwangulango, N., Meyer, B., Schwartz, M. K., Mbago, F., Sungula, M., & Caro, T. (2008). The woodland vegetation of Katavi-Rukwa ecosystem in western Tanzania. *Forest Ecology and Management, 255,* 3382-3395.

Barnes, M. E. (2001). Seed predation, germination and seedling establishment of *Acacia erioloba* in northern Botswana. *Journal of Arid Environments, 49,* 541-554.

Belbin, L. (1995). A multivariate approach to the selection of biological reserves. *Biodiversity and Conservation, 9,* 951-963.

Bendsen, H. (2002). *The dynamics of land use systems in Ngamiland: Changing livelihoods options and strategies.* Maun: University of Botswana Harry Openheimer Okavango Research Centre.

Bengtsson-Sjörs, K. (2006). *Establishment and survival of woody seedlingsin a semi-arid savanna in southern Botswana.* Minor Field Study 123. Uppsala: Uppsala University.

Ben-Shahar, R. (1993). Patterns of elephant damage to vegetation in northern Botswana. *Biological Conservation, 65,* 249-256.

Ben-Shahar, R. (1996a). Woodland dynamics under the influence of elephants and fire in northern Botswana. *Plant Ecology, 123,* 153-163.

Ben-Shahar, R. (1996b). Do elephants over-utilize mopane woodlands in northern Botswana? *Journal of Tropical Ecology, 12,* 505-515.

Ben-Shahar, R. (1998a). Changes in structure of the savannah woodlands in northern Botswana following the impacts of elephants and fire. *Plant Ecology, 136,* 189-194.

Ben-Shahar, R. (1998b). Elephant density and impact on Kalahari woodland habitats. *Transactions of the Royal Society of South Africa, 53,* 149-155.

Ben-Shahar, R., & Macdonald, D. (2002). The role of soil factors and leaf protein in the utilization of mopane plants by elephants in northern Botswana. *BMC Ecology, 2,* 3.

Birhane, E., Teketay, D., &Barklund, P. (2006). Actual and potential contribution of exclosures to enhance biodiversity of woody species in the drylands of Eastern Tigray. *Journal of Drylands, 1,* 134-147.

Bognounou, F., Tigabu, M., Savadogo, P., Thiombiano, A., Boussim, I. J., Óden, P.-C., & Guinko, S. (2010). Regeneration of five combretaceae species along a latitudinal gradient in Sahelo-Sudanian zone of Burkina Faso. *Annals of Forest Sciences, 67,* 306.

Bonyongo, C. M., Veenendaal, E. M., & Bredenkamp, G. (2000). Floodplain vegetation of seasonal floodplains in the Nxaraga Lagoon Area Okavango Delta Botswana. *South African Journal of Botany, 66,* 15-21.

Chimbari, M. J., Magole, L., Wiles, G., Dikgola, K., Kurugundla, N., Teketay, D., Ngwenya, B., Nyepi, M. S., Motsumi, S., Ama, K., Thakadu, O., & Chombo, O. (2009). *Application of the ecohealth approach to understand flood-recession (Molapo) farming in the context of hydro-climate variability and hydro-climate change in the Okavango Delta, Botswana.* Maun: University of Botswana, Harry Oppenheimer Okavango Research Centre.

Condit, R., Sukumar, R., Hubbell, S. P., & Foster, R. B. (1998). Predicting population trends from size distributions: A direct test in a tropical tree community. *American Naturalist, 152,* 495-509.

CSO (Central Statistics Office) (2011). *Botswana population and housing census.* Gaborone, Botswana.

DEA (2008). *Okavango delta management plan.* Gaborone: Department of Environmental Affairs (DEA).

Dovie, D. B. K., Witkowski, E. T. F., & Shackleton, C. M. (2008). Knowledge of plant resource use based on location, gender and generation. *Applied Geography, 28,* 311-322.

Ellery, K., & Ellery, W. (1997). *Plants of the Okavango Delta: A field guide.* Durban: Tsaro Publisher.

El-Sheikh, M. A. (2013). Population structure of woody plants in the arid cloud forests of Dhofar, southern Oman. *Acta Botanica Croatica, 72,* 97-111.

Eriksson, I., Teketay, D., & Granstrom, Λ. (2003). Response of plant communities to fire in an *Acacia* woodland and a dry Afromontane forest, southern Ethiopia. *Forest Ecology and Management, 177,* 39-50.

Faith, D. P., & Walker, P. A. (1996). Environmental diversity: On the best possible use of surrogate data for assessing the relative biodiversity of sets of areas. *Biodiversity and Conservation, 5,* 399-415.

FAO (2010). *Global forest resources assessment.* Rome: FAO.

Feeley, K. J., Davies, S. J., Noor, N. S., Kassim, A. R., & Tan, S. (2007). Do current stem size distributions predict future population changes? An empirical test of intraspecific patterns in tropical trees at two spatial scales. *Journal of Tropical Ecology, 23,* 191-198.

Fisaha, G., Hundera, K., & Dalle, G. (2013). Woody plants' diversity, structural analysis and regeneration status of Wof Washa natural forest, North-east Ethiopia. *African Journal of Ecology* (in press).

Gomez-Pompa, A., Whitmore, T. C., & Hadley, M. (1991). Paris: Rain Forest Regeneration and Management.

Harper, J. L. (1977). *Population biology of plants.* Lomdon: Academic Press.

Heinl, M., Sliva, J., & Tacheba, B. (2004). Vegetation changes after single fire-events in the Okavango Delta wetland, Botswana. *South African Journal of Botany, 70,* 695-704.

Heinl, M., Sliva, J., Murray-Hudson, M., & Tacheba, B. (2007). Post-fire succession on savanna habitats in the Okavango Delta wetland, Botswana. *Journal of Tropical Ecology, 23,* 705-713.

Heinl, M., Sliva, J., Tacheba, B., & Murray-Hudson, M. (2008). The relevance of fire frequency for the floodplain vegetation of the Okavango Delta, Botswana. *African Journal of Ecology, 46,* 350-358.

Helm, C. V., & Witkowski, E. T. F. (2012). Characterising wide spatial variation in population size structure of a keystone African savanna tree. *Forest Ecology and Management, 263,* 175-188.

Herrera, A. (2011). *Changes in spatial structure of woody savanna vegetation after 11 years of exclusion of large herbivores.* Minor Field Studies 162. Uppsala: Uppsala University.

Janzen, D. H. (1988). Tropical dry forests: The most endangered major tropical ecosystem. In E. O. Wilson (Ed.), *Biodiversity* (pp. 130-137). Washington: National Academic Press.

Kalema, V. N. (2010). *Diversity, use and resilience of woody plants in a multiple land-use equatorial African savanna, Uganda.* Ph.D. Thesis, Johannesbur: University of the Witwatersrand.

Käller, A. (2003). *Growth pattern and reproduction of woody vegetation in a semi-arid savanna in southern Botswana.* Minor Field Study 86. Uppsala: Uppsala University.

Kalwij, J. M., de Boer, W. F., Mucina, L., Prins, H. H. T, Skarpe, C., & Winterbach, C. (2010). Tree cover and biomass increase in a southern African savanna despite growing elephant population. *Ecological Applications, 20,* 222-233.

Kent, M., & Coker, P. (1992). *Vegetation description and analysis. A practical approach.* London: Belhaven Press.

Krebs, C. J. (1989). *Ecological methodology.* New York: Harper Collins Publishers.

Lamprecht, H. (1989). *Silviculture in the tropics: Tropical forest ecosystems and their tree species—Possibilities and methods for their long-term utilization.* Eschborn: Deutsche Gesellschaft für Technische Zusammenarbeit (GTZ) GmbH.

Leife, H. (2010). *Has woody vegetation in a semi-arid savanna changed after 11 years exclusion of large herbivores?* Minor Field Study 153. Uppsala: Uppsala University.

Liu, Q., & Brakenhielm, S. (1996). Variability of plant species diversity in Swedish natural forest and its relation to atmospheric deposition. *Academic Publishers, 125,* 63-72.

Louga, E. J., Witkowski, E. T. F., & Balkwill, K. (2004). Regeneration by coppicing (resprouting) of miombo (African savanna) trees in relation to land use. *Forest Ecology and Management, 189,* 23-35.

Luoga, E. J., Witkowski, E. T. F., & Balkwill, K. (2000). Differential utilisation and ethnobotany of trees in Kitulangalo forest reserve and surrounding communal lands, eastern Tanzania. *Economics Botany, 54,* 328-343.

Lykke, A. M. (1998). Assessment of species composition change in savanna vegetation by means of woody plants' size class distributions and local information. *Biodiversity and Conservation, 7,* 1261-1275.

Magurran, A. E. (2004). *Measuring biological diversity.* Malden and Oxford: Blackwell Publishing.

Makhabu, S. W. (2005a). *Interactions between woody plants, elephants and other browsers in the Chobe Riverfront, Botswana.* Doctoral Thesis, Oslo: Norwegian University of Science and Technology.

Makhabu, S. W. (2005b). Resource partitioning within a browsing guild in a key habitat, the Chobe Riverfront, Botswana. *African Journal of Ecology, 21,* 641-649.

Makhabu, S. W., Skarpe, C., Hytteborn, H., & Mpofu, Z. D. (2006). The plant vigour hypothesis revisited—How is browsing by ungulates and elephant related to woody species growth rate? *Plant Ecology, 184,* 163-172.

Margalef, R. (1972). Homage to Evelyn Hutchinson, or why is there an upper limit to diversity. *Transactions of the Connecticut Academy of Arts and Sciences, 44,* 211-235.

McCarthy, T. S., Bloem, A., & Larkin, P. A. (1998). Observations on the hydrology and geohydrology of the Okavango Delta, Botswana. *Journal of Geology, 101,* 101-117.

McLaren, K. P., McDonald, M. A., Hall, J. B., & Healey, J. R. (2005). Predicting species response to disturbance from size class distributions of adults and saplings in a Jamaican tropical dry forest. *Plant Ecology, 181,* 69-84.

Mengistu, T., Teketay, A., Hulten, H., & Yemshaw, Y. (2005). The role of enclosures in the recovery of woody vegetation in degraded dryland hillsides of central and northern Ethiopia. *Journal of Arid Environments, 60,* 259-281.

Mmolotsi, R. M., Obopile, M., Kwerepe, B. C., Sebolai, B., Rampart, M. P., Segwagwe, A. T., Ramolemana, G., Maphane, T. M., Lekorwe, L., & Kopong, I. (2012). Studies on Mukwa (*Pterocarpus angolensis* D. C.) dieback in Chobe Forest Reserves in Botswana. *Journal of Plant Studies, 1,* 154-157.

Moleele, N., Ringrose, S., Arnberg, W., Lunden, B., & Vanderpost, C. (2001). Assessment of vegetation indexes useful for browse (forage) prediction in semi-arid rangelands. *International Journal of Remote Sensing, 22,* 741-756.

Moleele, N., Ringrose, S., Matheson, W., & Vander Post, C. (2002). More woody plants? The status of bush encroachment in Botswana's grazing areas. *Journal of Environmental Management, 64,* 3-11.

Mosugelo, D. K., Stein, M., Ringrose, S., & Nellemann, C. (2002). Vegetation changes during a 36 year period in Northern Chobe National Park, Botswana. *African Journal of Ecology, 40,* 232-240.

Motsumi, S., Magole, L., & Kgathi, D. (2012). Indigenous knowledge and land use policy: Implications for livelihoods of flood recession farming communities in the Okavango Delta, Botswana. *Physics and Chemistry of the Earth, 50-52,* 185-195.

Mueller-Dombois, D., & Ellenberg, H. (1974). *Aims and methods of vegetation ecology.* New York: John Willey and Sons, Inc.

Murphy, P. G., & Lugo, A. E. (1986). Ecology of tropical dry forest. *Annual Review of Ecology, 17,* 67-88.

Mwavu, N. E., & Witkowski, E. T. F. (2008). Sprouting of woody species following cutting and tree-fall in a lowland semi-deciduous tropical rainforest, North-Western Uganda. *Forest Ecology and Management, 255,* 982-992.

Mwavu, N. E., & Witkowski, E. T. F. (2009a). Population structure and regeneration of multiple-use tree species in a semi-deciduous African tropical rainforest: Implications for primate conservation. *Forest Ecology and Management, 258,* 840-849.

Mwavu, N. E., & Witkowski, E. T. F. (2009b). Seedling regeneration, environment and management in a semi-deciduous African tropical rain forest. *Journal of Vegetation Science, 20,* 791-804.

Nangendo, G., Steege, H. T., & Bongers, F. (2006). Composition of woody species in a dynamic forest-woodland-savannah mosaic in Uganda: Implications for conservation and management. *Biodiversity and Conservation, 15,* 1467-1495.

Nduwamungu, J. (1997). *Tree and shrub diversity in Miombo Wood-

lands. A case study at SUA Kitulanghalo Forest Reserve, Morogoro, Tanzania. Master's Thesis, Morogoro: Sokoine University of Agriculture.

Neudeck, L., Avelino, L., Bareetseng, P., Ngwenya, B. N., Teketay, D., & Motsholapheko, M. R. (2012). The contribution of edible wild plants to food security, dietary diversity and income of households in Shorobe Village, northern Botswana. *Ethnobotanical Research & Applications, 10,* 449-462.

Norwegian Forestry Society (1992). *Chobe forest inventory and management plan.* Gaborone: Ministry of Agriculture.

Obiri, J., Lawes, M., & Mukolwe, M. (2002). The dynamics and sustainable use of highvalue tree species of the coastal Pondoland forests of the Eastern Cape Province, South Africa. *Forest Ecology and Management, 166,* 131-148.

Oosterbaan, R. J., Kortenhorst, L. F., & Sprey, L. H. (1986). *Development of flood-recession cropping in the molapo's of the Okavango Delta, Botswana.* Wageningen: International Institute for Land Reclamation and Improvement.

Pare, S., Savadogo, P., Tigabu, M., Odén, P.-C., & Ouadba, J. M. (2009). Regeneration and spatial distribution of seedling populations in Sudanian dry forests in relation to conservation status and human pressure. *Tropical Ecology, 50,* 339-353.

Peters, C. M. (1996). *The ecology and management of non-timber forest resources.* Washington: World Bank Technical Paper 322.

Poorter, L., Bongers, F., van Rompaey, A. R., & Klerk, M. D. (1996). Regeneration of canopy tree species at five sites in West African moist forest. *Forest Ecology and Management, 84,* 61-69.

Rampart, M. (2007). *Effects of fire and elephants on the growth of Pterocarpus angolensis (Mukwa) seedlings/saplings in the Chobe Forest Reserves (Botswana).* Master's Thesis, Bangor: University of Wales.

Rao, P., Barik, S. K., Pandey, H. N., & Tripathi, R. S. (1990). Community composition and tree population structure in a sub-tropical broad-leaved forest along a disturbance gradient. *Vegetatio, 88,* 151-162.

Ringrose, S. (2003). Characterisation of riparian woodlands and their potential water loss in the distal Okavango Delta, Botswana. *Applied Geography, 23,* 281-302.

Ringrose, S., & Matheson, W. (2001). Spatial characteristics of Riparian Woodlands in the Distal Okavango Delta. *Botswana Notes and Records, 33,* 101-114.

Ringrose, S., Chipanshi, A. C., Matheson, W., Chanda, R., Motoma, L., & Magole, I. (2002). Climate and human induced woody vegetation changes in Botswana and their implications for human adaptation. *Environmental Management, 30,* 98-109.

Ringrose, S., Lesolle, D., Botshoma, T., Gopolang, B., VanderPost, C., & Matheson, W. (1999). An analysis of vegetation cover components in relation to climatic trends along the Botswana Kalahari Transect. *Botswana Notes and Records, 31,* 33-52.

Ringrose, S., Matheson, W., & Vander Post, C. (1998). Analysis of soil organic carbon and vegetation cover trends along the Botswana Kalahari Transect. *Journal of Arid Environments, 38,* 379-396.

Robinson, J. A., Lulla, K. P., Robinson, J. A., Lulla, K. P., Kashiwagi, M., Suzuki, M., Nellis, M. D., Charles, Bussing, E., Long, W. J. L., & McKenzie, L. J. (2002). Conservation applications of astronaut photographs of earth: Tidal-flat loss (Japan), elephant effects on vegetation (Botswana), and seagrass and mangrove monitoring (Australia). *Conservation Biology, 15,* 876-884.

Rutina L. P. (2004). *Impalas in an elephant-impacted woodland: Browser-driven dynamics of the Chobe riparian zone, northern Botswana.* Ph.D. Thesis, Ås: Agricultural University of Norway.

Rutina, L. P., Moe, S. R., & Swenson, J. E. (2005). Elephant *Loxodonta africana* driven woodland conversion to shrubland improves dry-season browse availability for impalas *Aepyceros melampus. Wildlife*

Biology, 11, 207-213.

Sabiiti, E. N., & Wein, R. W. (1987). Fire and acacia seeds: A hypothesis of colonization success. *Journal of Ecology, 74,* 937-946.

Sano, J. (1997). Age and size distribution in a long-term forest dynamics. *Forest Ecology Management, 92,* 39-44.

Savadogo, P. (2007). *Dynamics of Sudanian savanna-woodland ecosystems in response to disturbances.* Doctoral Thesis, Umeå: Swedish University of Agricultural Sciences.

Savadogo, P., Tigabu, M., Sawadogo, L., & Odén, P.-C. (2007). Woody species composition, structure and diversity of vegetation patches of a Sudanian savanna in Burkina Faso. *Bois et Forêts des Tropiques, 294,* 7-20.

Schultka, W., & Cornelius, R. (1997). Vegetation structure of a heavily grazed range in northern Kenya: Tree and shrub canopy. *Journal of Arid Environments, 36,* 291-306.

Sekhwela, M. B. M. (2003). Woody vegetation resource changes around selected settlements along aridity gradient in the Kalahari, Botswana. *Journal of Arid Environment, 54,* 469-482.

Sekhwela, M. B. M., Yates, D., & Lamb, D. (2000). Woody vegetation structure and wood availability in arid and sem-arid Kalahari sand system in Botswana. In S. Ringrose and C. Raban (Eds.), *Towards Sustainable Natural Resource Management in the Kalahari Transect* (pp. 65-82). Gaborone: University of Botswana.

Senbeta, F., & Teketay, D. (2003). Diversity, community types and population structure of woody plants in Kimphee forest, a unique nature reserve in southern Ethiopia. *Ethiopian Journal of Biological Sciences, 2,* 169-187.

Setshogo, M. F. (2005). *Preliminary checklist of the plants of Botswana.* Pretoria and Gaborone: Southern African Botanical Diversity Network Report No. 37.

Setshogo, M. P., & Venter, F. (2003). *Trees of Botswana: Names and Distribution.* Pretoria: Southern African Botanical Diversity Network Report No. 18.

Shackleton, C. M. (1993). Demography and dynamics of the dominant woody species in a communal and protected area of the eastern Transvaal Lowveld. *South African Journal of Botany, 59,* 569-574.

Skarpe, C. (1990a). Structure of the woody vegetation in disturbed and undisturbed arid savanna. Botswana. *Vegetatio, 87,* 11-18.

Skarpe, C. (1990b). Shrub layer dynamics under different herbivore densities in an arid savanna, Botswana. *Journal of Applied Ecology, 27,* 873-885.

Skarpe, C. (1992). Dynamics of savanna ecosystems. *Journal of Vegetation Science, 3,* 293-300.

Sokpon, N., & Biaou, S. H. (2002). The use of diameter distributions in sustained-use management of remnant forests in Benin: Case of Bassila Forest Reserve in North Benin. *Forest Ecology and Management, 161,* 13-25.

Sop, T. K., Oldeland, J., Schmiedel, U., Ouedraogo, I., & Thiombiano. A. (2011). Population structure of three woody species in four ethnic domains of the sub-sahel of Burkina Faso. *Land Degradation & Development, 22,* 519-529.

Swaine, M. D., Lieberman, D., & Hall, J. B. (1990).Structure and dynamics of a tropical dry forest in Ghana. *Vegetatio, 88,* 31-51.

Tabuti, J. R. S. (2007). The uses local perceptions and ecological status of 16 woody species of Gadumire Sub-county Uganda. *Biodiversity and Conservation, 16,* 1901-1916.

Tacheba, B., Segosebe, E., Vanderpost, C., & Sebego, R. (2009). Assessing the impacts of fire on the vegetation resources that are available to the local communities of the seasonal wetlands of the Okavango, Botswana, in the context of different land uses and key government policies. *African Journal of Ecology, 47,* 71-77.

Diversity, Population Structure and Regeneration Status of Woody Species in Dry Woodlands Adjacent to
Molapo Farms in Northern Botswana

177

Teketay, D. (1996b). Germination ecology of twelve indigenous and eight exotic multipurpose leguminous species from Ethiopia. *Forest Ecology and Management, 80*, 209-223.

Teketay, D. (1997a). Seedling populations and regeneration of woody species in dry Afromontane forests of Ethiopia. *Forest Ecology and Management, 98*, 149-165.

Teketay, D. (1997b). Germination ecology of *Acacia negrii*, an endemic multipurpose tree from Ethiopia. *Tropical Ecology, 38*, 39-46.

Teketay, D. (1998). Soil seed bank at an abandoned Afromontane arable site. *Feddes Repertorium, 109*, 161-174.

Teketay, D. (2004-2005). Causes and consequences of dryland forest degradation in Sub-Saharan Africa. *Walia, 24*, 3-20.

Teketay, D. (2005a). Seed and regeneration ecology in dry Afromontane forests of Ethiopia: I. Seed production—Population structures. *Tropical Ecology, 46*, 29-44.

Teketay, D. (2005b). Seed and regeneration ecology in dry Afromontane forests of Ethiopia: II. Forest disturbance and succession. *Tropical Ecology, 46*, 45-64.

Teketay, D. (2011). Natural regeneration and management of *Podocarpus falcatus* (Thunb.) Mirb. in the Afromontane forests of Ethiopia. In: S. Günter, M. Weber, B. Stimm and R. Mosandi (Eds.), *Silviculture in the tropics* (pp. 325-336). London and New York: Springer-Verlag Berlin Heidelberg.

Teketay, D., & Granström, A. (1995). Soil seed banks in dry Afromontane forests of Ethiopia. *Journal of Vegetation Science, 6*, 777-786.

Teketay, D., & Granström, A. (1997). Seed viability of Afromontane tree species in forest soils. *Journal of Tropical Ecology, 13*, 81-95.

Teketay. D. (1996a). *Seed ecology and regeneration in dry Afromontane Forests of Ethiopia*. Doctoral Thesis, Umeå: Swedish University of Agricultural Sciences.

Tesfaye, G., Teketay, D. & Fetene, M. (2002). Regeneration of fourteen tree species in Harenna forest, southeastern Ethiopia. *Flora, 197*, 461-474.

Tesfaye, G., Teketay, D., Fetene, M., & Beck, E. (2010). Regeneration of seven indigenous tree species in a dry Afromontane forest, southern Ethiopia. *Flora, 205*, 135-143.

Vanderpost, C. (2009). *Molapo farming in the Okavango Delta. fact sheet 7/2009*. Maun: Harry Oppenheimer Okavango Research Centre, University of Botswana.

Venter, S. M., & Witkowski, E. T. F. (2010). Baobab (*Adansonia digitata* L.) density, size-class distribution and population trends between four land-use types in northern Venda, South Africa. *Forest Ecology and Management, 259*, 294-300.

West, A. G., Midgley, J. J., & Bond, W. J. (2000). Regeneration failure and potential importance of human disturbance in a subtropical forest. *Applied Vegetation Science, 3*, 223-232.

Whelan, R. J. (2001). *The ecology of fire*. Cambridge: Cambridge University Press.

Wilson, B. G., & Witkowski, E. T. F. (2003). Seed banks, bark thickness and change in age and size structure (1997-1999) of the African savanna tree, *Burkea Africana. Plant Ecology, 167*, 151-162.

Witkowski E. T. F., & O'Connor, T. G. (1996). Topo-edaphic, floristic and physiognomic gradients of woody plants in a semi-arid African savanna woodland. *Vegetatio, 124*, 9-23.

Witkowski, E. T. F., & Garner, R. D. (2000). Spatial distribution of soil seed banks of three African savanna woody species at contrasting sites. *Plant Ecology, 149*, 91-106.

Woldemariam, T., Teketay, D., Edwards S., & Olsson, M. (2000). Woody plant and avian species diversity in a dry Afromontane forest on the central plateau of Ethiopia: Biological indicators for conservation. *Ethiopian Journal of Natural Resources, 2*, 255-293.

Worku, A., Teketay, D., Lemenih, M., & Fetene, M. (2012). Diversity, regeneration status and population structure of gum and resin producing woody species in Boarana, Southern Ethiopia. *Forests, Trees and Livelihoods*, 1-12.

Zahabu, E. (2001). *Impact of charcoal extraction to the miombo woodlands: The case of Kitulangalo area, Tanzania*. Master's Thesis, Morogoro: Sokoine University of Agriculture.

Zar, J. H. (1999). *Biostastical analysis* (4th ed.). Upper Saddle River, NJ: Prentice Hall.

Zegeye, H., Teketay, D., & Kelbessa, E. (2006). Diversity, regeneration status and socio-economic importance of the vegetation in the islands of Lake Ziway, south-central Ethiopia. *Flora, 201*, 483-498.

Zegeye, H., Teketay, D., & Kelbessa, E. (2011). Diversity and regeneration status of woody species in Tara Gedam and Abebaye forests, northwestern Ethiopia. *Journal of Forestry Research, 22*, 315-328.

Zida, D., Savadogo, L., Tigabu, M., Tiveau, D., & Oden, P. C. (2007). Dynamics of sapling population in savanna woodlands of Burkina Faso subjected to grazing, early fire and selective tree cutting for a decade. *Forest Ecology and Management, 243*, 102-115.

27

Survival, Growth and *Orygmophora mediofoveata* Shoot Borer Attack of *Nauclea diderrichii* Progenies Established in Three Ecological Zones in Ghana

Paul P. Bosu[1*], Stephen Adu-Bredu[1], Yaovi Nuto[2], Kouami Kokou[3]

[1]Forestry Research Institute of Ghana, Council for Scientific and Industrial Research, Kumasi, Ghana

[2]Département de Zoologie, Faculté des Sciences, Université de Lomé, Lomé, Togo

[3]Laboratoire de Botanique et Ecologie Végétale, Faculté des Sciences, Université de Lomé, Lomé, Togo

Nauclea diderrichii is a tropical African hardwood species and a suitable candidate for plantation development. However, attack by the *Orygmophora mediofoveata*, Hamps shoot borer threatens establishment of the species in plantations. A genotype * environment assessment of 15 *N. diderrichii* progenies from Ghana and Togo was conducted in the Wet Evergreen, Moist Semi-deciduous and Dry Semi-deciduous forest zones. Progeny performance (Attack intensity, survival and growth) varied significantly between sites, and marginally within sites after 2.7 years. Overall, incidence of shoot borer attack was lower at the wet zone than at the moist or dry zones. Percent survival was higher at the wet (79.5%) than at the moist (50.8%) or dry (55.0%) forest zones. Mean height across the 15 progenies was 5.40 m, 4.30 m, and 2.73 m at the wet, dry and moist forests, respectively. Similarly, mean diameter was 5.31 cm, 4.58 cm, and 2.83 cm at the wet, dry and moist zones, respectively. The relatively low growth rate recorded at the moist zone was attributed to the paucity of soil conditions at the experimental site. Three wet forest zone progenies (BS9, BS3 and BS2) and two moist forest zone progenies (BE2 and GA1) performed better than average and have been recommended for planting.

Keywords: Genotype * Environment Assessment; *Nauclea diderrichii*; *Orygmophora mediofoveata*; Growth Rates; Survival Rate; Ghana; Togo

Introduction

Nauclea diderrichii (*Sarcocephalus diderrichii* De Wild) is a tropical African hardwood species belonging to the family Rubiaceae. The species is widely distributed across tropical Africa, from Liberia eastward through the Congo Basin to Uganda and Angola. It is a moderately fast-growing species, with fairly high density timber and durable wood. The tree grows up to about 60 m in height, with straight, cylindrical bole clear to 30 - 40 m, and trunk diameter ranging from 1.0 - 2.5 m. Its natural habitat is subtropical or tropical moist lowland forests. As a sun-loving species, the plant regenerates abundantly in gaps and openings and is often almost gregarious in the transition zone between freshwater swamps and lowland forests (Hawthorne, 1995). Young trees are often found in secondary bushy growth in humid areas.

The straightmonopodial growth habit of this species has promoted interest in it for the production of transmission poles, veneers and timber for heavy construction, flooring and furniture. In recent years, *N. diderrichii* has been planted in high rainfall areas (2000 - 4500 mm) throughout West Africa, but especially in south-east Nigeria. The biophysical limits for the growth of *Nauclea diderrichii* are; altitude 0 - 500 m; mean annual rainfall 1600 - 3000 mm; and mean annual temperature 24°C - 30°C. It does not grow well on excessively wet soils or on lateritic ones that dry out completely in the dry season.

In Ghana, the species is found in both the deciduous and evergreen forest zones. It is found at constant low densities and is never very abundant (Hawthorne, 1995). *Nauclea* is considered vulnerable (1994 IUCN threat category) due to its excessive exploitation (Hawthorne, 1995). It has been awarded a scarlet star for Ghana which means that it is common but under profound pressure from heavy exploitation (Hawthorne, 1995). In year 2000, an opportunity was created to increase the planting of *N. diderrichii* in Ghana when the species was selected as one of the five priority indigenous species for the national forest plantation development project. Selection of the five priority species was based on several factors including fast growth rate and low susceptibility to pest attacks. However, the vulnerability of *Nauclea* to opepe shoot borer *Orygmophora mediofoveata* (Lepidoptera: Noctuidae) was grossly underestimated. Not long after the project was started, considerable *O. mediofoveata* damage was recorded in several nurseries in the Eastern and

*Corresponding author.

Ashanti Regions, including two owned by the Forestry Re-search Institute (FORIG) at Mesewam and Fumesua near Kumasi (Bosu, et al., 2004).

Orygmophora mediofoveata attacks in *N. diderrichii* plantations were observed in Nigeria as far back as the 1930s. However, the identity of the borer was not discovered until 1962 when Eidt (1965a) reared the moth for the very first time. The damage caused by the shoot borer is principally through stunting of the trees in nursery transplant beds, and is rare in seed beds. Attack results in the rapid formation of a callus tissue over the injured parts, and may lead to mortality in heavy multiple attacks. Parry (1956) noted that "attack by *Orygmophora mediofoveata* was not severe enough to discourage use of the tree in pure plantations in Nigeria, but in Ghana, Opepe is so badly damaged that it is an unreasonable risk as a plantation crop".

The life history of *Orygmophora mediofoveata* is generally unknown, particularly in the egg stage and the first instar larvae (Eidt, 1965b). The larvae which attack and damage the plants are grub-like and morphologically quite unusual. The larva is short and stout; the head is partly withdrawn into the prothorax and the legs and prolegs are well developed. Fully grown larvae of the ultimate instar are about 14mm long, and average head width is about 1.5 mm. Early instar larvae are translucent and appear greenish because of the plant tissue in the gut. Ultimate instar larvae are deep red on the dorsum but remain green on the venter. They infest the terminal shoots, boring in the last two or three internodes, and preferring the more apical shoot. They do not girdle the shoots, but bore in the pith and produce galleries several inches long. In the cause of tunneling, the larvae eject dark brown frass which accumulates in the leaf axils. This tunneling can reveal their presence. Pupation occurs within the galleries although there is no cocoon. The pupal period lasts about 3 weeks. However, the length of a generation is unknown and has been estimated to be about three or four months.

To minimize the impact of the shoot borer in *N. diderrichii* plantations, mixed-species planting trials were conducted, which showed some promise (Addo-Danso, et al., 2012). However, planting genetic strains inherently resistant to *O. mediofoveata* could be a more effective way of minimizing the impact of this endemic pest on its *N. dedirrichii* host. In this study, we examined the susceptibility of fifteen *N. diderrichii* progenies from Ghana and Togo to *O. mediofoveata* shoot borer attack and its impact on the survival and growth of the plant in three forest zones.

Materials and Methods

Seed Collection

Seeds were collected from trees in the *N. diderrichii* distribution range in Ghana and Togo. In Ghana, the seeds were obtained from three mother trees in the Wet Evergreen Forest zone (WEF), eight trees in the Moist Semi-deciduous Forest zone (MSF) and one tree from the Dry Semi-deciduous Forest zone (DSF). Seeds from Togo, were collected from threetrees in the Plain of Litimé (**Table 1**).

Seeds of all 15 progenies were sent to Mesewamnursery where they were processed for germination and seedling production. As *N. diderrichii* seeds are very small and photoblastic, seed germination was done in plastic bowls (45 cm diameter and 12 cm deep) placed in a shade house of about 50% radiation and watered regularly until germination occurred. Seeds took an average of 6 - 8 weeks to germinate, and were then transferred to plastic bags until used.

Establishment of G * E Plots

Standard genotype by environment (G * E) experiments involving 15 *Nauclea diderrichii* progenies were established in three ecological zones namely, WEF, MSF and DSF zones. The WEF plot was established inside the Nueng Forest Reserve near Benso in the Western Region (5°16'N, 1°89'W). The MSF plot was established at the FORIG nursery area at Mesewam (6°44'N, 1°30'W), near Kumasi in the Ashanti Region. The DSF plot was established inside the Afram Headwaters Forest Reserve near Abofour, also in the Ashanti Region (7°10'N, 1°40'W). On each site, the plots were established using four blocks each consisting of 15 progenies, with each progeny replicated 10 times per block, using a Randomized Complete Block Design (RCBD). All three trials were established in April 2008. The plots were maintained by regular weeding, and monitored for a period of three months during which beating-up was done with seedlings from the original stock.

Assessment of Plots

Assessments of the plots for incidence of the shoot borer attack, seedling survival, and growth rates (height and diameter) were carried out between 4 - 6 months intervals, from August 2008 to December 2010. Except for the December 2010 (age 2.7 years) assessment at which diameter was measured at breast height (dbh = 1.3 m), all previous measurements were carried

Table 1.
Characteristics of ecological zones and number of accessions collected.

Forest zone	Rainfall (mm)	Country	Population	Population code	No. of progenies
Moist/Wet evergreen	1750 - 2000	Ghana	Benso	BS	3
Moist semi-deciduous, southeast	1500 - 1800	Ghana	Amantia	AM	2
Moist semi-deciduous, southeast	1200 - 1800	Ghana	Begoro	BE	2
Moist semi-deciduous, northwest	1200 - 1800	Ghana	Gambia village	GA	4
Moist semi-deciduous, plateau region	1200 - 1800	Togo	Badou	TG	3
Dry semi deciduous	1250 - 1500	Ghana	Berekum	BR	1

out at 10 cm above the soil level. Shoot borer attack intensity was ranked on a scale of 1-5; with 1 indicating no visible damageand 5 indicating severely damaged.

Data Analyses

Two factorial analyses of variance were used to estimate average differences in the variables (height, diameter and attack) using progeny and site as the main factors. Percentage data (survival) were arcsine transformed to conform to normality prior to analysis of variance.

Family heritability was estimated based on variance components obtained from analysis of variance as described by Zobel and Talbert (1991). The least square mean values were transformed to percentage deviation from the trial mean and were further multiplied by the heritability to provide the predicted family value also known as genetic gain (G) as: $G = h^2S$, where S is selection differential or deviation from trial mean (Ofori et al., 2007).

Results

Attack

Orygmophora shoot borer attack was recorded at all the three sites four months after the establishment of the plots. However, incidence of attacks were generally low and not statistically significant among progenies or across sites. After 12 months, attack rates were largely unchanged at the moist and dry forest sites, whiles only marginal increase in attack was recorded at the wet site. After 20 months (1.7 years), differences in the incidence of attack began to show among progenies, sites, and between the interaction of progeny and site (**Table 2**). Across progenies, attack was generally lower at the wet site than at either the dry or moist sites (**Figure 1**). Progenies with low in-

Table 2.
Results of two-way analyses of variance of *Orygmophora mediofoveata* shoot borer attack, percent survival, height and diameter of 15 *Nauclea diderrichii* progenies established in three forest zones in Ghana.

	Sources	*F*	*P*
Attack[*]	Site	102.37	<.0001
	Progeny	1.53	0.0945
	Site[*]progeny	1.95	0.0022
%survival	Site	12.240	<.0001
	Progeny	2.05	0.0183
	Site[*]progeny	1.12	0.3283
Height	Site	335.22	<.0001
	Progeny	1.87	0.0254
	Site[*]progeny	0.75	0.8217
Diameter	Site	153.05	<.0001
	Progeny	2.46	0.002
	Site[*]progeny	1.14	0.2835

[*]Assessment carried out at 1.7 years.

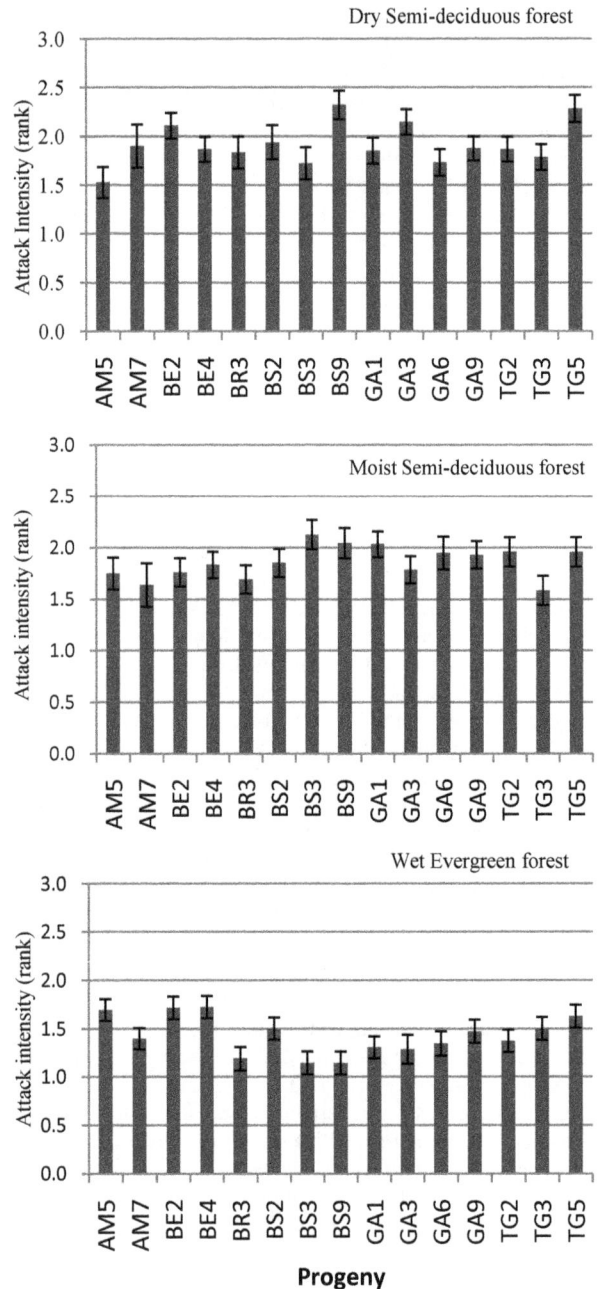

Figure 1.
Orygmophora mediofoveata shoot borer attack of 15 *Nauclea diderrichii* progenies 1.7 years after planting in three forest zones in Ghana.

cidence of attack at the wet site were BR3, BS3 and BS9, whiles TG5, AM5, BE2 and BE4 sustained significantly higher incidence of attack. At the dry forest site, AM5, BS3, GA6 and TG3 experienced lower incidence of attack whiles TG5, GA3 and BS9 recorded higher rates of attack. However, incidence of attack did not vary significantly among progenies at the moist site.

Survival

The survival rates of seedlings were generally high for all 15 progenies and at all three sites four months after planting ($P > 0.05$). Overall survival across progenies was 96.3% at the wet-site, 90.3% at the moist site and 88.8% at the dry site. However, differences in survival rates were observed among sites and among progenies at plantation age 2.7 years (**Table 2**). Mean percent survival was significantly higher at the wetsite (79.5%) compared to the moist (50.8%) or dry (55.0%) forest sites. Average overall survival across all three sites was 61.7%. At the wet site, all except AM7 recorded higher than average percent survival. At the dry site, five progenies (BE2, BE4, GA1, GA9 and TG2) were higher than average. However, only two progenies (BE2 and GA3) recorded higher than average percent survival at the moist site (**Figure 2**).

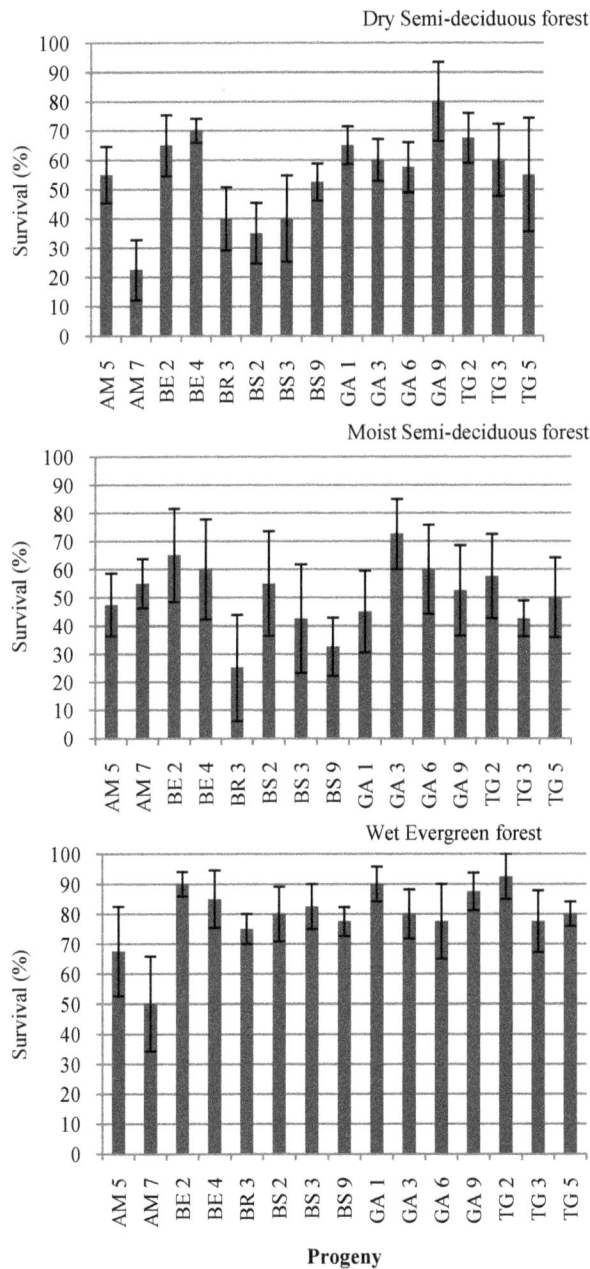

Figure 2.
Percent survival of 15 *Nauclea diderrichii* progenies 2.7 years after planting in three forest zones in Ghana.

Growth

Height: Significant differences in mean height were observed among progenies, sites, and the interaction of progeny and site in the early stage of the plantation (4 months). Across progenies, mean height was significantly higher at the wet site (39.40 ± 1.12 cm) compared to themoist (22.57 ± 0.61 cm) or dry (24.55 ± 0.23 cm) forest sites. Within sites, however, height varied only at the moist (F = 6.129, $P < 0.0001$) and dry (F = 5.413, $P < 0.0001$) sites.

At one year, differences in height were observed among progenies, sites, and between the interaction of progeny and site (**Table 2**). Mean height across all 15 progenies was highest at the wet site (74.62 ± 4.11 cm), followed by the dry site (57.40 ± 3.14cm), and the moist site (34.43 ± 2.96 cm). Mean differences in height of the 15 progenies varied somewhat randomly, from site to site. The pattern in height growth observed during the 4th and 12th months remained largely unchanged at the last assessment (age 2.7 years), although by this time the progeny-progeny differences were not significant at any of the three sites. Mean height across sites were; wet (540.34 ± 6.20 cm), dry (430.05 ± 7.83 cm), and moist (273.11 ± 8.17 cm). At age 2.7 years, the following progenies were among the fastest growing in the plots: wet site (BS2, BS3, BS9 and TG2), dry site (AM7, BS2, and BS9), and moist site (BS9, BS3, and GA3) (**Figure 3**).

Diameter

Diameter of seedlings measured 10 cm from above the soil level after one year varied significantly among sites. Mean diameter was 1.43 ± 0.02 cm at the wet site, which was significantly higher than that at the moist (1.16 ± 0.28 cm) or dry (0.92 ± 0.27 cm) forest sites. A similar trend was observed when diameter was measured at breast height (dbh = 1.3 m) at age 2.7 years (**Figure 4**). Mean diameter was highest at the wet site (5.31 ± 0.21 cm), followed by the dry site (4.58 ± 0.30 cm) and moist site (2.83 ± 0.32 cm). All progenies except BR3 recorded high growth in diameter at the wet site. At the dry site, AM5, BE2 and BR3 recorded relatively low growth indiameter. Three progenies BS3, BS9 and GA1 recorded moderately higher diameter growth than the others at the moist site.

Heritability and Genetic Gain in Height at the Moist Semi-Deciduous Forest Zone

Heritability for total height at the Moist Semi-deciduous Forest Zone was 0.438, indicating that approximately 44% of the variation observed in height growth at the MSF site was under genetic control. The variation in genetic gain from the site ranged from 28.36% (BS9) above mean performance to -20.08% (TG 5) below mean performance (**Table 3**).

Discussion

Shoot Borer Damage

None of the 15 progenies assessed escaped *O. mediofoveata* attack during the period of the study. In other words, none exhibited complete resistance or immunity to the shoot borer. However, it was clear from the results that significant variability to *O. mediofoveata* shoot borer attack occurs among the progenies, which were strongly influenced by habitat or site factors. While infestation was recorded at all the three sites during

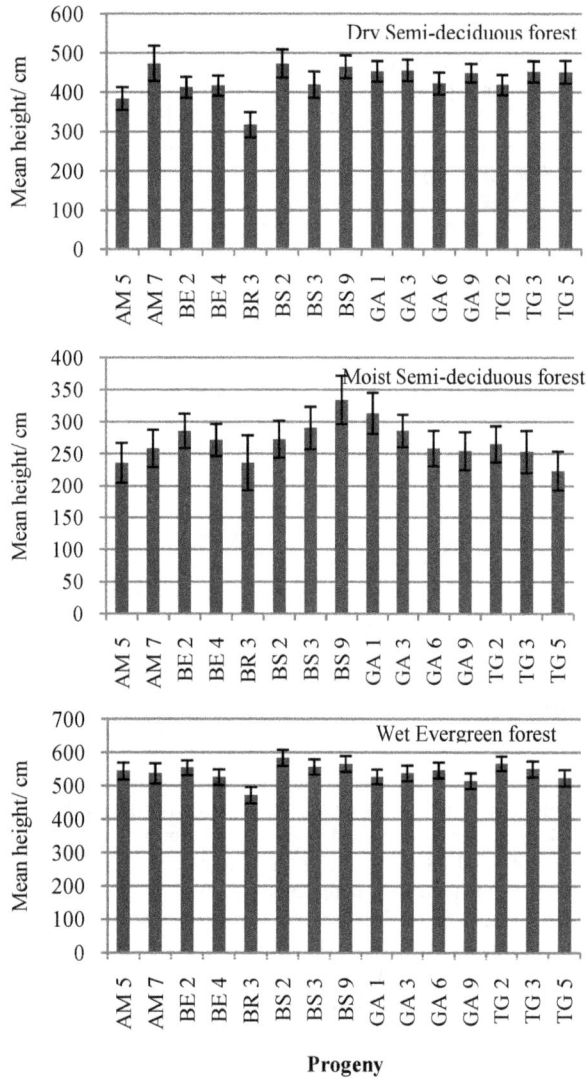

Figure 3.
Mean heights of 15 *Nauclea diderrichii* progenies 2.7 years after planting in three forest zones in Ghana.

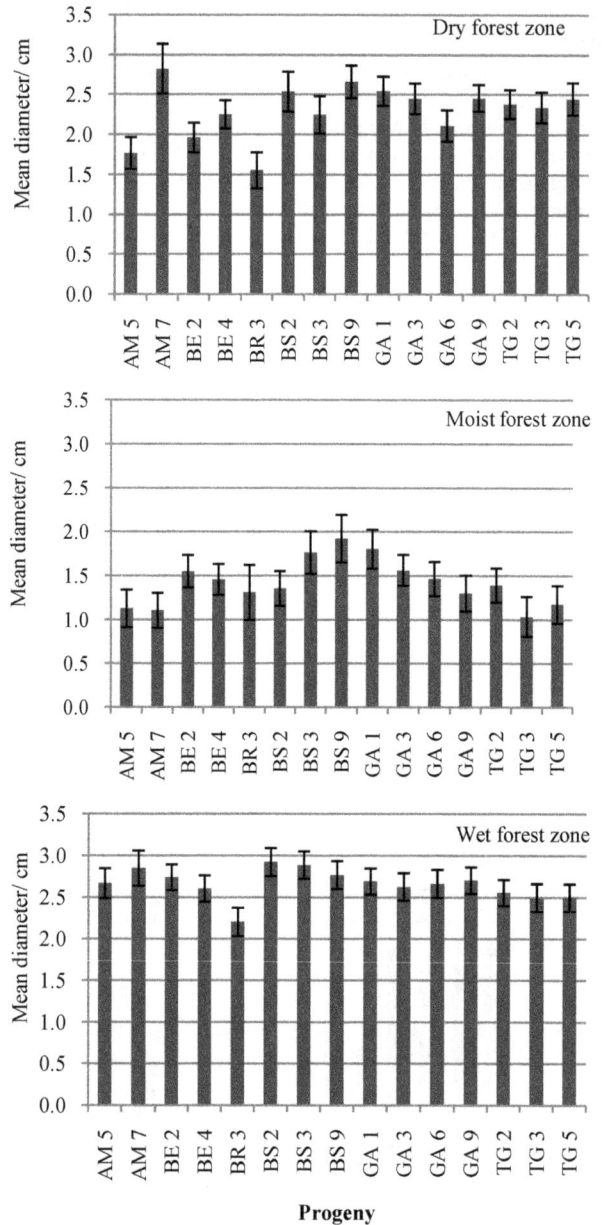

Figure 4.
Mean diameter (at breast height) of 15 *Nauclea diderrichii* progenies 2.7 years after planting in three forest zones in Ghana.

assessment in the fourth month, it was not until the 20th month that clear-cut differences were noticed. For example, while attacks were marginally lower at the Moist Semi-deciduous and Dry Semi-deciduous Forest zones at the 4th and 12th month, the reverse was observedduring the peak of the infestation at 20 months. The higher initial attacks at the Wet Evergreen Forest zone might be due to the presence of an older *N. diderrichii* plantation established in 1972 and located at about 100 m away from the experimental plot. This plantation was very likely the source of *O. mediofoveata* population which facilitated an early colonization of the plots at the WEF site as compared to the remaining two sites.

Shoot borer attacks were observed in the plantation after the peak infestation at 1.7 years, however, the impact of the attack did not show on the trees. The occurrence of fresh attacks were evident but visually this did not appear to substantially impact on the growth of the plant. This is quite contrary to the situation with, for example, the mahogany shoot borer (*Hypsipyla robusta*) attack which is a serious pest of species of the Meliaceae

in Ghana and many tropical countries (Ofori et al., 2007, Opuni-Frimpong et al., 2008; Bosu & Nkrumah, 2011). *Orygmophora medofoveata* damage did not cause *N. diderrichii* to develop profuse epicormic branching as is often the case with *H. robusta* attack of the Meliaceae. Although epicormic branches occurred in some cases, there was usually a clearly distinguishable lead terminal that suppressed the growth of the remaining shoots as the tree grew. The worst case scenarios recorded in the plantation were just a dominant stem and a smaller or poorly growing stem. In general, the damage levels observed did not cause significant mortalities to the seedlings. Healthy and vigorously growing seedlings tended to recover (coppice) from attack and subsequent dieback (withering of the shoot) when growing conditions were optimum, thus mortality

Table 3.
Selection differential and genetic gain in total height growth (cm) for 15 *Nauclea diderrichii* progenies at Mesewam nursery in the Moist Semi-deciduous Forest zone.

Progeny	Mean height	Deviation	% Deviation	% gain	Rank
BS 9	333.62	3.34	24.020	28.36	1
GA 1	312.78	3.13	16.274	19.21	2
BS 3	290.29	2.90	7.916	9.34	3
GA 3	285.69	2.86	6.204	7.32	4
BE 2	285.58	2.86	6.162	7.27	5
BS 2	272.50	2.73	1.301	1.54	6
BE 4	271.37	2.71	0.880	1.04	7
TG 2	265.00	2.65	−1.487	−1.76	8
AM 7	258.41	2.58	−3.937	−4.65	9
GA 6	258.33	2.58	−3.965	−4.68	10
GA 9	254.05	2.54	−5.558	−6.56	11
TG 3	252.94	2.53	−5.970	−7.05	12
AM 5	235.63	2.36	−12.404	−14.64	13
BR 3	235.50	2.36	−12.454	−14.70	14
TG 5	223.25	2.23	−17.007	−20.08	15

due to *O. mediofoveata* shoot borer attack was lower than expected.

However, the impact of the damage to *N. diderrichii* seedlings at the nursery stage appeared much higher and devastating than were observed in the field. It appears that bigger and sturdier seedlings used for outplanting in the field make the seedlings more tolerant to shoot borer damage than smaller and more tender seedlings that are frequently attacked at the nursery. By maintaining seedling vigour, minimizing shock during transportation from the nursery to field sites, and planting at the beginning of the rainfall season (April-May), the impact of *O. mediofoveata* attack could be substantially reduced in plantations.

Survival

The percentage survival of 88.8% - 96.3% recorded four months after establishment is encouraging for a native timber species growing under pressure of its primary insect pest. In a study to evaluate the growth of *N. diderrichii* in pure and mixed species trials conducted at the BiaTano Shelterbelt forest reserve in Ghana (Moist Semi-deciduous Forest zone), Addo-Danso et al. (2012) recorded an initial (after 6-month) survival of 70.8% in monoculture plots, 63.0% in a two-species mixed plots, and 38.9% in a 4-species mixed plots. In the same study, overall survival of Nauclea in the plantation was 40% in the monoculture after 24 months (2 years) and remained largely unchanged after 36 months (3 years). Compared to the present study, survival of the worst performing progenies were better than those reported by Addo-Danso et al., (2012).

Growth

As Nauclea is a wet/moist forest species, we expected that overall performance (resistance to attack by the shoot borer, survival, and growth rates) of the progenies would be better at the wet (WEF), followed by the moist (MSF) site, before the dry (DSF) zones. Our hypothesis was partially supported, in that performance was better at the WEF. However, between the moist and dry forest zones, average performance was better at the DSF site than the MSF site (**Table 4**). It seems that the poor survival and growth at the MSF was due largely to poor site factors that prevailed in the location where the plots were established. Besides, having been heavily cropped for many years, the soil in this location was mostly clayey and liable to flooding during periods of heavy rainfall.

Differences in growth (height and diameter) were recorded at various stages, however, at the last assessment at age 2.7 years, progenies BS9, BS3, BS2 and GA1 were amongthose with the highest overall height and diamter growth rates in the plantations. However, the Bensoprogenies (BS9, BS3 and BS2) were also those with the lowest overall survival rates among the 15 progenies. Perhaps, the high level of mortality afforded the surviving seedlings less competition for growth. The growth rates achieved under the pressent study compares favourably well with previous studies of Nauclea in Ghana and elsewhere in West Africa. In Ghana, Addo-Danso et al., (2012) recorded overall height growth of 1.9 m in monoculture Nauclea plots at the Bia-Tano shelterbelt after 24 months, 2.8 m at 36 months and 6.8 m at 60 months. The corresponding diameters of these heights were 3.8 cm, 4.1 cm and 9.7 cm, respectively. It is worth noting that all 15 progenies assessed in this study achieved a mean height of more than 3.5 m and diameter more than 3.5 cm at 32 months (2.7 years), both of which are greater than what was achieved at 36 months by Addo-Danso et al., (2012). Heights and diameter in mixed plots were lower than in the monoculture plots. In Nigeria, Onyekwelu (2007) reported a mean total height of 9.0 m height and diameter of 9.6 cm for a 5-year old Nauclea plantation in the Omo forest reserve. Mean tree diameter at breast height (dbh), total height and standbole volume ranged from 9.6 to 29.3 cm; 9.0 to 23.6 m and 23.27 to 535.52 m³/ha, respectively from plantations ranging from 5 - 30 years of age. Also in Nigeria, Fawape et al., (2001) recorded mean height of 14.07 m and 0.29 m dbh in a 20-year-old even-aged stand of *Nauclea diderrichii*in the Akure forest reserve located in the humid rainforest zone of Ondo State.

Conclusion

None of the 15 progenies was distinctly different from the others in any measure during the entire evaluation period. Rather, site-site differences were clear. It appears therefore that good or suitable site factors, especially the soil and water, are the important factors to be considered when establishing plantations of *N. diderrichii*. It will be important to ensure that seedlings used are sturdy and the plantation is properly managed. In addition, it is also important that progenies planted are not far away from the ecological zones where they were obtained, such as planting Dry Semi-deciduous progenies in the Wet Evergreen Forest zone and vice versa.

The low growth rate recorded at the Mesewam nursery area (MSF) though unexpected provided insight as to which of the progenies could be best suited for planting under harsh or

Table 4.
Mean overall performance of 15 *Nauclea diderrichii* progenies established in three forest zones in Ghana at 2.7 years.

Site	Height ± S.E (cm)	Diameter ± S.E (cm)	Survival ± S.E (%)	*Attack ± S.E. (rank)
Dry Semi-deciduous forest zone DSF (Abofour)	430.78 ± 7.8 a	14.38 ± 0.34 a	55.00 ± 3.09 a	1.92 ± 0.03 a
Moist Semi-deciduous forest zone–MSDF (Mesewam)	273.28 ± 8.1 b	8.93 ± 0.35 b	50.83 ± 3.68 a	1.86 ± 0.03 a
Moist/Wet Evergreen forest zone (Benso)	540.20 ± 6.2 c	16.67 ± 0.67 a	79.50 ± 2.44 b	1.44 ± 0.01 b

*Based on assessment carried out at 1.7 years.

stressed environmental conditions. As observed (**Table 3**), the Benso, Gambia and Begoro progenies were the most suitable under the prevailing poor soil condition. Indeed, the Benso (WEF) progenies BS9, BS3 and BS2 came close to what may be described as best performing progenies of the study. These progenies were obtained from an existing *N. diderrichii* (establish in 1972) plantation located in the Nueng Forest where wet forest zone trial was conducted. The origin of the 1972 plantation is unknown, but it appears that they were carefullly selected for planting.

We recommend that the following progenies BS9, BS3, BS2, GA1, GA3 and BE2 should be considered for plantation establishment. With the distribution range of *N. diderrichii* spanning across the African continent, it is recommended that future studies should consider a range-wide progeny and provenance assessments.

Acknowledgements

This study was conducted with funds from the African Forestry Research Network (AFORNET) Grant No. 252005. Messrs. Elvis E. Nkrumah,K. Prempeh Bandoh and Emmanuel A. Manu provided technical support for the field work.

REFERENCES

Addo-Danso, S. D., Bosu, P. P., Nkrumah, E. E., Pelz, D. R., Coke, S. A., & Adu-Bredu, S. (2012). Survival and growth of *Naucleadiderrichii* (De Wild.) and *Pericopsiselata* (Harms) in monoculture and mixed-species plots in Ghana. *Journal of Tropical Biology, 24,* 37-45.

Bosu, P. P., & Nkrumah, E. E. (2011). Companion planting of insect repellent plants with *Khayaivorensis* and its impact on growth and *Hypsipyla* shoot borer attack of the host species. *Ghana Journal of Forestry, 27,* 40-51.

Bosu, P. P., Adu-Bredu, S., & Nkrumah, E. E. (2004). Observations of insect pest activities within selected nurseries in Ashanti, Ghana. In: J. R. Cobbinah, D. A. Ofori, & P. P. Bosu (Eds.), *Pest management in tropical plantations* (pp. 154-162). International Workshop Proceedings, Edited Powerpoint Presentations, Forestry Research Institute of Ghana.

Eidt, D. C. (1965a). The Opepe shoot borer, *Orygmophora mediofoveata* Hmps. (Lepidoptera: Noctuidae), a pest of *Nauclea diderrichii* in West Africa Commonwealth. *Forestry Review, 44,* 123-125.

Eidt, D. C. (1965b). Description of the larva of *Orygmophora mediofoveata* Hampson (Lepidoptera: Noctuidae). *Canadian Entomology,* 612-617.

Fawupe, J. A., Onyekwelu, J. C., & Adekunle, V. A. J. (2001). Biomass equations and estimation for *Gmelina arborea* and *Nauclea diderrichii* stands in Akure forest reserve. *Biomass and Bioenergy, 21,* 401-405.

Hawthorne, W. D. (1995). Ecological profiles of Ghanaian forest trees. *Tropical Forestry Papers,* 25 p.

Ofori, D. A., Opuni-Frimpong, E., & Cobbinah, J. R. (2007). Provenance variation in *Khaya* species for growth and resistance to shoot borer *Hypsipylarobusta*. *Forest Ecology and Management, 242,* 438-444.

Onyekwelu, J. C. (2007). Growth, biomass yield and biomass functions for plantation-grown *Naucleadiderrichii* (de wild) in the humid tropical rainforest zone of south-western Nigeria. *Bioresource Technology, 98,* 2679-2687.

Opuni-Frimpong, E., Karnosky, D. F., Storer, A. J., Abeney, E. A., & Cobbinah, J. R. (2008). Relative susceptibility of four species of African mahogany to the shoot borer *Hypsipylarobusta* (Lepidoptera: Pyralidae) in the moist semideciduous forest of Ghana. *Forest Ecology and Management, 255,* 313-319.

Parry, M. S. (1956). Tree planting practices in tropical Africa. FAO Forestry Development Paper No. 8, 298 p.

Zobel, B., & Talbert, J. (1991). *Applied forest tree improvement.* IL: Waveland Press, Inc.

Permissions

The contributors of this book come from diverse backgrounds, making this book a truly international effort. This book will bring forth new frontiers with its revolutionizing research information and detailed analysis of the nascent developments around the world.

We would like to thank all the contributing authors for lending their expertise to make the book truly unique. They have played a crucial role in the development of this book. Without their invaluable contributions this book wouldn't have been possible. They have made vital efforts to compile up to date information on the varied aspects of this subject to make this book a valuable addition to the collection of many professionals and students.

This book was conceptualized with the vision of imparting up-to-date information and advanced data in this field. To ensure the same, a matchless editorial board was set up. Every individual on the board went through rigorous rounds of assessment to prove their worth. After which they invested a large part of their time researching and compiling the most relevant data for our readers. Conferences and sessions were held from time to time between the editorial board and the contributing authors to present the data in the most comprehensible form. The editorial team has worked tirelessly to provide valuable and valid information to help people across the globe.

Every chapter published in this book has been scrutinized by our experts. Their significance has been extensively debated. The topics covered herein carry significant findings which will fuel the growth of the discipline. They may even be implemented as practical applications or may be referred to as a beginning point for another development. Chapters in this book were first published by Scientific Research Publishing Inc.; hereby published with permission under the Creative Commons Attribution License or equivalent.

The editorial board has been involved in producing this book since its inception. They have spent rigorous hours researching and exploring the diverse topics which have resulted in the successful publishing of this book. They have passed on their knowledge of decades through this book. To expedite this challenging task, the publisher supported the team at every step. A small team of assistant editors was also appointed to further simplify the editing procedure and attain best results for the readers.

Our editorial team has been hand-picked from every corner of the world. Their multi-ethnicity adds dynamic inputs to the discussions which result in innovative outcomes. These outcomes are then further discussed with the researchers and contributors who give their valuable feedback and opinion regarding the same. The feedback is then collaborated with the researches and they are edited in a comprehensive manner to aid the understanding of the subject.

Apart from the editorial board, the designing team has also invested a significant amount of their time in understanding the subject and creating the most relevant covers. They scrutinized every image to scout for the most suitable representation of the subject and create an appropriate cover for the book.

The publishing team has been involved in this book since its early stages. They were actively engaged in every process, be it collecting the data, connecting with the contributors or procuring relevant information. The team has been an ardent support to the editorial, designing and production team. Their endless efforts to recruit the best for this project, has resulted in the accomplishment of this book. They are a veteran in the field of academics and their pool of knowledge is as vast as their experience in printing. Their expertise and guidance has proved useful at every step. Their uncompromising quality standards have made this book an exceptional effort. Their encouragement from time to time has been an inspiration for everyone.

The publisher and the editorial board hope that this book will prove to be a valuable piece of knowledge for researchers, students, practitioners and scholars across the globe.

List of Contributors

Muhammad Yusuf, Sudirman Baco and Muhammd Nasir Karim
Department of Animal Production, Faculty of Animal Science, Hasanuddin University, Makassar, Indonesia

Sudirman Baco, Muhammad Yusuf, Basit Wello and Muhammd Hatta
Department of Animal Production, Faculty of Animal Science, Hasanuddin University, Makassar, Indonesia

Wayne A. Geye and Keith D. Lynch
Division of Forestry, Kansas State University, Manhattan, USA

Bajrang Singh and Kripal Singh
National Botanical Research Institute, Council of Scientific and Industrial Research, Rana Pratap Marg, Lucknow, India

Karunakar Prasad Tripathi
Dolphin (PG) Institute of Biomedical and Natural Sciences, Manduwala, Dehradun, India.

Md. Nazmus Sadath
Chair of Forest and Nature Conservation Policy, Georg August University Goettingen, Goettingen, Germany
Forestry and Wood Technology Discipline, Khulna University, Khulna, Bangladesh

Max Krott and Carsten Schusser
Chair of Forest and Nature Conservation Policy, Georg August University Goettingen, Goettingen, Germany

Yozo Yamada
Graduate School of Bio-Agricultural Sciences, Nagoya University, Nagoya, Japan

Efi Yuliati Yovi
Faculty of Forestry, Bogor Agricultural University, Bogor, Indonesia

Dianne Staal Wästerlund
Department of Forest Resource Management, Swedish University of Agricultural Sciences, Umeå, Sweden

John J. Garland
Garland & Associates, Waldport, USA

Janusz M. Sowa
Department of Forest and Wood Utilization, University of Agriculture in Krakow, Krakow, Poland

Azamal Husen
Department of Biology, Faculty of Natural and Computational Sciences, University of Gondar, Gondar, Ethiopia

Guozhong Lin, Xiaorong Wen, Chunguo Zhou and Guanghui She
College of Forest Resources and Environment, Nanjing Forestry University, Nanjing, China

Philip Beckschäfer, Philip Mundhenk and Christoph Kleinn
Chair of Forest Inventory and Remote Sensing, Georg-August-Universität Göttingen, Göttingen, Germany

Yinqiu Ji
State Key Laboratory of Genetic Resources and Evolution, Kunming Institute of Zoology, Chinese Academy of Sciences, Kunming, China

Douglas W. Yu
State Key Laboratory of Genetic Resources and Evolution, Kunming Institute of Zoology, Chinese Academy of Sciences, Kunming, China
School of Biological Sciences, University of East Anglia, Norwich, UK

Rhett D. Harrison
Key Laboratory for Tropical Forest Ecology, Xishuangbanna Tropical Botanical Garden, Chinese Academy of Sciences, Menglun, China

Ouahiba Meddour-Sahar, Rachid Meddour and Arezki Derridj
Facultés Sciences Agronomiques et Biologiques, Department of Sciences Agronomiques B.P., Université Mouloud Mammeri de Tizi-Ouzou, Tizi-Ouzou, Algeria

Raffaella Lovreglio
Department of Agriculture, University of Sassari, Sassari, Italy

Vittorio Leone
Department of Crop Systems, Forestry and Environmental Sciences, University of Basilicata, Potenza, Italy

Nathsuda Pumijumnong and Paramate Payomrat
Faculty of Environment and Resource Studies, Mahidol University, Nakhon Pathom, Thailand

Henrique O. Sawakuchi and Maria Victoria R. Ballester
Center of Nuclear Energy in Agriculture, University of São Paulo, Piracicaba, Brazil

Manuel Eduardo Ferreira
Social-Environmental Studies Institute/LAPIG, Federal University of Goiás, Goiânia, Brazil

Sabrina Palanti and Elisabetta Feci
CNR IVALSA, Istituto per la Valorizzazione del Legno e delle Specie Arboree, Sesto Fiorentino, Italy

Sayumi Kosaka and Yozo Yamada
Graduate School of Bio-Agricultural Sciences, Nagoya University, Nagoya, Japan

Sabrina Palanti
CNR IVALSA, Consiglio Nazionale delle Ricerche Istituto per la Valorizzazione del Legno e delle Specie Arboree, Sesto Fiorentino, Italy

T. D. Reid and R. L. H. Essery
School of Geosciences, The University of Edinburgh, Edinburgh, UK

Wayne A. Geyer and Keith D. Lynch
Division of Forestry, Kansas State University, Manhattan, USA

Peter Schaefer
Department of Plant Science, South Dakota University, Vermillion, USA

William R. Lovette
Nebraska Forest Service, Lincoln NE 68583 Nebraska Forest Service, Lincoln, USA

Daniabla Natacha Edwige Thiombiano and Issaka Joseph Boussim
Laboratoire de Biologie et d'Ecologie Végétales, Université de Ouagadougou, Ouagadougou, Burkina Faso

Niéyidouba Lamien
Département de Production Forestière, Institut de l'Environnement et de Recherches Agricoles, Koudougou, Burkina Faso

Ana M. Castro-Euler
Instituto Estadual de Floresta, Embrapa Amapa Rodovia Juscelino Kubitschek, Macapa, Brazil

Barbara Vinceti
Bioversity International, CGIAR, Headquaters: Via Dei TreDenari, Rome, Italy

Dolores Agundez
INIA-CIFOR, Dpto Sistemas y Recursos Forestales, Carretera de la Coruña, Madrid, Spain

Gerardo P. Reyes
Department of Biological Sciences, Centre for Forest Research, University of Quebec in Montreal, Montreal, Canada
Faculty of Interdiscipliniary Studies, Lakehead University, Orillia, Canada

Daniel Kneeshaw
Department of Biological Sciences, Centre for Forest Research, University of Quebec in Montreal, Montreal, Canada

Louis de Grandpré
Department of Biological Sciences, Centre for Forest Research, University of Quebec in Montreal, Montreal, Canada
Natural Resources Canada, Canadian Forest Service, Laurentian Forestry Centre, Ste-Foy, Canada

Alex C. Londeau and Thomas J. Straka
School of Agricultural, Forest, and Environmental Sciences, Clemson University, Clemson, USA

Xizi Zhang
International Department, The Affiliated High School of South China Normal University, Guangzhou, China

Ping Zhou, Weiqiang Zhang and Yongfeng Wang
Department of Forest Ecology, Guangdong Academy of Forestry, Guangzhou, China

Weihua Zhang
Department of Forest Breeding and Silviculture, Guangdong Academy of Forestry, Guangzhou, China

Saroj Panthi
Department of National Park and Wildlife Conservation, Dhorpatan Hunting Reserve, Baglung, Nepal

Sher Singh Thagunna
Department of National Park and Wildlife Conservation, Api-Nampa Conservation Area, Darchula, Nepal

Noël H. Fonton, Gilbert Atindogbé, Brice T. Missanon and Belarmain Fadohan
Laboratory of Study and Research in Applied Statistics and Biometry, University of Abomey-Calavi, Abomey-Calavi, Benin

Arcadius Y. Akossou
Faculty of Agronomy, University of Parakou, Parakou, Benin

Philippe Lejeune
Unit of Forest and Nature Management, Gembloux Agro-Bio Tech, University of Liege, Gembloux, Belgium

Yozo Yamada and Sayumi Kosaka
Graduate School of Bio-Agricultural Sciences, Nagoya University, Nagoya, Japan

Nina Yulianti and Hiroshi Hayasaka
Graduate School of Engineering, Hokkaido University, Sapporo, Japan

Alpon Sepriando
Meteorological, Climatology and Geophysical Agency of Central Kalimantan, Palangkaraya, Indonesia

John Neelo, Wellington Masamba and Keotshephile Kashe
Okavango Research Institute, University of Botswana, Maun, Botswana

Demel Teketay
Department of Crop Science and Production, Botswana College of Agriculture, Gaborone, Botswana

Paul P. Bosu and Stephen Adu-Bredu
Forestry Research Institute of Ghana, Council for Scientific and Industrial Research, Kumasi, Ghana

Yaovi Nuto
Département de Zoologie, Faculté des Sciences, Université de Lomé, Lomé, Togo

Kouami Kokou
Laboratoire de Botanique et Ecologie Végétale, Faculté des Sciences, Université de Lomé, Lomé, Togo

www.ingramcontent.com/pod-product-compliance
Lightning Source LLC
Chambersburg PA
CBHW050455200326
41458CB00014B/5192